Tubercular Drug Delivery Systems

Ranjita Shegokar • Yashwant Pathak
Editors

Tubercular Drug Delivery Systems

Advances in Treatment of Infectious Diseases

Springer

Editors
Ranjita Shegokar
CapnoPharm GmbH
Tuebingen, Germany

Yashwant Pathak
Taneja College of Pharmacy
University of South Florida
Tampa, FL, USA

Faculty of Pharmacy
Airlangga University
Surabaya, Indonesia

ISBN 978-3-031-14102-7 ISBN 978-3-031-14100-3 (eBook)
https://doi.org/10.1007/978-3-031-14100-3

This Springer imprint is published by the registered company Springer Nature Switzerland AG
The registered company address is: Gewerbestrasse 11, 6330 Cham, Switzerland

Preface

An article published in the *Washington Post* in 2016 stirred an intense discussion between the public, scientific communities, and health authorities. How many diseases are precisely known to humankind? At the moment, scientists estimate the presence of more than 10,000 human diseases and only fewer available treatments, that too for major diseases.[1]

In 2022, the scenario is not very different considering the deliberate speed of academic/industry research, economic ups and downs, tougher regulatory policies, complex clinical trial setups, and the impact of the Covid-19 pandemic in slowing processes and businesses, and changing world political dynamics and policies. The same question on "availability of effective treatment" is valid now and maybe even after next two to three decades.

Diseases can be genetic or caused by environmental factors (mainly known as infectious diseases). Human infectious diseases are typically classified according to the source of infection as anthroponoses (human–human transmission), zoonoses (animal–human transmission), and sapronoses (abiotic decaying substrate–human transmission). These infectious diseases contribute to the enormous financial burden on a country's economy. By 2001, around 1415 species of organisms had been recorded to be pathogenic to humans, which mainly comprised of bacteria, viruses/prions, fungi, protozoa, and helminths.

This book is a trivial attempt to compile all possible and available information on etiology, pathology, current therapy options available of wide spectrum of diseases, the role of drug delivery sciences, advances in new techniques, diagnostic tools, and new drug research of various infectious diseases.

A total **four volumes** are compiled to accommodate the vast available information:

Volume 1 – *Malarial Drug Delivery Systems* (MDDS)
Volume 2 – *Tubercular Drug Delivery Systems* (TDDS)
Volume 3 – *Viral Drug Delivery Systems* (VDDS)
Volume 4 – *Infectious Disease Drug Delivery Systems* (IDDS)

[1] Are there really 10,000 diseases and just 500 'cures'? – The Washington Post https://www.orpha.net/

Volume 1: MDDS

Malaria is a disease caused by the parasite Plasmodium. The parasite spreads to humans through the bites of infected mosquitoes causing high fever, nausea, vomiting, diarrhea, body pain, rapid heart rate, and shaking chills. Each year, millions of people get infected by malaria, and many hundred-thousand people die. Some of the most significant risk-prone areas include Sub-Saharan Africa, South and Southeast Asia, Pacific Islands, Central America, and Northern South America. Treatment of malaria mainly comprises the most common antimalarial drugs like chloroquine and primaquine. In the case of drug resistance, artemisinin-based combination therapies (ACTs) are preferred. ACT is an amalgamation of two or more drugs that work against the malaria parasite using a different mechanism of action.

Volume 2: TDDS

Tuberculosis (TB) is a potentially severe infectious disease that affects the lungs and, in some cases, the kidney, spine, and brain. The *mycobacterium* causes tuberculosis via air route. As a result, two TB-related scenarios are possible: latent TB infection (LTBI) and TB disease. If not treated properly, TB disease can be fatal. TB bacteria usually grow in the lungs (pulmonary TB). The typical test used to diagnose TB is the Mantoux tuberculin skin test (TST). The medications used to treat latent TB infection include Isoniazid, Rifapentine, and Rifampin. Classically, the patient may undergo several treatment regimens (1st/2nd/3rd line) recommended as per disease condition and health policy of that specific country. TB treatment can take 4, 6, or 9 months depending on the regimen.

Volume 3: VDDS

Viruses are very tiny infectious germs, which cause infectious diseases such as the common cold, flu, and wart, and severe illnesses such as HIV/AIDS, Ebola, and Covid-19 (which caused the recent pandemic where millions of people lost their lives). They invade living, normal cells and use those cells as host. Depending upon the type of virus, the target body cells are different. Virus infections and diseases are categorized under ten other groups, that is, contagious, respiratory, gastrointestinal, exanthematous, hepatic, transmission, cutaneous, hemorrhagic, neurologic, and rest of the viruses not in these categories. All viruses have a protein coat and a core of genetic material, either RNA or DNA; unlike bacteria, viruses can't survive without a host. The diagnosis of viral diseases/infections can be performed by viral culture, serological tests, virus antigen detection, and viral nucleic acid or antibody detection. Treatment of viral diseases/infections depends on the type of viral infection.

Antibiotics do not work for viral infections. FDA has already approved several antiviral medicines for the treatment of certain illnesses.

Volume 4: IDDS

Each infectious disease has its specific signs and symptoms. Diagnosis of infectious diseases needs lab testing. Samples of body fluids, for example, blood, urine, and saliva, can reveal evidence of the particular microbe that is causing the illness. While imaging scans using X-rays, computerized tomography, and magnetic resonance imaging can help pinpoint disease states. Often, local tissue biopsies provide helpful information on the state of infection and adverse observations of disease (if any). This volume is focused on diagnosis, detection, disease models, the link between two or multiple infectious diseases, and vaccine development for the treatment of infectious diseases

This book series compiles all the new treatment avenues that have been explored to treat **malaria, tuberculosis, viral infections, and other infectious diseases like Ebola and hepatitis**. It covers various aspects of drug delivery advances for disease targeting, new drug molecules, analysis of currently ongoing clinical trials, vaccine development, and availability of disease models to evaluate drug performance. Dedicated chapters are included on herbal treatment opportunities for each disease. In addition, readers can refer to information on global disease health scenarios, cellular pathophysiology, and drug resistance, and full coverage on polymeric nanoparticles, solid lipid nanoparticles, dendrimers, liposome, and micro/nanoemulsions as drug delivery carriers.

Experts from all over the world have shared their knowledge to generate this one-stop resource. This book series is destined to fill the knowledge gap through information sharing and organized research compilation from the diverse expert area of pharma, medicine, clinical, chemist, and academics to fulfill following specific objectives:

- To discuss opportunities and challenges in the treatment of infectious diseases
- Enlist current efforts by researchers and experts
- Facilitate the insight and knowledge sharing
- Highlight innovative, cutting-edge micro and nanotechnology research
- Establish collaborations between academic scientists and industrial and clinical researchers

In summary, we are sure this book series will provide you great insights into drug delivery sciences (conventional, micro-nanomedicines, and upcoming drug delivery trends) along with updates on clinical and chemical drug research for the treatment of infectious diseases.

Tuebingen, Germany — Ranjita Shegokar
Tampa, FL, USA — Yashwant Pathak

Contents

Global Health and Tuberculosis; Past, Present, and Future

Suryaveer Sankineni, Sarika Chauhan, Ranjita Shegokar, and Yashwant Pathak

Abstract As we study tuberculosis as one of the leading causes of death from a single infection, it is important to understand the origins and interaction of newly developing drug-resistant strains of *Mycobacterium tuberculosis*. Health organizations across the globe have been combating TB for decades and have seen incredible successes in the quality of patient health. However, mitigating the spread of TB has been a constant struggle. Across Africa and a majority of Asia, there has been a significant decline in TB incidence, but due to poor execution of preventative methods and the rise of multidrug-resistant strains, these regions are nowhere near total eradication. On the other hand, the Middle East has seen a significant decrease in TB death and incidence in the past few decades making the steadiest progress toward eradication. The few incidences of TB in these regions have been the result of immigrations. With the inception of Coronavirus, many global initiatives have redirected resources and personnel, effectively stalling the progression of TB prevention. Combatting this disease can help us streamline healthcare interventions for TB and many other infectious diseases.

Keywords Tuberculosis · *Mycobacterium tuberculosis* · Global · Drug resistant · Transmission · Elimination/eradication · Treatment · Infectious disease

S. Sankineni · S. Chauhan
Taneja College of Pharmacy, University of South Florida, Tampa, FL, USA
e-mail: sankineniveer@gmail.com

R. Shegokar
CapnoPharm GmbH, Tuebingen, Germany
e-mail: ranjitashegokar@gmail.com

Y. Pathak (✉)
Taneja College of Pharmacy, University of South Florida, Tampa, FL, USA

Faculty of Pharmacy, Airlangga University, Surabaya, Indonesia

R. Shegokar, Y. Pathak (eds.), *Tubercular Drug Delivery Systems*,
https://doi.org/10.1007/978-3-031-14100-3_1

1 What Is Tuberculosis

Tuberculosis (TB) is the second leading cause of death via a single infectious agent (after COVID-19) and comes thirteenth in overall cause of death worldwide [1]. Infecting approximately ten million people worldwide in 2020, the chronic inflammatory disease is airborne, meaning it is spread between humans through the inhalation of air infected with TB pathogens [3]. The aforementioned pathogen is *Mycobacterium tuberculosis (M. tuberculosis)*. The pathogen is able to adapt effectively within a host and can be fatal if not treated properly. *M. tuberculosis* resides in granulomas of immune cells, compromising the host's immunity as they are infected. TB can manifest in one of two forms, pulmonary and extrapulmonary TB. The disease mainly affects the lungs but can occur in other places like the kidneys, lymph nodes, and brain [2]. Although anyone can develop the disease, it is more prevalent among adult males [3].

Latent tuberculosis infection (LTBI) occurs in people with *M. tuberculosis* present in their bodies, but not at levels where transmission to others is possible [2]. The infection begins as a result of extracellular bacilli consumption by macrophages. Usually, the immune system is able to limit and cease the rapid multiplication of the bacilli, preventing the infection's progression to disease. It is important to note, however, that anyone with LTBI can develop TB. Some individuals are at greater risk of their TB diagnoses progressing to LTBI. Human immunodeficiency virus (HIV) is the foremost risk factor which facilitates LTBI infection. In 2020, 214,000 out of the 1.5 million people that died from TB had HIV (define) [1]. Those infected with HIV are approximately 18 times more likely to develop TB than those without HIV. Children under the age of 5 are also at a greater risk of developing the disease. Other risk factors include diabetes mellitus, renal failure, cancers of the lung or neck, smoking and drug abuse, malnutrition, and those going through immunosuppressive therapy [2].

The pathogen has been associated with multiple comorbidities and health complications such as autoimmune diseases, pulmonary complications, and metabolic syndromes [4]. Further research on *M. tuberculosis* in relation to these complications and comorbidities would help more develop effective treatments for TB and its side effects.

2 The History of the Outbreak

TB is one of the oldest and most infectious diseases scientists have encountered; its cause was unknown until the discovery of *M. tuberculosis*, by Robert Koch in 1882 [3]. Evidence of the etiological agent *M. tuberculosis* infecting humans has been found in skeletons dating back as far as 4000 years [5].

The first documented evidence detailing similar symptoms to modern TB was found in India dating back approximately 3300 years ago [6]. Other written forms of documentation of TB-like symptoms can be found in ancient Hebrew texts; passages from the Old Testament use Hebrew words to describe diseases of *wasting*. Evidence suggests the most common disease resulting in body deterioration would have been TB [7]. Even ancient Greece observed their fair share of a disease they knew by the name Phthisis, which we can deduce as being TB. Hippocrates was able to accurately describe the symptoms we currently equate to TB along with its effects on an individual's lung health. Surviving the ancient timeline, TB remained prevalent and documented through the Middle Ages; during these times, it was primarily documented as *Scrofula* also known as the *king's evil*. At the time it was widely believed that the touch of a royal could cure such an illness [8]. Around the eighteenth century, TB had reached epidemic levels, with a mortality rate nearing 900 deaths per 100,000 individuals per year, with a large proportion of the deceased including younger people [6].

The height of the industrial revolution brought poor working conditions, air quality, and malnutrition; all these factors exacerbated the risk factors for deadly TB infections. Due to lesser infections among those of higher social classes, TB soon received the nomenclature and stigma of the *poor man's disease* creating a greater obstacle for preventative or curative practices. Near the beginning of the nineteenth-century talks of these various historical diseases of *wasting*, *Phasis??*, and *Scrofula*, all being forms of a singular disease arose across various scientific communities.

This discovery followed hundreds of years and scientists' worth of efforts dedicated to the cure and prevention of what we know as TB.

3 Treatments over the Time

Despite being both preventable and treatable, TB has shown its resilience over multiple centuries, being at the forefront of global health issues. After peaking in the nineteenth century, the incidence of TB began its slow but steady decline, beginning with the sanatorium movement originating with Hermann Brehmer. Brehmer opened his high-altitude sanatorium in Poland focusing on high airflow living conditions paired with low-intensity exercise and a healthy diet [9]. His efforts were superseded by Edward Livingston Trudeau, who established the incredibly successful Adirondack Cottage Sanatorium in New York [10]. Due to the extensive research, he conducted on the effectiveness of sanatorium treatment programs, and being the first to cultivate the tubercle bacillus in the United States, he was widely appreciated as the pioneer of modern TB treatment programs [11]. Many began to follow in Trudeau's footsteps, and from his opening of the Adirondack Cottage Sanatorium in 1884–1953, more than 800 institutions were functional with the occupant capabilities to support more than 100,000 patients [9]. During most patients' stays in sanatoriums, there was a great emphasis on bed rest; surprisingly this practice was incredibly successful for various reasons. In the late 1950s, cardiologist William (Bill) Dock found that the recumbent position for those suffering from tuberculosis allowed for increased blood and lymph flow to the upper regions of the lungs and equalized hypertension with the lower regions [12]. By the end of the 1960s, more effective and consistent chemotherapy programs began to emerge as scientists and doctors understood more about this disease.

The chemical battle against TB began with the discoveries of Streptomycin (SM) and para-aminosalicylic acid (P.A.S.). SM treatment greatly benefited patient mortality in the short term; however, long-term studies found an increase in SM resistance resulting in death further down the line [14]. When used in conjunction with one another, these medications proved to be reasonably effective with P.A.S. treatment significantly reducing SM resistance. In 1952, an incredibly effective replacement for SM had been synthesized, Isoniazid (INH). Since its discovery, INH has served as the main anti-TB chemical, due to its capability in preventing antibiotic resistances while maintaining low toxicity, all at the fraction of the cost [13]. Studies conducted by the British Medical Research Council concerning the development of drug resistance showed results stating that the majority of TB cases tended to only show resistance to one of the three available strains of treatment [14]. These studies gave rise to the most effective methods of chemotherapy of this time. Patients were treated with varying combinations of SM, P.A.S., and INH depending on drug resistance, age, and other health factors. However, the downside to the application of chemotherapy lay within the necessity for continuous use of such medications. After successful discharge from the hospital, a patient is expected to continue the daily oral intake of small quantities of SM and INH; failure to complete the proper cycle of chemotherapy medication resulted in the resurgence of extremely drug-resistant infection in the afflicted patients, more often than not ending in death [14]. The necessary treatments required expensive year-long hospital stays inevitably leaving poorer countries with no accessible forms of treatment.

More modern programs came to fruition following the late 1960s primarily to address the issue of cost efficiency. Studies at the Tuberculosis Chemotherapy Center in India found that at-home chemotherapy programs could be just as effective as the in-patient alternative [15]. These substitute programs allowed for cheaper and mass access regardless of available hospital and sanatoria beds. Decreasing treatment times was also a significant advancement that would make treatment programs more adaptable to any lifestyle. Clinical studies conducted on TB at Cornell University employed pyrazinamide (PZA) to effectively kill any leftover bacteria following heavy TB treatment in mice [16]. In other news, Rifampicin (RMP) was identified as an accelerant in the elimination of TB bacilli in the lungs of mice. The addition of RMP and PZA to existing treatment plans involving SM and INH was found to drastically reduce the opportunity for relapse while only consisting of a six-month chemotherapy timeline [17]. These chemotherapy programs began to experience vast testing in the following years with varying combinations of existing drugs in an effort to create the most cost- and medically effective treatment programs (Fig. 1).

In 2020, approximately 5.8 million people developed TB, a large decrease from 2019 which had shown an approximation of 7.1 million new cases. This however is no cause for celebration, as disruptions due to Covid-19 are hypothesized as being the cause for such a drastic decrease in TB reporting [3]. Despite the World Health

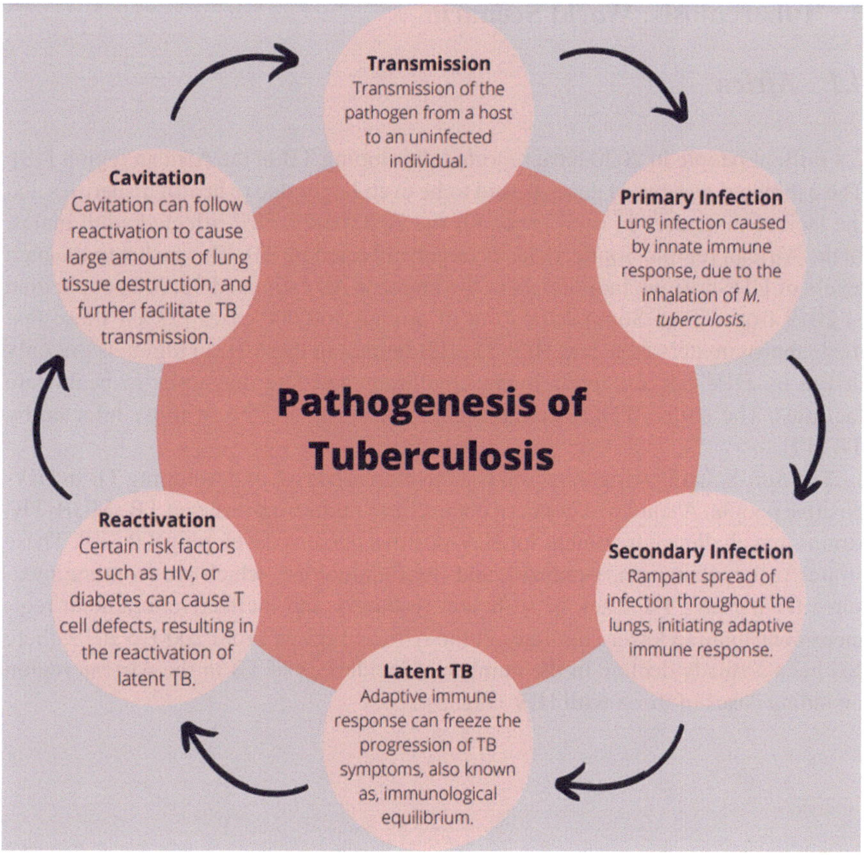

Fig. 1 Pathogenesis of tuberculosis [30]

Organization's (WHO) estimates for TB eradication, the timeline has been set back exponentially due to the unexpected complications which arose from the Covid-19 pandemic. Incomplete TB treatments and mistreatment have led to the rise of multidrug-resistant TB (MDR). With more and more complications presenting themselves at every corner, TB eradication seems to be steadily creeping further and further away.

4 Tuberculosis: World Scenario

4.1 Africa

2.4 million people in 2020 have reported developing TB in the African region [18]. The estimated number of cases is said to be even larger due to health disparities and the lack of diagnoses in rural areas. Of the 2020 cases, 58% affected adult males. In the African region, South Africa is largely affected by TB. This is due to the high levels of HIV-positive individuals in the country. An estimated 58,000 people died in 2019 from TB in South Africa out of around 360,000 cases [3]. Of those that died, approximately 62% had HIV. The TB burden in the African region is not only driven by HIV but also poor living conditions and late diagnoses in healthcare facilities. The entire WHO African region accounts for 25% of those infected by TB [19].

The Sub-Saharan African region is at an increased risk of developing TB in HIV-positive people. A source of concern comes from multidrug-resistant TB (MDR-TB) strains that challenge treatment for HIV-positive persons developing TB [20]. These strains resist rifampicin, isoniazid, and fluoroquinolone, which are frontline measures for treating TB. Lack of sufficient resources and ineffective treatment regimens contribute to lower cure rates in the African region. From 2000 to 2019, there has been a steady decline in the number of incidences of TB in the African region including cases of those with HIV (Fig. 2).

	Patients *n*	**Relapses** %
SM+INH	112	29
SM+INH+thioacetazone	104	22
SM+INH+PZA	153	8
SM+INH+RMP	152	3

PZA = pyrazinamide; RMP = rifampicin; SM = streptomycin; INH = isoniazid

Fig. 2 Effect of the addition of thioacetazone (control), PZA, or RMP to a basic 6-month SM + INH regimen on the relapse rate in a multicenter East African regimen study [14] (*PZA* pyrazinamide, *RMP* rifampicin, *SM* streptomycin, *INH* isoniazid)

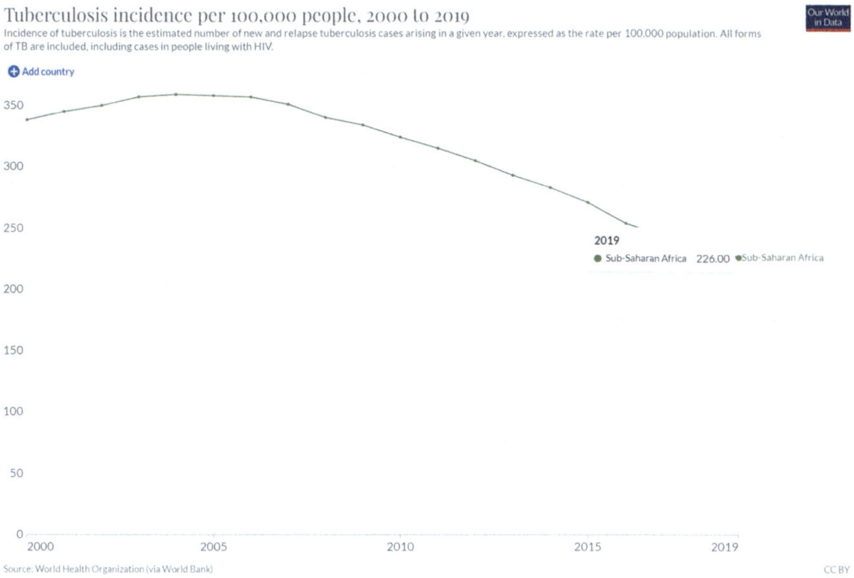

"Incidence of Tuberculosis (per 100,000 people)". Published online at OurWorldInData.org. Retrieved from: 'https://ourworldindata.org/grapher/Tuberculosis-incidence-per-100000-people' [Online Resource]

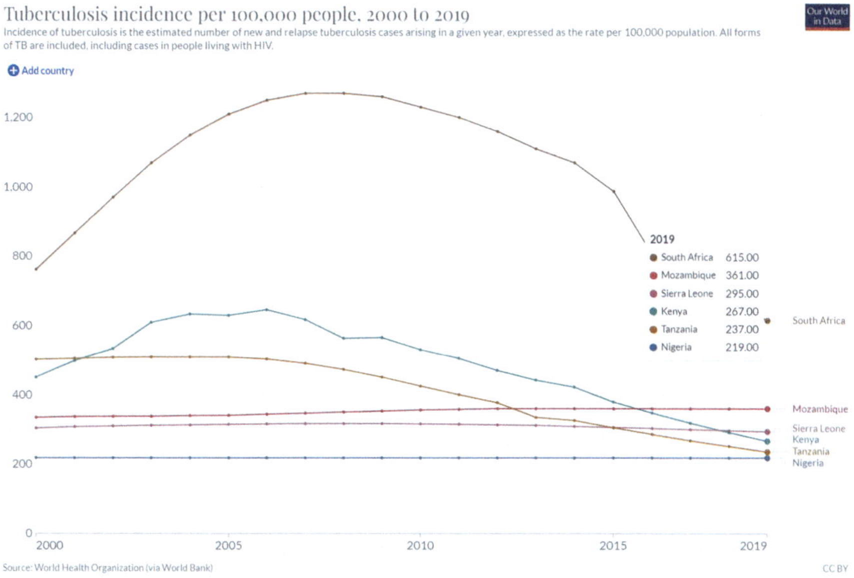

"Incidence of Tuberculosis (per 100,000 people)". Published online at OurWorldInData.org. Retrieved from: 'https://ourworldindata.org/grapher/Tuberculosis-incidence-per-100000-people' [Online Resource]

Other than South Africa, the country with the largest incidence in Africa, Mozambique, Kenya, and Tanzania have decreased incidences from 2000 to 2019. Inadequacies in testing for drug-resistant TB and confirmed diagnoses have led to ineffective treatment. Those under the age of 15 are a missing subgroup of those with TB that contribute to larger treatment gaps in children [3]. There needs to be a sustained focus on treating missing subgroups to facilitate the further identification of cases in the region and the hopeful eradication of the disease.

4.2 Western Pacific/East Asia

The Western Pacific contributed approximately 1.8 million cases of TB in 2019 [21]. Of the 1.8 million individuals infected, 833,000 cases stemmed from China, 170,000 from Vietnam, and 59,000 from the Philippines. These three epidemic countries contributed to 95% of all the cases in this region, and more than 90,000 deaths resulted from TB within the past year. Of the reported cases, approximately more than 100,000 were found to be infected by multidrug-resistant TB [22].

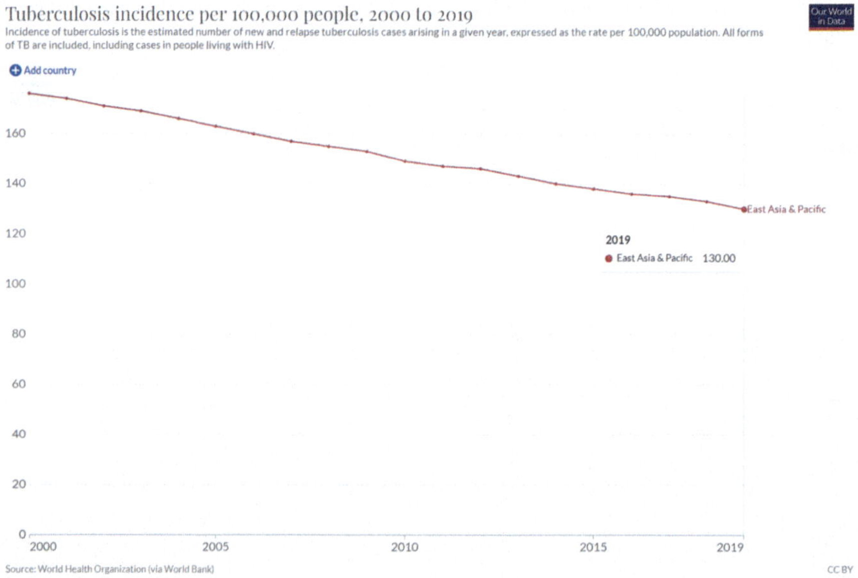

"Incidence of Tuberculosis (per 100,000 people)". Published online at OurWorldInData.org. Retrieved from: 'https://ourworldindata.org/grapher/Tuberculosis-incidence-per-100000-people' [Online Resource]

All these numbers are an improvement to years past, moving from approximately 2.3 million cases (greater than 135 cases per 100,000 population) recorded in 2000 to about 130 cases per 100,000 population in 2019. However, the success rate of treatment programs has remained stagnant for many years due to such long-term monetary and commitment necessities. Even the incidence statistics seem to be nearing a plateau and the emergence of the covid-19 global pandemic only created more complications in the TB detection and reporting systems. As for preventative tactics, only 11% of recorded children under the age of 5 had taken any preventative form of TB treatment [22]. Due to these countries' low resources for the detection of TB, low levels of preventative participation, and the necessity for long-term treatment programs, the region continues to suffer from both undetected and drug-resistant forms of TB [23].

4.3 Southern Asia

TB in South-East Asia carries an unreasonable proportion of burden in terms of incidence and mortality in comparison to any other region in the world. With approximately 4.3 million cases in 2019 and 630,000 deaths, this region was responsible for more than 50% of the entire world's TB deaths.

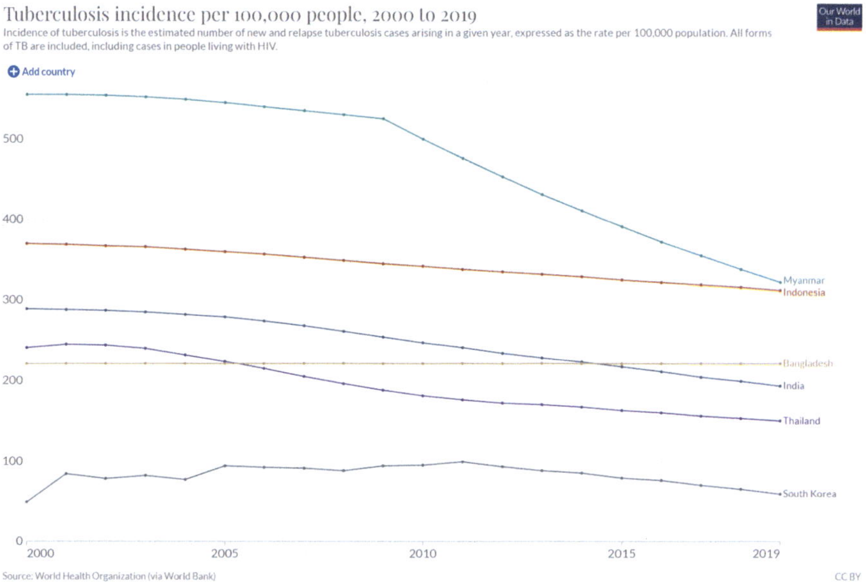

"Incidence of Tuberculosis (per 100,000 people)". Published online at OurWorldInData.org. Retrieved from: 'https://ourworldindata.org/grapher/Tuberculosis-incidence-per-100000-people' [Online Resource]

The most susceptible countries within this region include India, Bangladesh, Indonesia, the Democratic People's Republic of Korea, Thailand, and Myanmar [24]. All of these countries have experienced minimal decreases in incidence and death rates in the past recorded years but have not seen any significant values displaying steady progression toward eradication. Many anti-TB programs have been enacted in the past in hopes of speeding up the eradication process, for example, the Indian government's National Tuberculosis Programme started in 1962, and was replaced soon after by the Revised National Tuberculosis Control Programme (RNTCP) [25]. This organization and many other projects all shared similar goals to increase detection rates, treatment accessibility, and treatment completion. Nonetheless, early-stage and minor successes can be identified in large-scale incidence numbers, but as it still stands numerous cases go undetected or undergo improper treatment resulting in resistance, or worse, death.

More recently, countries that are suffering from similar complications regarding the treatment of TB are being faced with a new obstacle, extensively drug-resistant TB (XDR-TB). XDR-TB currently has no known cure, and various test results display that it is completely unaffected by any forms of pre-existing treatment/chemical therapy. Individuals undergoing treatment for this incredibly dangerous form of TB show no signs of progression regardless of undergoing years' worth of treatment and nearly all cases resulted in death [26].

4.4 *The Middle East*

The eastern Mediterranean region pales in comparison to most other regions, making up only about 8% of the world's TB incidence. Pakistan is the largest contributor to the incidence rates of this location with a total of 5.8% of world incidence, more than 50% of the region itself [3].

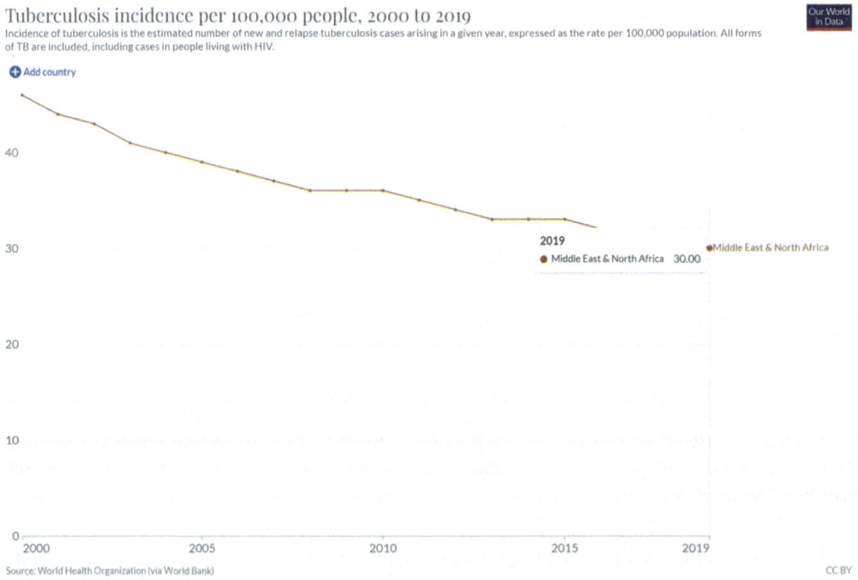

"Incidence of Tuberculosis (per 100,000 people)". Published online at OurWorldInData.org. Retrieved from: 'https://ourworldindata.org/grapher/Tuberculosis-incidence-per-100000-people' [Online Resource]

The main disease transmission factor in this region stems from the large number of refugees often traveling across borders in unsuitable living conditions, a great compounding factor for the transmission of TB [27]. Despite having to manage extraneous factors, most Middle Eastern countries experience relatively low incidence rates of TB, and the region as a whole has dropped its rates significantly in the past 19 years to approximately 30 cases per 100,000 population.

5 Conclusion and Future Trends

Although the world has known of and studied *M. tuberculosis* for over a century, the disease continues to persist as one of the largest leading causes of death from single infection. Second only to Covid-19, TB infected approximately ten million individuals in 2020 [3]. Despite being curable and preventable, global health organizations across the world have been relatively unsuccessful in combatting this disease. The contributing factors: expensive medical programs, little outreach for detection across rural communities (who are suffering the most), and the emergence of new drug-resistant strains of the bacterium due to incomplete treatment.

Smaller, less afflicted regions such as the Middle East and a few others across the globe are successfully nearing less than 10 cases per 100,000 population (the burden which the WHO considers to be the pre-elimination stage) [3]. However, this entails that the countries in this stage continue identifying latent TB infection, individuals who are susceptible to relapse, and the potential emergence of drug-resistant strains of TB [27].

The African region has seen both sides of the spectrum; they are the second largest contributor to global TB cases at nearly 2.4 million in 2020, but also is home to multiple countries that have met the global milestone. Kenya, Namibia, Zimbabwe, and the United Republic of Tanzania had all reached the global milestone of an overall 20% decrease in TB incidence since 2015 [3].

The Southern Asia region provides the greatest obstacle in TB eradication reporting 4.3 million cases in 2019, accounting for more than 50% of the worldwide reported cases [24]. Coupled with the recent discovery of XDR-TB, for which there is no developed/reliable cure or treatment at the current moment, high-risk regions are in more danger than ever before [28]. Nevertheless, organizations such as the WHO are continually creating and implementing new methods such as the END TB Strategy. The END TB Strategy focuses on increasing detection, treatment coverage, and completion across numerous countries in this region. These efforts will hopefully combat the growing circumstance of drug-resistant TB in order to push toward large-scale prevention and eventually eradication.

All in all, 2020 has disastrously impacted the global effort against TB; the Covid-19 global pandemic has unfortunately exacerbated all the complications public health organizations have been working to combat. In 2020, more access to TB diagnostic and treatment services has been disrupted and certain resources have even been repurposed to cater to individuals affected by Covid-19. As a result of this disruption, more people have died from TB complication, more cases have gone undetected, and more cases were left untreated. The WHO projected that TB deaths would decrease up to 35% by 2020 but currently is only a quarter of the way there [29]. Total TB international funding has taken an enormous blow as countries began diverting resources to the prevention and treatment of Covid-19. All these compounding factors set back the global effort to end TB by many years. Organizations like the WHO are taking immediate action to redistribute and provide essential resources toward TB detection and treatment with the goal to eliminate TB by 2030.

References

1. World Health Organization: Tuberculosis. https://www.who.int/news-room/fact-sheets/detail/Tuberculosis
2. Centers for Disease Control and Preventions: Core curriculum on Tuberculosis: what the clinician should know. https://www.cdc.gov/tb/education/corecurr/index.htm
3. WHO. Global tuberculosis report 2021. Geneva: WHO Press; 2021.
4. Chai Q, Zhang Y, Liu CH. *Mycobacterium tuberculosis*: an adaptable pathogen associated with multiple human diseases. Front Cell Infect Microbiol. 2018;8:158. Published 2018 May 15. https://doi.org/10.3389/fcimb.2018.00158.
5. Mac Donald EM, Izzo AA. Tuberculosis vaccine development. In: Ribbon W, editor. Tuberculosis-expanding knowledge. In Tech; 2015.
6. Barberis I, Bragazzi NL, Galluzzo L, Martini M. The history of tuberculosis: from the first historical records to the isolation of Koch's bacillus. J Prev Med Hyg. 2017;58(1):E9–E12.
7. Daniel VS, Daniel TM. Old testament biblical references to tuberculosis. Clin Infect Dis. 1999;29(6):1557–8. https://doi.org/10.1086/313562.
8. Murray JF, Rieder HL, Finley-Croswhite A. The King's evil and the Royal Touch: the medical history of scrofula. Int J Tuberc Lung Dis. 2016;20(6):713–6. https://doi.org/10.5588/ijtld.16.0229.
9. Murray JF, Schraufnagel DE, Hopewell PC. Treatment of tuberculosis. A historical perspective. Ann Am Thorac Soc. 2015;12(12):1749–59. https://doi.org/10.1513/AnnalsATS.201509-632PS.
10. Shampo MA, Kyle RA, Steensma DP, Edward L. Trudeau–founder of a sanatorium for treatment of Tuberculosis. Mayo Clin Proc. 2010;85(7):e48. https://doi.org/10.4065/mcp.2010.0379.
11. Roberts CS. Trudeau at Lake Saranac. In: Walker HK, Hall WD, Hurst JW, editors. Clinical methods: the history, physical, and laboratory examinations. 3rd ed. Boston: Butterworths; 1990. Available from: https://www.ncbi.nlm.nih.gov/books/NBK699/.
12. Murray JF. Bill dock and the location of pulmonary tuberculosis: how bed rest might have helped consumption. Am J Respir Crit Care Med. 2003;168:1029–33.
13. Donald PR, Sirgel FA, Botha FJ, et al. The early bactericidal activity of isoniazid related to its dose size in pulmonary tuberculosis. Am J Respir Crit Care Med. 1997;156(3 Pt 1):895–900. https://doi.org/10.1164/ajrccm.156.3.9609132.
14. Crofton J. Chemotherapy of pulmonary tuberculosis. Br Med J. 1959;1(5138):1610–4. https://doi.org/10.1136/bmj.1.5138.1610.
15. Tuberculosis Chemotherapy Centre. A concurrent comparison of home and sanatorium treatment of pulmonary tuberculosis in South India. Bull World Health Organ. 1959;21(1):51–144.

16. Mccune RM Jr, Mcdermott W, Tompsett R. The fate of mycobacterium tuberculosis in mouse tissues as determined by the microbial enumeration technique. II. The conversion of tuberculous infection to the latent state by the administration of pyrazinamide and a companion drug. J Exp Med. 1956;104(5):763–802. https://doi.org/10.1084/jem.104.5.763.
17. Mitchison D, Davies G. The chemotherapy of tuberculosis: past, present and future. Int J Tuberc Lung Dis. 2012;16(6):724–32. https://doi.org/10.5588/ijtld.12.0083.
18. Tbfacts.org Information about Tuberculosis: TB Statistics – 2020 – deaths, case notifications. https://tbfacts.org/tb-statistics/
19. Chakaya J, Khan M, Ntoumi F, et al. Global tuberculosis report 2020 – reflections on the global TB burden, treatment and prevention efforts. Int J Infect Dis. 2021;113(1):S7–S12.
20. Torpey K, Agyei-Nkansah A, Ogyiri L, et al. Management of TB/HIV co-infection: the state of the evidence. Ghana Med J. 2020;54(3):186–96. https://doi.org/10.4314/gmj.v54i3.10.
21. World Health Organization: Ending Tuberculosis in the Western Pacific by 2030. https://www.who.int/westernpacific/news-room/feature-stories/item/ending-Tuberculosis-in-the-western-pacific-by-2030
22. World Health Organization: Draft Regional Framework to End TB in the Western Pacific (2021–2030). https://www.who.int/docs/default-source/wpro%2D%2D-documents/regional-committee/session-72/wpr-rc72-07_Tuberculosis_annex.pdf?sfvrsn=79905e00_21
23. van Maaren PJ. Fighting the tuberculosis epidemic in the Western Pacific region: current situation and challenges ahead. Kekkaku. 2010;85(1):9–16.
24. World Health Organization: Tuberculosis in South-East Asia region. https://www.who.int/southeastasia/health-topics/Tuberculosis
25. Singh S, Kumar S. Tuberculosis in India: road to elimination. Int J Prev Med. 2019;10:114. Published 2019 Jun 12. https://doi.org/10.4103/ijpvm.IJPVM_492_17.
26. Velayati AA, Farnia P, Masjedi MR. The totally drug resistant tuberculosis (TDR-TB). Int J Clin Exp Med. 2013;6(4):307–9. Published 2013 Apr 12
27. Ahmad S, Mokaddas E, Al-Mutairi NM. Prevalence of tuberculosis and multidrug resistant tuberculosis in the Middle East region. Expert Rev Anti-Infect Ther. 2018;16(9):709–21. https://doi.org/10.1080/14787210.2018.1519393.
28. Centers for Disease Control and Preventions: Extensively drug-resistant tuberculosis (XDR_TB). https://www.cdc.gov/tb/publications/factsheets/drtb/xdrtb.htm
29. World Health Organization: Tuberculosis deaths rise for the first time in more than a decade due to the Covid-19 pandemic. https://www.who.int/news/item/14-10-2021-Tuberculosis-deaths-rise-for-the-first-time-in-more-than-a-decade-due-to-the-covid-19-pandemic
30. Ernst J. The immunological life cycle of tuberculosis. Nat Rev Immunol. 2012;12:581–91.

Tuberculosis: Cellular Understanding of Disease

Nana Mireku-Gyimah and Nana Ama Mireku-Gyimah

Abstract Tuberculosis is undoubtedly a public health concern. An understanding of the cellular make-up of the disease can help in its mitigation by several ways aimed at providing new preventive and therapeutic approaches. The chapter thus deliberates on some of the cellular mechanisms involved in the pathogenesis of tuberculosis (Mechanism of Mtb-host interaction, phagocytosis and immune cells, inhibition of phagolysosome maturation, inhibition of phagolysosome acidification, inhibition of ubiquitination) as well as some forms of tuberculosis (pulmonary, miliary, spinal (Pott's) tuberculosis, and tuberculous lymphadenitis).

Keywords *Mycobacterium tuberculosis* · Cellular mechanisms · Tuberculosis

1 Introduction

Tuberculosis (TB) is an ancient ailment that has plagued the world dating as far back as the Predynastic era. The disease has been documented to have been long recognized in Egyptian mummies by paleopathological changes [1]. Over the past decades, stringent measures including the availability of a vaccine and anti-TB medications [2] have been developed and implemented to arrest the global burden of TB [3]. This contagious airborne disease, caused by one of a group of closely related bacteria, termed the *Mycobacterium tuberculosis* (Mtb) complex, however, remains among the leading killer diseases worldwide. It is responsible for an

N. Mireku-Gyimah (✉)
Public Health Department, Kwesimintsim Government Hospital, Takoradi, Ghana
e-mail: nmireku-gyimah@gmail.com

N. A. Mireku-Gyimah
Department of Pharmacognosy and Herbal Medicine, University of Ghana, Accra, Ghana

R. Shegokar, Y. Pathak (eds.), *Tubercular Drug Delivery Systems*,
https://doi.org/10.1007/978-3-031-14100-3_2

estimated 1.5 million fatalities annually [4]. The discovery of the etiological agent Mtb, by Robert Koch over 50 years after the disease was first described, is arguably among the most important events in history [5]. Mtb is the 13th leading cause of death worldwide, and was previously noted to be deadlier than any other single infectious disease only to become second place to COVID-19 in recent times [4, 6]. In 2020, ten million men, women, and children were infected with TB even though the disease is preventable and treatable. Between 2015 and 2020 only an 11% reduction out of the 20% targeted reduction milestone in the End TB Strategy was realized [4]. TB funding for low- and middle-income countries, who contribute more than 95% of the world's TB infections, has consistently remained inadequate. A projected 13 billion USD is required for the diagnosis, management, and prevention of the disease in 2022 alone [4]. The disproportionate distribution of TB among low- and middle-income countries is highlighted in its dominance among the poor, homeless, and malnourished. The dearth of knowledge in the mechanisms of TB virulence, emergence of multiple drug resistance (MDR) strains of Mtb as well as comorbidity with human immunodeficiency virus/acquired immunodeficiency syndrome (HIV/AIDS) which causes more severe disease due to immune suppression has given rise to new complications in the control of TB [3]. The grave impact of TB on public health can therefore be glossed over. This chapter thus throws light and sets the tone for in-depth discussions on the cellular understanding of the disease TB.

2 Mtb Complex

The Mtb complex consists of seven common genetically closely related mycobacteria namely *Mycobacterium tuberculosis, Mycobacterium bovis, Mycobacterium africanum, Mycobacterium microti, Mycobacterium canettii, Mycobacterium pinipedii, and Mycobacterium caprae.* Other relatively newer discovered species include *Mycobacterium orygis, Mycobacterium mungi, Mycobacterium suricattae,* and dassie bacillus [7–10].

The first three, *Mycobacterium tuberculosis, Mycobacterium bovis,* and *Mycobacterium africanum* characteristically, are able to cause disease in humans. *Mycobacterium tuberculosis* and *Mycobacterium africanum* are transmitted only through the respiratory route and they do not infect animals. *Mycobacterium bovis* which is also airborne, however, is known to also cause infection through the oral route when high titers of the pathogen penetrate the gastrointestinal tract after infected milk has been ingested. *Mycobacterium bovis* is zoonotic with cattle as its reservoir. The live-attenuated form of *Mycobacterium bovis* is employed in the BCG vaccine for immunization against TB [8]. Table 1 [7–17] gives a summary of the characteristics of the various species of the *Mycobacterium* complex.

Table 1 Characteristics of species of the *Mycobacterium* complex

Mycobacterium species	Organisms they infect	Route of transmission	Other
Mycobacterium tuberculosis	Humans	Respiratory/airborne	Main causative agent for human TB
Mycobacterium africanum	Humans	Respiratory/airborne	*M. africanum* lineages are exclusive to Western Africa
Mycobacterium bovis	Humans Animals	Respiratory/airborne Oral	Used in BCG (Bacillus Calmette–Guérin) vaccine for immunization against TB Infects a wide range of animal species, both domestic and wildlife It is zoonotic, with cattle as the reservoir
Mycobacterium microti	Animals Humans	Skin wounds Oral Aerogenic infection	Infects small rodents The field vole (*Microtus agrestis*) is the maintenance host Fewer cases of human infection reported
Mycobacterium canettii	Humans	Route of transmission not clearly known	Also known as smooth tubercle bacilli Initially isolated among people groups in the horn of Africa region
Mycobacterium pinipedii	Animals	Possibly airborne	Infects pinnipeds (Walrus, True seals, Fur seals, Sea lions) Zoonotic
Mycobacterium caprae	Humans Goats	Possibly by contact	Primarily found in Europe Causes less number of human TB Detected in other animals such as sheep and grey wolves
Mycobacterium orygis	Animals Humans	Possibly by contact	Infects various species Reported cases of transmission from humans to animals and from animals to humans
Mycobacterium mungi	Banded mongooses (*Mungos mungo*)	Hypothesized to occur via environmental contamination Causative agent infecting via erosions of the nasal planum	Endemic to Southern Africa
Mycobacterium suricattae	Meerkats (*Suricata suricattae*)	Respiratory Oral (ingestion of bacteria from bite wounds or discharging skin wounds)	Endemic to Southern Africa
Dassie bacillus	Rock hyraxes (*Procavia capensis*)	Respiratory	Endemic to Southern Africa

3 Initial Transmission of Mtb

Infection with Mtb is initially established in humans mainly by droplet nuclei production through coughing, sneezing, singing, or speaking from an infected person into human air; aerolization from medical procedures such as bronchoscopy; or nosocomial handling of infected tissue and secretions. There is then subsequent invasion of the lungs of a healthy person where the pathogen exerts its virulence factors against the immune system of the host [3, 8, 18]. Infection may also be established via the digestive tract or damaged skin and mucous membranes [2].

There is the need for the bacteria to replicate within human cells and disseminate to other persons in order to propagate the disease process. Infection with Mtb may cause active infection in an otherwise healthy person, but the pathogen is known to be warded off and eliminated in 10% of people. The pathogen may also escape and become dormant during the infection process in a latent infection. In this situation, an active infection can be induced with a decline in immunity [2]. How Mtb is able to thrive in the cells of some individuals while other individuals are able to ward off the infection has however not been completely understood [6].

The nuclei droplets responsible for the initiation of infection typically range between the sizes of 0.65 μm to 7.0 μm [19]. These droplets which may contain 1–3 bacteria [20] plays a role in the pathogenesis of the disease. Smaller droplets are able to transit pass the nasopharyngeal and tracheobronchial regions into the lower respiratory tract while larger particles will usually become trapped in the oropharynx and upper airway. Studies have shown that nuclei droplet size is inversely related to the burden of infection [20]. In pre-millennial studies conducted in rabbits, large aerosolized particles containing large quantities of bacteria were observed to be trapped in the upper airway and caused little or no disease, while contrastingly, smaller aerosolized particles containing fewer numbers of bacteria traversed to the lower airway and caused significant progressive infection [20]. In the upper airway, Mtb may cause localized TB of the oropharynx and cervical lymph nodes. Once in the distal airways, Mtb invades the alveolar epithelial cells where they are phagocytosed by resident macrophages, dendrites, and neutrophils. Mtb may also invade non-myelocytic cells according to some studies [6].

4 Mechanism of Mtb-Host Interaction

Pathogen-associated molecular patterns (PAMP) present on the surface of the bacteria are recognized by pattern recognition receptors (PRR) on host cells such as Toll-like receptors (TLR), scavenger receptors (SR), Fc receptors (FcR), complement receptors (CR), C-type lectin receptors (CTL), and cytosolic DNA sensors. This recognition commences the interaction between Mtb and phagocytic cells. Following the stimulation of PRR, bacterial phagocytosis and stimulation of immune response occur. Also, several cellular processes such as apoptosis, antigen processing, inflammasome activation, phagosome maturation, and autophagy are activated [21–23].

For a clearer understanding of how Mtb infects and invades the host innate immune system, one ought to have a look at the nature and structure of the bacteria and thus understand what aids it to thrive as a pathogen in the host system. Mtb is a member of the phylum *Actinobacteria.* It is a non-spore-forming, aerobic bacilli characterized by an impermeable, hydrophobic cell envelope with a number of components. These components comprise the capsule, the outer mycomembrane, the peptidoglycan layer, and the inner plasma membrane. Different secretion systems including the twin arginine translocation (TAT) pathway, general secretion (SEC) pathway, and ESX system are employed by Mtb to enable protein transport across its unique envelope [18, 21]. These pathways play a significant role in the virulence and pathogenesis of Mtb. While the TAT and SEC pathways are common to other bacteria, ESX system is specific to *Mycobacteria* [24].

A recent study by Tucci et al. (2020) discovered 1314 proteins in an in vitro culture filtrate of Mtb. They further found that these proteins were important for detoxification, adaptation, virulence, cell wall, and cell process [25]. From this work, it is apparent that Mtb secretes a lot of proteins although proteins secreted in vitro may not be a true representation of what happens in vivo during pathogenesis. These proteins play important roles in host immune modulation including enzymatic activity, DNA modification, epigenetic modulation, immunomodulation, regulation of cell survival, modulation of free radicals and cytokines, regulation of inflammation, and inhibition of phagosome–lysosome fusion and antigen presentation [24].

The interaction between TLR and Mtb has been shown to be responsible for the primary activation of phagocytic cells. The mannose receptor, a member of the CTL family, is known to recognize mannose molecules and glycoproteins on the surface of the bacteria, and its stimulation results in the production of anti-inflammatory cytokines and non-activation of oxidative responses. Mincle, Dectin 1 and 2, as well as dendritic cell immunoactivating factor are other members of the CLR family which have been implicated in Mtb-host interaction although the exact mechanisms by which they do so have not been completely elucidated [21]. FcR and CR are expressed prominently on the surface of macrophages and have been shown to be essential in the phagocytosis of Mtb by macrophages. SR are expressed on all macrophage surfaces, and the macrophage receptor with collagenous (MARCO) which is the most researched SR has been shown to cooperate with TLR to induce the secretion of pro-inflammatory cytokines [21]. Also, cytosolic DNA sensors recognize bacterial DNA in the cytosol, a mechanism that is dependent on the rupture of phagosomes induced by the ESX system.

Myeloid differentiation factor 88 (MyD88) is a TLR-adaptor protein involved in the innate immunity response. It mediates several biologically important signal transduction pathways including inducing regulatory signals to prevent excessive inflammation and cellular damage in the lungs. MyD88 plays a crucial role in promoting full-scale activation of macrophages by IFN-γ. Indeed, experiments with mice deficient in MyD88 protein showed that the murine models were much more susceptible to granulomatous pulmonary Mtb infection [26]. The impact of

TLR-mediated pathogen recognition and MyD88-dependent signaling on anti-mycobacterial host responses was determined experimentally using genetically modified TLR2/4/9 triple- and MyD88 k.o. mice. After induction of Mtb infection, the researchers revealed that there was little impact on the induction and expression of Th1 immune responses by TLR-mediated pattern recognition and MyD88-dependent adaptor signaling. They also observed that MyD88 shaped anti-bacterial effector mechanisms in macrophages in response to IFN-c independently of its function as a TLR signal transducer. Their work did "not only underscore the dual function of MyD88 as a TLR-coupled signaling adaptor on the one hand and a master regulator of macrophage activation on the other, they also assign critical relevance for *in vivo* protection against TB only to the latter" [27].

4.1 Phagocytosis and Immune Cells

In the lungs, the bacteria are delivered to the alveoli which consist of type I and II epithelial cells with immune cells including alveolar macrophages, dendritic cells, as well as neutrophils. Following phagocytosis of Mtb by these immune cells, the pathogen deploys various virulence factors which enable it to evade being killed by macrophages and also to replicate within the phagosomes. Again, the pathogen is able to modulate intracellular macrophage signaling so as to modify the cytokine environment leading to a reduction in the effectiveness of the host immune response. This creates an environment that tolerates Mtb and enables the pathogen to thrive within cells [21]. The innate immune response and subsequently the adaptive immune response are established progressively in the lungs, leading to the formation of granuloma. This constitutes an organized structure with centralized macrophages surrounded by different cells including neutrophils, fibroblasts, T-lymphocytes, and giant cells [21, 28]. In vitro studies in mice corroborate that phagocytosed Mtb induces a local inflammatory response that attracts immune cellular aggregates consisting of several cell types. This multicellular aggregation results in the formation of the granulomatous lesion, the cardinal pathological insult of TB. The granuloma has been shown to provide an enabling environment for the propagation of Mtb although how this is so is not clearly understood [6]. The establishment of an immune balance, which is determined by the pathogen-host interaction, ultimately determines whether the granuloma will recoil into silent infection or disseminate into other systems. Mtb may conceal itself within the caseous necrotic centers of granulomatous lesions or host cells and remain for long periods in latent TB. Where there is an anomaly in the immune balance, active TB may ensue as the host is unable to arrest the infection [21].

4.2 Inhibition of Phagolysosome Maturation

After internalization and subsequent phagosome formation of Mtb by macrophages, which are the first line of the host defense system, sequential fusions immediately occur rendering it microbicidal in a process termed maturation. The maturation of phagosomes is actively regulated by proteins called Rab GTPases (Rab) which sequentially take the phagosome through early to the end stages of maturation where there is lysosomal fusion called phagolysosomes which contributes to clearance of the pathogen using the hydrolytic enzymes contained in the lysosomes [21]. The structural characteristics and composition of the mycelium of Mtb enable the bacteria to inhibit the maturation process and survive the harsh environment posed to it. Bacterial proteins prevent vacuolar accumulation of ATP and GTP leading to a reduction in pH which interferes with the maturation process. Also, by increasing the expression of the protein coronin 1 in the host-phagocyte membrane, Mtb inhibits the formation of lysosomes and thus prevents the formation of phagolysosomes in the maturation process. The cytokine interferon, IFN α, and pro-inflammatory transcription factor, NF-κB, have also been implicated in the inhibition of maturation [2, 21].

4.3 Inhibition of Phagolysosome Acidification

As maturation occurs, the pH of phagosome rapidly drops as a result of high activity of vesicular proton-pump ATPase. This acidification of phagosomes is essential for the clearance of the pathogen-containing phagolysosome. This is because an acidic pH is a prerequisite for optimal activity of the hydrolytic enzymes contained in the lysosomes as well as formation of reactive oxygen species which also contribute to pathogen clearance [21]. The structure of the bacterial cell wall inhibits acidification of phagosomes. Tyrosine kinase A, an Mtb protein, has also been shown to play a key role in acid inhibition [2].

4.4 Inhibition of Ubiquitination

The host ubiquitin system, inclusive of which are ubiquitin Ligase Parkin, Ubiquilin1, and SMURF1, is known to contribute to the recognition and elimination of intracellular Mtb via a ubiquitin-mediated autophagy [21]. Several studies have shown that Mtb interferes with the activities of the ubiquitin system, although the exact mechanisms by which this is done are not completely established [23]. A recent study has provided further insight into how Mtb suppresses host immunity. The study showed that a core subunit of the anaphase-promoting complex/cyclosome E3 ubiquitin ligase ANAPC2 has been determined to suppress the expression of proinflammatory cytokines. It is further reported to do so by interacting with the mycobacterial protein Rv0222 and helping in the attachment of lysine-11-linked ubiquitin chains to lysine 76 of Rv0222. It should be noted that Rv0222 plays an important role for the virulence of Mtb. Genetic mutations of the ubiquitination site on the mycobacterial protein Rv0222 were thus noted to reduce the virulence of the mycobacterial while impairing the inhibition of proinflammatory responses. Furthermore, the inhibitory effect of Rv0222 on proinflammatory responses was eliminated upon inhibition of ANAPC2 by specific short hairpin RNA [29]. Mtb is also known to induce apoptosis and autophagy, although the mechanisms by which this is done are also not clearly understood [2].

5 Classification of Tuberculosis

There are two main ways by which TB can be classified.

1. According to location:
   (a) TB that occurs in the lung is termed pulmonary TB [30, 31].
   (b) TB that occurs outside the lungs is termed extrapulmonary TB [30, 31].
2. According to Mtb and host interactions:
   (a) TB is termed primary infection if a previously uninfected individual is infected on exposure to Mtb. Primary TB occurs mainly in children but it is increasingly being reported in immunocompromised patients (HIV). Mediastinal or hilar lymph node enlargement is often observed in primary TB as well as other diseases such as lymphoma, testicular carcinoma, small cell lung cancer, and Whipple and Crohn's diseases. However, a positive tuberculin skin test, history of exposure to TB, and a focal area of parenchymal consolidation in primary TB can be used to confirm primary TB infection [32].
   (b) Post-primary disease refers to TB infection which develops after a latent period of about 2 years after primary infection either by reactivation or reinfection. Mtb infection can lie dormant for decades in most healthy individuals (90–95%) [33, 34]. It can occur as a result of delayed-type hypersensitivity/adaptive immunity. It usually occurs in adults [32, 35].

5.1 *Pulmonary and Extrapulmonary TB*

Mtb has the tendency to infect many organs of the body, sometimes infecting more than one organ at a time. When the lungs are implicated, the disease is termed pulmonary TB whereas extrapulmonary TB refers to Mtb infection involving other parts of the body other than the lungs. Several extrapulmonary TB occurs. They include miliary, spinal, abdominal, peritoneal, genitourinary, bone/joint, pleural, renal, and lymphatic TB. The risk factors for the development of extrapulmonary TB differ from country to country depending also on the patient's comorbidities and demographic, geographical, ethnic, social, and economic factors [36–38]. Depending on the site of infection (type of organ (s) infected), different signs and symptoms may be exhibited.

5.1.1 Pulmonary TB

TB of the lungs (pulmonary tuberculosis) is the commonest form of TB. Upon initial infection with Mtb, either one of a combination of these four presentations of primary pulmonary TB may be seen. They include parenchymal disease, lymphadenopathy, pleural effusions, and miliary TB. The parenchymal manifestation shows Ghon lesions in the lungs, especially where ventilation is best: in the subpleural and mostly in the upper part of the lower lobe and lower part of the middle or upper lobe, visible by chest x-ray [39].

Typically, infected people will present with weight loss, general malaise, fever, chronic coughing which worsens with time, fatigue, and night sweats [31, 40].

A number of studies have pointed out that a previous history of pulmonary TB is a risk factor for lung cancer, airflow obstruction, and chronic obstructive pulmonary disease (COPD) [41]. A study in South Korea demonstrated that pulmonary Tb was indeed a predisposing factor for future lung cancer.

Diagnosis of TB can be by various ways. In diagnosing, care should be critically taken not to overlook latent TB which may present with little to no symptoms. These include [31, 35, 42, 43]:

1. The tuberculin skin test: It is still a widely used test for diagnosing TB. It should be noted however that in immunosuppressed patients, it is less sensitive [35]. Also, a positive test may be obtained from a patient with latent TB infection and not actually from the symptoms he or she may be showing. The test thus lacks sensitivity and specificity.
2. Acid-fast-bacilli smear and culture test: The suspected patient is made to produce sputum. This is then cultured for the presence of Mtb. It is a specific test for the presence of Mtb. However, a negative culture test does not rule out active TB.
3. Hematological investigations: Analyses of the blood of a pulmonary TB patient can show the presence of elevated leukocytes, leukocytosis, anemia (normochromic, normocytic), mild monocytosis or eosinophilia as well as elevated ferritin, lactate dehydrogenase, platelet, and phosphatase counts. These results are not specific only to pulmonary TB. For example, patients with leukemia also will present with high leukocyte counts.
4. Radiography (Chest x-rays): Chest x-rays typically show Ghon focus either with hilar adenopathy and bilateral infiltrates or not in primary lung TB while upper lobe cavitary lesion is observed in post-primary pulmonary TB. The accuracy of chest x-rays for diagnosing TB is marred by underlying airway diseases such as

COPD and lung cancer. Also, it becomes difficult to interpret the chest x-rays in the presence of immunosuppression.

5. Pathological assessments: Tissues such as the lungs can be subjected to pathological investigations to rule out pulmonary TB. The presence of many gray–white areas of caseation and multiple areas of softening and cavitation with fluids in the cavities usually responding to the acid-fast-bacilli smear and culture test confirms pulmonary TB.
6. Active case-finding (ACF) or systematic screening for TB: This type deals with subjecting individuals to routine/ regular check-ups for lung TB. It may be included in annual family or workplace routine checks, antenatal care, visa applications to certain countries depending on the region the applicant is from, school evaluations, HIV volunteer testing, HIV clinics, immunosuppressed patients, and community-based active case findings. The idea is to identify and initiate treatment quickly for a TB patient.

5.1.2 Extrapulmonary TB

Miliary

It is a disseminated form of TB which gives a miliary pattern (numerous tiny millet-like spots in the lungs) on x-ray. It is described as both pulmonary and extrapulmonary TB. This type of TB affects the lungs and can spread to the liver, bone marrow, meninges of the brain and spinal cord, and or the pericardium. The disease is thus sometimes described as disseminated TB. It commonly occurs in children but can occur in adults too [44–46].

The disease presents in two forms: acute and cryptic miliary TB. Whereas the acute miliary TB is marked by the presence of fever and commonly affects people of age 40 and below, cryptic miliary TB typically presents with no fever and occurs in the elderly, 60 years and above. Aside these differences, cryptic miliary TB is also marked by weight loss and other symptoms which resemble that of metastatic cancer. The diagnosis of miliary TB is thus difficult and requires multi-faceted clinical investigations [44, 47]. Again, miliary TB can occur after a person suffers a disease that weakens the immune system such as in the case of measles [46]. A recent case study has shown the rare association of disseminated TB with Coombs-positive autoimmune hemolytic anemia as well as the rare presence of autoimmune hemolytic anemia and hypoadrenalism in the same TB patient [48].

The infection can also involve the skin where the appearance of erythematous macules or papules (tuberculosis miliaria cutis) is visible. This type is commonly seen in patients with AIDS. Thoracic transverse myelopathy can occur when the infection involves the CNS leading to tubercular meningitis, motor or sensory abnormality, and headaches secondary to tubercular meningitis with or without tuberculoma formation. The patient can be seen to have motor and sensory abnormalities as well as the general symptoms associated with Pott's disease (see section "Spine") if there is neurological involvement. If the musculoskeletal system is affected, conditions such as septic arthritis, osteomyelitis, and bursitis can be seen in the patient. If the abdominal quarters are involved in the infection, the following manifestations suggestive of abdominal TB involving the liver, intestines, and or peritoneal cavities would be observed – nausea, vomiting, fever, stomach pain in the right upper quadrant, generalized fatigue (hepatic); micro and macronutrient deficiencies, irregular bowel movements, subacute to acute intestinal obstruction, fever (intestinal); fatigue, abdominal pain, ascites (peritoneal/peritonitis) [44].

The disseminating nature of the disease within the host, coupled with the fact that it is insidious makes the diagnosis difficult. Different clinical investigations would have to be performed to correctly diagnose miliary TB. Miliary TB is highly fatal and should be suspected in TB cases and treated with immediate effect should it be diagnosed. As Vohra and Dhaliwal [44] puts it "a multi-pronged approach comprising of meticulous history taking, thorough clinical examination, radiological, and laboratory investigation is required for the early diagnosis and adequate treatment" of miliary TB.

A chest radiograph showing classical miliary pattern of diffuse small nodules and or bilateral diffuse reticulonodular lung lesions on a background of miliary mottling, clinical signs including weight loss and fever at night as well as the detection of Mtb in the patient either by microbiology, histopathology, and or cytopathology should lead the physician to query miliary TB. It should be noted that tuberculin skin test shows anergy in miliary TB than in pulmonary TB [18, 44, 49]. It has also been noted that blood neutrophil-lymphocyte ratio levels can be used to determine the prognosis of miliary Mtb in patients. Elevated levels of neutrophil-lymphocyte ratio lead to a bad prognosis with the likelihood of the development of acute respiratory distress syndrome. Han and colleagues suggested that levels greater than 5 needed immediate investigation and treatment [50]. Other factors that have been used as a marker for bad prognosis of miliary TB include old age, consciousness disturbance, and high BUN levels [49]. Also, initial chest CT scan for miliary TB can be missed which could make diagnosis and subsequent treatment take a longer duration of time [51].

Spine (Pott's disease)

TB of the spine otherwise known as Pott's disease or tuberculous spondylitis occurs when Mtb infects the spinal cord. It is a typical manifestation of extrapulmonary TB and can occur through the spread of the infection from the lungs. It has been described as the most destructive form of TB. Classically, this infection leads to spinal deformities with the upper lumbar and lower thoracic spine being mostly affected. It involves the destruction of the intervertebral disk space and the adjacent vertebral bodies. Pott's disease is also seen to cause the collapse of the spinal elements, and anterior wedging leading to the characteristic angulation and gibbus (anterior collapse of the spine with destruction of vertebral bodies) as well as progressive kyphotic deformity. This extrapulmonary TB infection generally leads to functional impairment with paraplegia being it most injurious complication. Generally, the disease is a slow-progressing one but with harmful effects [52].

Back pain, usually localized to the site of infection, is the most prominent sign and symptom of spinal TB. Others include local tenderness, night sweats, muscle stiffness, muscle spasms, and a gradual creeping in of cold abscesses as well as spinal deformity and neurological deficits. It should be noted that where the tuberculous vertebral lesion occurs will denote the spinal deformity [52–54].

Diagnosis of spinal TB involves physical examination of the back for deformities, examining the spine for non-contiguous vertebral lesions, microbiological examination using acid-fast bacilli to determine the presence of Mtb in samples obtained through biopsy, polymerase chain reaction, and neuroimaging. Some neuroimaging techniques employed to diagnose the disease include [52, 55]:

1. Magnetic resonance imaging (MRI), which is more sensitive and specific for spinal TB to tell whether the disease is spreading to soft tissues.
2. Computed tomography (CT), to show disco-vertebral lesions and paravertebral abscesses.
3. Plain vertebral radiograph, to point out rarefaction of the vertebral end plates, loss of disk height, osseous destruction, new-bone formation, soft-tissue abscess, and cold abscesses.
4. Conventional x-ray, which visualizes the disease in the underlying vertebrae by showing widening of the superior mediastinum in antero-posterior x-ray and increased prevertebral soft tissue shadow with anterior convexity of tracheal shadow in lateral x-rays of the upper dorsal spine. It is important to have a chest x-ray to rule out the possibility of coexisting pulmonary infection in a patient with Pott's disease.

For the various diagnostic methods, a decision as to the one to select lies on the fastest, accurate, and cost-effective method. Acid-fast bacilli stain is time consuming and shows low sensitivity and specificity. Again false-negatives of biopsies are a common occurrence in microbial assessments. Polymerase chain reaction is reported to have high specificity and sensitivity and provides quick results [52].

Lymphatic (Tuberculous Lymphadenitis)

TB in the lymph node is known as tuberculous lymphadenitis. It presents as a gradual increase in swelling that is painless of one or more lymph node, and these symptoms usually last from weeks to months. As compared to other forms of extra pulmonary TB, tuberculous lymphadenitis is commonly found in children and women [56].

A review on tuberculous lymphadenitis pointed out cervical lymph as the commonest followed by mediastinal, axillary, mesenteric, hepatic portal, perihepatic, and inguinal lymph nodes [57]. Interestingly, another literature review and meta-analysis on the epidemiologic and clinical characteristics of head and neck TB found that cervical lymph nodes (87.9%) were the commonest followed by larynx

(8.7%) with the involvement of other head and neck regions being the least (3.4%) [58].

The diagnosis of tuberculous lymphadenitis is mostly by culture or polymerase chain reaction demonstration of Mtb in an affected lymph. Culture remains the gold standard for diagnosis. Culture can and may take 2–4 weeks to yield results. A positive acid-fast bacilli stain test indicates the presence of Mtb. The test is specific for Mtb [59].

Other diagnostic methods include taking into consideration clinical symptoms presented by the patient plus fine-needle aspiration, cytology, and excisional biopsies [58].

Fine-needle aspiration involves taking some lymph node tissue for investigations. The procedure is less complicated, much less painful, and yields rather small tissues for examinations. On the other hand, excisional biopsies require surgically removing a lymph node. This method although provides enough lymph node tissues for examinations, the procedure is invasive, can leave visible permanent scars, and can cause permanent destruction to some nerves as in the case of cervical excising cervical posterior triangle [60]. Excisional biopsy however has the highest sensitivity and produces rapid symptomatic response. It is mostly recommended when the patient presents with multiple affected lymph nodes.

6 Organizations Involved in TB Research

A number of research institutions, universities, and organizations around the world are dedicated to understanding the TB disease. Active research is ongoing in a bid to get the cellular and pathophysiological understanding of the disease, as well as, to use the knowledge acquired in developing newer and efficient diagnostic tools, treatments, and treatment regimens. Table 2 summarizes a non-exhaustive list of some of these organizations around the globe.

7 Conclusion

Tuberculosis is undoubtedly a public health concern. Getting the understanding of the cellular makeup of the disease can help in its mitigation in many ways aimed at providing new preventive and therapeutic approaches. These may include developing newer drug molecules, developing targeted drug delivery systems as well as truncating the progression of the bacteria in the biological system. The chapter has discussed some of the cellular mechanisms involved in the pathogenesis of tuberculosis as well as some forms of tuberculosis (pulmonary, miliary, spinal tuberculosis, and tuberculous lymphadenitis) with the aim of shedding more light on the subject of tuberculosis.

Table 2 Organizations involved in TB research

Country	Location	Name of organization	Website
Australia	University of Sydney	The World Health Organization (WHO) Collaborating Centre for Tuberculosis	https://www.sydney.edu.au/medicine-health/our-research/research-centres/who-collaborating-centre-on-tuberculosis.html
China		The National Center for TB Control and Clinical Medicine, Center for Disease Control (CDC)	https://www.cdc.gov/globalhealth/countries/china/default.htm
Ghana	University of Ghana	The Noguchi Memorial Institute for Medical Research (NMIMR)	https://www.noguchimedres.org/index.php/bacteriology
	Kwame Nkrumah University of Science and Technology (KNUST)	Kumasi Centre for Collaborative Research in Tropical Medicine (KCCR)	https://kccr-ghana.org/
	Komfo Anokye Teaching Hospital (KATH), Department of Medicine	Chest clinic	http://www.kathhsp.org/directorates/medicine/
India		India TB Research Consortium	https://itrc.icmr.org.in/
		The Indian Council of Medical Research (ICMR) – National Institute for Research in Tuberculosis	https://nirt.res.in/
		National Institute of Tuberculosis and Respiratory Diseases	http://www.nitrd.nic.in/index.aspx?Lang_Id=1
Japan		The Research Institute of Tuberculosis, Japan Anti-Tuberculosis Association (RIT/JATA)	https://jata.or.jp/english/
Kenya		Kenya Tuberculosis Research and Training Program	https://trtc.uw.edu/research/kenya-research-and-training-program
Switzerland	Swiss Tropical and Public Health Institute	Tuberculosis Research Unit	https://www.swisstph.ch/en/about/mpi/tuberculosis-research/

Country	Location	Name of organization	Website
The Netherlands		TuBerculosis Vaccine Initiative	https://www.tbvi.eu/
South Korea		International Tuberculosis Research Center (ITRC)	http://www.itrc.kr/en/
		The Korean Academy of Tuberculosis and Respiratory Disease	http://eng.lungkorea.org/
Sweden	Karolinska Institutet	Centre for Tuberculosis Research	https://ki.se/en/gph/centre-for-tuberculosis-research
United Kingdom (UK)	The VALIDATE Network (Vaccine development for complex intracellular neglected pathogens)	Tuberculosis	https://www.validate-network.org/pathogens/tuberculosis#collapse3343376
United States of America (USA)	University of California San Francisco	Curry International Tuberculosis Center (CITC)	https://www.currytbcenter.ucsf.edu/
	The State University of New Jersey	Global Tuberculosis Institute at Rutgers, The State University of New Jersey	https://globaltb.njms.rutgers.edu/
	The University of Texas Health Science Center at Tyler	Heartland National Tuberculosis Center	https://www.heartlandntbc.org/
	University of Florida	Southeastern National Tuberculosis Center	https://sntc.medicine.ufl.edu/home/index#/
	Johns Hopkins University	The Center for Tuberculosis Research	http://tbcenter.jhu.edu/
		Tuberculosis Trials Consortium (TBTC)	https://www.cdc.gov/tb/topic/research/tbtc/default.htm?
	Vanderbilt University School of Medicine	Vanderbilt Tuberculosis Center	https://www.vumc.org/tb-center/welcome
	University of Washington	Tuberculosis Research and Training Centre	https://trtc.uw.edu/

(continued)

Table 2 (continued)

Country	Location	Name of organization	Website
	1. Emory University, Atlanta (TBRU ASTRa)	(TBRU-N) Tuberculosis Research Units Network	1. TBRU ASTRa (http://tbru.emory.edu/index.html)
	2. Boston Medical Center, Boston (The Boston University (BU) – Rutgers University (RU) Tuberculosis Research Unit (TBRU))		2. BU-RU TBRU (https://www.bumc.bu.edu/tbru/10-2/)
	3. Brigham and Women's Hospital, Boston (The Tuberculosis Research Unit for Lipidomic, Immunologic, Metabolomic, and Allelic Associations (LIMAA), TBRU LIMAA)		3. TBRU LIMAA (https://projects.iq.harvard.edu/tbru/home)
	4. Weill Cornell Medicine and New York-Presbyterian (NYP)/Weill Cornell Medical Center (The Division of Pulmonary and Critical Care Medicine)		4. https://medicine.weill.cornell.edu/divisions-programs/infectious-diseases/research/tuberculosis
Worldwide	Worldwide	The International Union Against Tuberculosis and Lung Disease	https://theunion.org/

References

1. Donoghue HD, Lee OY-C, Minnikin DE, Besra GS, Taylor JH, Spigelman M. Tuberculosis in Dr Granville's mummy: a molecular re-examination of the earliest known Egyptian mummy to be scientifically examined and given a medical diagnosis. Proc R Soc B Biol Sci. 2010;277(1678):51–6.
2. Zhai W, Wu F, Zhang Y, Fu Y, Liu Z. The immune escape mechanisms of *Mycobacterium tuberculosis*. Int J Mol Sci. 2019;20(2):340.
3. Sun J, Champion PA, Bigi F. Cellular and molecular mechanisms of *Mycobacterium tuberculosis* virulence. Front Cell Infect Microbiol. 2019;9:331.
4. WHO. Global tuberculosis report 2020: executive summary. 2020.
5. Zuñiga J, Torres-García D, Santos-Mendoza T, Rodriguez-Reyna TS, Granados J, Yunis EJ. Cellular and humoral mechanisms involved in the control of tuberculosis. Clin Dev Immunol. 2012;2012:193923.
6. Bussi C, Gutierrez MG. *Mycobacterium tuberculosis* infection of host cells in space and time. FEMS Microbiol Rev. 2019;43(4):341–61.
7. Van Soolingen D, Van Der Zanden AG, De Haas PE, Noordhoek GT, Kiers A, Foudraine NA, et al. Diagnosis of *Mycobacterium microti* infections among humans by using novel genetic markers. J Clin Microbiol. 1998;36(7):1840–5.
8. ATS. Diagnostic standards and classification of tuberculosis in adults and children. Am J Respir Crit Care Med. 2000;161(4):1376–95.
9. Clarke C, Van Helden P, Miller M, Parsons S. Animal-adapted members of the *Mycobacterium tuberculosis* complex endemic to the southern African subregion. J S Afr Vet Assoc. 2016;87(1):1322.
10. Kanabalan RD, Lee LJ, Lee TY, Chong PP, Hassan L, Ismail R, et al. Human tuberculosis and *Mycobacterium tuberculosis* complex: a review on genetic diversity, pathogenesis and omics approaches in host biomarkers discovery. Microbiol Res. 2021;246:126674.
11. Kipar A, Burthe SJ, Hetzel U, Rokia MA, Telfer S, Lambin X, et al. *Mycobacterium microti* tuberculosis in its maintenance host, the field vole (*Microtus agrestis*):characterization of the disease and possible routes of transmission. Vet Pathol. 2014;51(5):903–14.
12. Orgeur M, Brosch R. Evolution of virulence in the *Mycobacterium tuberculosis* complex. Curr Opin Microbiol. 2018;41:68–75.
13. Dawson KL, Bell A, Kawakami RP, Coley K, Yates G, Collins DM. Transmission of *Mycobacterium orygis* (*M. tuberculosis* complex species) from a tuberculosis patient to a dairy cow in New Zealand. J Clin Microbiol. 2012;50(9):3136–8.
14. Supply P, Brosch R. The biology and epidemiology of *Mycobacterium canettii*. In: Gagneux S, editor. Strain variation in the *Mycobacterium tuberculosis* complex: its role in biology, epidemiology and control. Cham: Springer; 2017. p. 27–41.
15. Vidal E, Grasa M, Perálvarez T, Martín M, Mercader I, Pérez de Val B. Transmission of tuberculosis caused by *Mycobacterium caprae* between dairy sheep and goats. Small Rumin Res. 2018;158:22–5.
16. Yeboah-Manu D, de Jong BC, Gehre F. The biology and epidemiology of *Mycobacterium africanum*. In: Gagneux S, editor. Strain variation in the *Mycobacterium tuberculosis* complex: its role in biology, epidemiology and control. Cham: Springer; 2017. p. 117–33.
17. Asante-Poku A, Yeboah-Manu D, Otchere ID, Aboagye SY, Stucki D, Hattendorf J, et al. *Mycobacterium africanum* is associated with patient ethnicity in Ghana. PLoS Negl Trop Dis. 2015;9(1):e3370.
18. Walter N, Daley CL. Chapter 31 – Tuberculosis and nontuberculous mycobacterial infections. In: Spiro SG, Silvestri GA, Agustí A, editors. Clinical respiratory medicine. 4th ed. Philadelphia: W.B. Saunders; 2012. p. 383–405.
19. Fennelly KP, Jones-López EC. Quantity and quality of inhaled dose predicts immunopathology in tuberculosis. Front Immunol. 2015;6:313.

20. Cambier CJ, Falkow S, Ramakrishnan L. Host evasion and exploitation schemes of *Mycobacterium tuberculosis*. Cell. 2014;159(7):1497–509.
21. Queval CJ, Brosch R, Simeone R. The macrophage: a disputed fortress in the battle against *Mycobacterium tuberculosis*. Front Microbiol. 2017;8:2284.
22. Liu CH, Liu H, Ge B. Innate immunity in tuberculosis: host defense vs pathogen evasion. Cell Mol Immunol. 2017;14(12):963–75.
23. Chai Q, Wang L, Liu CH, Ge B. New insights into the evasion of host innate immunity by *Mycobacterium tuberculosis*. Cell Mol Immunol. 2020;17(9):901–13.
24. Pal R, Bisht MK, Mukhopadhyay S. Secretory proteins of *Mycobacterium tuberculosis* and their roles in modulation of host immune responses: focus on therapeutic targets. FEBS J. 2022;289:4146–71.
25. Tucci P, Portela M, Chetto CR, González-Sapienza G, Marín M. Integrative proteomic and glycoproteomic profiling of *Mycobacterium tuberculosis* culture filtrate. PLoS One. 2020;15(3):e0221837.
26. Cervantes JL. MyD88 in *Mycobacterium tuberculosis* infection. Med Microbiol Immunol. 2017;206(3):187–93.
27. Hölscher C, Reiling N, Schaible UE, Hölscher A, Bathmann C, Korbel D, et al. Containment of aerogenic *Mycobacterium tuberculosis* infection in mice does not require MyD88 adaptor function for TLR2, −4 and −9. Eur J Immunol. 2008;38(3):680–94.
28. Ramakrishnan L. Revisiting the role of the granuloma in tuberculosis. Nat Rev Immunol. 2012;12(5):352–66.
29. Wang L, Wu J, Li J, Yang H, Tang T, Liang H, et al. Host-mediated ubiquitination of a mycobacterial protein suppresses immunity. Nature. 2020;577(7792):682–8.
30. WHO. Global Tuberculosis Control 2008: surveillance, planning, financing. Geneva: World Health Organization. p. 2008.
31. Hoffmann CJ, Churchyard GJ. Chapter 29 – pulmonary tuberculosis in adults. In: Schaaf HS, Zumla AI, Grange JM, Raviglione MC, Yew WW, Starke JR, et al., editors. Tuberculosis. Edinburgh: W.B. Saunders; 2009. p. 332–41.
32. Silva CIS, Müller NL. Case 70. In: Silva CIS, Müller NL, editors. The teaching files: chest. Philadelphia: W.B. Saunders; 2010. p. 140–1.
33. Gideon HP, Flynn JL. Latent tuberculosis: what the host "sees"? Immunol Res. 2011;50(2):202–12.
34. Letang E, Ellis J, Naidoo K, Casas EC, Sánchez P, Hassan-Moosa R, et al. Tuberculosis-HIV co-infection: progress and challenges after two decades of global antiretroviral treatment roll-out. Archivos de Bronconeumologia. 2020;56(7):446–54.
35. Goldman L, Schafer A. Goldman-Cecil medicine. Philadelphia: Elsevier Inc.; 2020. p. 2458–63.
36. Abdallah TEM, Mohmmed Toum FE, Bashir OH, Mansoor TI, Yuosif MM, Awad-Elseed Elkhawad M, et al. Epidemiology of extra pulmonary tuberculosis in eastern Sudan. Asian Pac J Trop Biomed. 2015;5(6):505–8.
37. Khan AH, Sulaiman SAS, Laghari M, Hassali MA, Muttalif AR, Bhatti Z, et al. Treatment outcomes and risk factors of extra-pulmonary tuberculosis in patients with co-morbidities. BMC Infect Dis. 2019;19(1):691.
38. Namperumalsamy P, Rathinam SR. Mycobacterial infections. Retina Elsevier. 2013:1473–8.
39. Delgado BJ, Bajaj T. Ghon complex StatPearls [Internet] 2021.
40. Muntean PE. A 50-year-old male with fever, cough, dyspnoea, chest pain, weight loss and night sweats. Breathe (Sheff). 2018;14(1):43–8.
41. Ravimohan S, Kornfeld H, Weissman D, Bisson GP. Tuberculosis and lung damage: from epidemiology to pathophysiology. Eur Respir Rev. 2018;27(147):170077.
42. Dacso CC. Skin testing for tuberculosis. 3rd ed. Boston: Butterworths; 1990.
43. Burke RM, Nliwasa M, Feasey HR, Chaisson LH, Golub JE, Naufal F, et al. Community-based active case-finding interventions for tuberculosis: a systematic review. Lancet Public Health. 2021;6(5):e283–e99.

44. Vohra S, Dhaliwal HS. Miliary Tuberculosis. 10. In: StatPearls [Internet]. Treasure Island (FL): StatPearls Publishing; 2022 Jan–. PMID: 32965971.
45. Ribeiro S, Trabulo D, Cardoso C, Oliveira A, Cremers I. Disseminated tuberculosis in an immunocompetent patient: the answer is in the liver. GE Port J Gastroenterol. 2015 Dec 3;23(4):208–213.
46. James W, Elston D, Treat J, Rosenbach M, Neuhaus I. Mycobacterial diseases. Andrews' diseases of the skin. 13th ed. Philadelphia: Elsevier; 2020.
47. Albrecht H. Miliary tuberculosis in critical care. In: Infectious diseases in critical care medicine. 3rd ed. New York: M. Dekker; 2009. p. 420–31.
48. Venkitakrishnan R, Augustine J, Paul M, Abraham L. Disseminated tuberculosis associated with autoimmune haemolytic anaemia and adrenal deficiency: a rare association. BMJ Case Rep. 2022;15(3):e249277.
49. Wakamatsu K, Nagata N, Kumazoe H, Honjyo S, Hara M, Nagaoka A, et al. Prognostic factors in patients with miliary tuberculosis. J Clin Tuberc Other Mycobact Dis. 2018;12:66–72.
50. Han Y, Kim SJ, Lee SH, Sim YS, Ryu YJ, Chang JH, et al. High blood neutrophil-lymphocyte ratio associated with poor outcomes in miliary tuberculosis. J Thorac Dis. 2018;10(1):339–46.
51. Choe J, Jung KH, Park J-H, Kim S-H, Kim MY. Clinical and radiologic characteristics of radiologically missed miliary tuberculosis. Medicine. 2021;100(8):e23833.
52. Garg RK, Somvanshi DS. Spinal tuberculosis: a review. J Spinal Cord Med. 2011;34(5):440–54.
53. Hidalgo JA. Pott disease. Department of Internal Medicine, Division of Infectious Disease Peru. 2016(cited 5 December 2011) available from: http://medscape.com
54. Carvalho P, Fernandes C, Cunha S, Teixeira C. Is it back pain or Pott's disease? Cureus. 2022;14(3):e22834.
55. Rajasekaran S, Soundararajan DCR, Shetty AP, Kanna RM. Spinal tuberculosis: current concepts. Glob Spine J. 2018;8(4_suppl):96S–108S.
56. Gupta SP, Arara A, Bhargava DK. An unusual presentation of oesophageal tuberculosis. Tuber Lung Dis. 1992;73(3):174–6.
57. Mohapatra PR, Janmeja AK. Tuberculous lymphadenitis. J Assoc Physicians India. 2009;57(6):585–90.
58. Qian X, Albers AE, Nguyen DTM, Dong Y, Zhang Y, Schreiber F, et al. Head and neck tuberculosis: literature review and meta-analysis. Tuberculosis. 2019;116:S78–88.
59. Fontanilla J-M, Barnes A, von Reyn CF. Current diagnosis and management of peripheral tuberculous lymphadenitis. Clin Infect Dis. 2011;53(6):555–62.
60. Knox J, Lane G, Wong JSJ, Trevan PG, Karunajeewa H. Diagnosis of tuberculous lymphadenitis using fine needle aspiration biopsy. Intern Med J. 2012;42(9):1029–36.

Metal Nanoparticles in Tuberculosis

Nilayan Guha and Md. Kamaruz Zaman

Abstract Tuberculosis (TB) is a bacterial illness that affects a number of human organs, predominantly the lungs but also the liver, spleen, and spine. It causes fever, fatigue, and a chronic cough, among other symptoms, and can be fatal if not treated effectively. Every year, ten million people contract active tuberculosis, with an estimated 1.3 million deaths. Oral administration of a combination of first-line anti-TB medicines for at least 6 months is recommended by current treatment guidelines. Patient compliance is poor due to extensive treatment times and poor pharmacokinetics, as well as side effects of drugs, which have accelerated the emergence of multidrug-resistant (MDR) organisms. All of this, together with the scarcity of novel TB drugs to treat MDR-TB and shorten typical treatment times, has underlined the need for new targeted drug delivery methods. In this regard, there has recently been a focus on nanotechnology to construct organic or/and metal, bi-metal nanoparticles loaded with TB drugs for targeted delivery via the inhaled route to improve their efficacy. This chapter identifies recent studies that have employed metal nanoparticles to provide a reliable diagnostic system and an inhaled drug delivery system to more effectively treat TB.

Keywords Metal composites · Nanoparticles · Targeted delivery · Tuberculosis

1 Introduction

Tuberculosis (TB) is a serious global health issue, with one-third of the world's population afflicted with *Mycobacterium tuberculosis* (M.tb). In 2018, around ten million new cases of TB were reported, with 1.3 million fatalities [1–3]. Despite the availability of vaccines and antibiotics today, and the fact that early detection and treatment of persons with TB save millions of lives each year, the "end" of TB as an epidemic and a major public health problem remains a pipe dream for most

N. Guha · M. K. Zaman (✉)
Department of Pharmaceutical Sciences, Dibrugarh University, Dibrugarh, Assam, India
e-mail: kzaman71@dibru.ac.in

R. Shegokar, Y. Pathak (eds.), *Tubercular Drug Delivery Systems*,
https://doi.org/10.1007/978-3-031-14100-3_3

countries. To close the vast and persistent gaps in TB diagnosis and treatment, many approaches such as radiometric detection, immunoassays such as enzyme-linked immunospot, polymerase chain reaction, and TB rapid culture detection systems have been developed [4]. However, because these approaches require complex instrumentation and highly qualified technical staff, they are relatively expensive and centralized in large stationary laboratories, making it critical to develop a portable, fast, and highly sensitive real-time system for accurate diagnosis and screening of tuberculosis infection in a timely manner.

Nanotechnology sparked the development of novel molecular nanodiagnostic tools with greater sensitivity, specificity, and speed at reduced costs as an enticing alternative to traditional procedures. In fact, a range of nanomaterials-based biosensors, such as metal nanoparticles (MeNPs), have been created for the detection of tuberculosis [5, 6]. The term "nano-biosensor" refers to a small analytical instrument that combines a biorecognition element with a physicochemical nanometric materials-sized transducer [7]. Different chemical or physical methods immobilize the biorecognition element, which can be an enzyme, antigen-antibody, nucleic acid, entire cell, etc., onto the transducer, which can then precisely quantify the emerging signals [8].

Treatment of individuals with multidrug-resistant tuberculosis (MDR-TB) and extensively drug-resistant TB strains has become a significant concern [9]. Patients with tuberculosis require long-term therapy with multidrug regimens. Some of these multidrug regimens interact with other antibacterial medications, making drug toxicity more likely [10]. Rifampicin has been shown to cause liver harm in rats [11], and various antitubercular drugs can cause hepatotoxicity [12]. As a result, uncomplicated, long-term, and effective antitubercular medication regimens are always needed to treat TB. Different metallic nanoparticles (MeNPs) such as silver, gold, iron, gallium, zinc, zinc oxide, magnesium oxide, and titanium dioxide have been discovered to prevent tuberculosis [13]. Researchers have focused on developing co-delivery mixed MeNPs with antitubercular antibiotics encapsulated in non-toxic and biodegradable polymers in recent years [14, 15]. The harmful effects of MeNPs on THP-1 and normal human lung cells (MCF-7 cell lines) should, of course, be taken into account. MeNPs produce various toxins in human tissues and cell cultures, according to studies, which can enhance oxidative stress and the generation of inflammatory cytokines. MeNPs could play a role in cell apoptosis. MeNPs have the ability to enter the cell membrane bilayer, mitochondria, and nucleus. As a result, they may damage mitochondria and trigger DNA mutations. Toxicity of MeNPs is determined by their size, dimensions, chemical composition, shape, surface structure, surface charge, density, and solubility [16]. According to studies, the initial concentration of MeNPs has a significant impact on their toxicity toward human cells, particularly in THP-1 cell lines [17].

The main goal of this chapter is to introduce the critical role of MeNPs as a diagnostic tool for TB as well as a treatment agent with anti-TB activity that can be changed and designed further to have a more potent and efficacious impact on reducing the current TB morbidity rate.

2 Pathogenesis and Immunology of TB

Inhalation of droplets produced by a person with active tuberculosis starts the initial stage of tuberculosis. These droplets can linger in the air for prolonged periods of time. A single droplet breathed could be enough to induce sickness. The majority of droplets land in the upper respiratory tract, killing the germs, but a few make it deeper down. The bacteria reach the alveoli in the lungs, where they are phagocytozed by alveolar macrophages. Mannose receptors, Toll-like receptor 2 (TLR2) and Toll-like receptor 4 (TLR4), surfactant protein A receptors, CD14, scavenger receptors, complement receptors, and immunoglobulin receptors are all involved in the absorption process [18]. Sometimes macrophages fail to eliminate bacteria because the bacteria's chemicals inactivate them or because M.tb inhibits phagosome-lysosome fusion mechanisms, avoiding low pH exposure and hydrolytic surrounds of phagolysosomes [19].

Mycobacterium multiplies in the macrophage in the second stage, eventually triggering its lysis. Cellular damage occurs as a result, attracting inflammatory cells and blood monocytes to the location. Monocytes become macrophages and try to kill the microbe that has been consumed by the macrophages and has grown inside the phagocyte. Due to the bacterial burden, these macrophages lyse and perish once more [20]. The third stage begins 2–3 weeks following infection. T cells acquire immunity, and lymphocytes migrate to the infection site. T cells are stimulated when mycobacterial antigens are presented to them, resulting in the release of interferon and other cytokines. IL-12, TNF-α, IL-8, and other pro-inflammatory cytokines are secreted by macrophages when they are activated by interferon. M.tb's rapid expansion slows at this point, and the host cell develops cell-mediated immunity. Because of the high lipid composition of the mycobacterial cell wall, those outside of cells are immune to antibody-activated complement assault. Much of the pathophysiology of tuberculosis is caused by cell-mediated immunity. When activated macrophages release lytic enzymes, reactive intermediates, and different cytokines, tissue injury can occur. The immune system, specifically macrophages, will cage the germs into tubercles at this point. The environment between these structures is anoxic and acidic, preventing mycobacteria from growing. Anoxic and acidic conditions exist between these structures, preventing mycobacteria from growing. Latency, which is one of the hallmarks of tuberculosis, is the equilibrium between the host and the mycobacterium. Many circumstances, including starvation, immunological suppression, steroid usage, or HIV infection, might cause the tubercles to dissolve in the fifth and final stage. Tubercle centers may liquefy for unknown reasons, giving an excellent growth substrate for the bacterium, which now proceeds to develop rapidly in the extracellular fluid. The enormous quantity of bacteria and the immunological reaction to them finally cause necrosis and formation of a hole in the lung tissue surrounding the tubercles [21]. At stage three, most tuberculosis infections are over. It is well known that CD4+ (helper) and CD8+ (cytotoxic) T cells are involved in a cell-mediated immune response and that both play an important role in TB protection. CD4+ (helper) T cells boost macrophage antibacterial activity by producing cytokines such as interferon-γ (IFN-γ) and TNF, whereas CD8+ cells kill infected macrophages and probably Mtb by releasing cytotoxic mediators such as perforins, granzymes, and granulysin [22]. Despite our improved understanding of the immune response to M.tb, the sort of immunological response required for successful immunization generated by vaccination remains unknown [23].

3 Role of Macrophages

Understanding the intracellular adaptation processes that allow tuberculosis to survive in macrophages is critical for developing effective treatments. M.tb has adaptive mechanisms that make it difficult for the human immune system, particularly macrophages, to combat it. These methods will be explained in more detail later.

3.1 Primary Phagosome-Lysosome Fusion Prevention

M.tb can interfere with primary endosome transformation and phagosome maturation, causing fusion with lysosomes to be delayed or inhibited [24]. M.tb can also prevent primary endosomes from transforming into phagolysosomes by lowering proton ATPase levels inside the endosomes. M.tb can prevent the inducible nitric oxide synthase (iNOS) from joining [25, 26]. According to studies, tuberculosis can also remove phosphatidylinositol 3-phosphate [27].

3.2 Escape of M.tb into the Cytosol

By shattering the endocytic vesicle wall and accessing the cytosol, tuberculosis can avoid destruction in phagolysosomes. The ability of M.tb in the cytosol to escape the endocytic vesicle is critical for intra-macrophage survival [28]. Some of the H37Rv strains of tuberculosis escaped from phagolysosomes and entered the cytosol, according to a study [16]. *M. marinum* can also exit its phagolysosome and move around using the motive force of actin via Arp2/3 complex-mediated actin rearrangement that is dependent on WASP activation [29, 30]. In addition, proteins released by tuberculosis, such as ESAT-6, play a key role in pathogenicity. ESAT-6, alone or in combination with its chaperone CFP-10 (ESAT-6: CFP-10), is required for phagosome escape [31]. Increased replication rates in the cell cytoplasm are caused by these proteins [32]. Furthermore, because of its lipid-rich architecture, the mycobacterial cell envelope is more likely to be mistaken for a phagosome membrane [33]. The presence of cytosolic M.tb in some preparations could be explained by the role of host triacylglycerol in phagosome formation and persistence in granulomatous lesions [34].

3.3 *Non-phagocytic Internalization of M.tb in Macrophages*

Non-phagocytic mechanisms for internalization of tuberculosis in macrophages involve interactions between tuberculosis and the macrophage membrane, resulting in the production of vesicles. Internalization of macrophages is accomplished by linking M.tb with lipid rafts and receptors that promote non-phagocytic endocytosis [35]. The synthetic antimicrobial polymers were found to be able to trigger membrane lysis and bind to the genomic material of mycobacteria, resulting in mycobacterial cell death, as well as destroy intracellular mycobacteria effectively without causing any harm to mammalian cells, according to studies. Synthetic antimicrobial polymers were found to have clathrin-independent penetration, survive hydrolytic lysosomal breakdown, and successfully kill intracellular bacteria in a study [36]. M.tb can also boost the expression of anti-apoptotic molecules such as Bcl2 by interacting with macrophage apoptotic pathways [37].

4 Interaction of Metal Nanoparticles and Macrophages

Macrophages can ingest opsonized NPs, modified mannose, IgG, and a variety of complements larger than 500 nm, including TB derived through phagocytic pathways [38, 39]. Macro-pinocytosis is an actin-dependent pinocytic mechanism that allows macrophages to ingest extracellular fluid droplets within enormous vacuoles generated by the fusion of an extended plasma membrane with a non-extended plasma membrane. It is possible that the macro-pinocytosis route is more likely to be used for swallowing agglomerated particles. In other words, macrophages can ingest agglomerated particles, ligand-modified NPs, viral NPs, and polyethylene glycol NPs via macro-pinocytosis pathways. Specific receptors, such as clathrin-mediated, caveolin-mediated, or clathrin/caveolin-mediated pinocytosis, have been shown to allow ligand-modified NPs and some viral NPs to enter macrophages [40]. MeNPs and dendrimers with diameters of 4–10 nm have been found to directly permeate macrophages. THP-1 cells exposed to Ag and ZnO NPs might permeate the cell bilayer membrane and enter the cell, according to transmission electron microscopy photos from a prior study. Figure 1[41] depicts various internalization mechanisms for MeNPs.

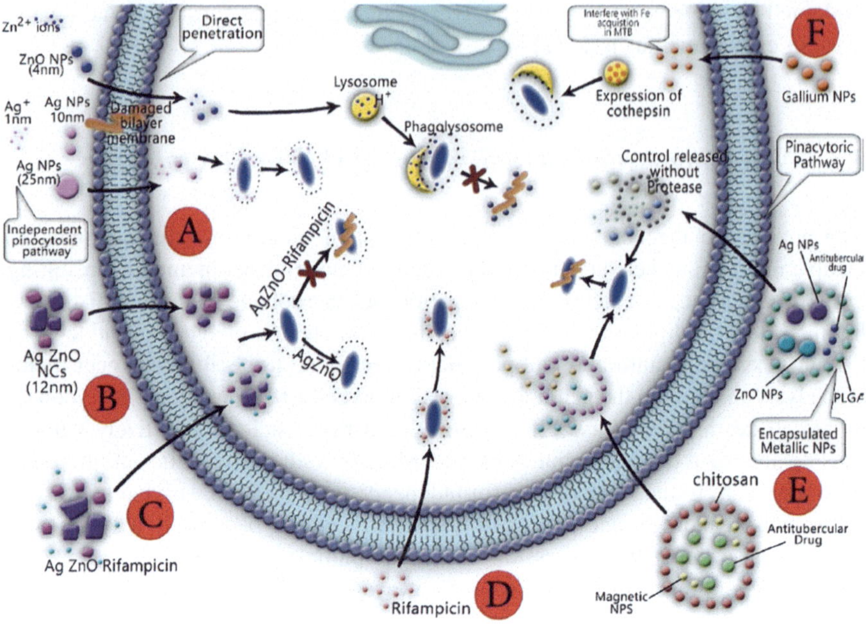

Fig. 1 Colloidal ZnO nanoparticles (NPs) can pass right through the bilayer membrane and end up in lysosomes. (**a**) Lysosomes with ZnO NPs are integrated into infected phagosomes and kill M.tb; colloidal Ag NPs alone are ineffective. (**b**) Compared to Ag/ZnO nanocrystals in a combination. (**c**, **d**) Ag/ZnO-rifampicin can prevent TB from entering the phagosome. (**e**) Endocytosis transports encapsulated magnetic NPs and antibiotic-loaded polymers to macrophages, which then release the NPs and antibiotics into the cytosol. M.tb can be killed in macrophages using a combination of magnetic NPs and antitubercular NPs. (**f**) The macrophages infected with M.tb expressed cathepsin D, a protein that is involved in macrophage activation

5 Diagnostic Applications

The use of nanomaterials, particularly nanoparticles with unique features, in biosensor design resulted in more simple, quick, sensitive, and hybrid nano-biosensor platforms with synergetic properties and functionalities [42]. Due to their high specific surface, which allows for the immobilization of an expanded number of bioreceptor units, the intelligent usage of such nano-objects resulted in demonstrably superior performances with increased sensitivities and lowered detection limits of several orders of magnitude. Nanoparticles also have chemical, physical, and electrical properties that are distinct from bulk materials, which were exploited to create novel and improved biosensing devices, due to their small size (usually in the range of 1–100 nm). Many different types of nanoparticles of various sizes and compositions are now commercially available, making their use in the detection of infectious diseases, particularly tuberculosis, easier. The many forms of metal nanoparticles-based DNA biosensors for the detection and identification of tuberculosis are summarized in Table 1.

Table 1 List of various metal composites used for the detection and identification of TB

Detection method	Nano metal composite	Analyzed samples	Limit of detection	Time of detection	References
Colorimetric sensing	Gold	DNA	0.75 μg	2 h	[45]
	Gold	DNA	50 fmol μL^{-1}	3 h	[46]
	Gold	178 bp of IS6110 insertion element	5 pg	1 h	[47]
	Gold	DNA	90 ng	30 min	[48]
	Gold	rpoB gene	N/A	N/A	[49]
	Gold	123 bp of IS6110 insertion element of dsDNA	1.95×10^{-2} ng mL^{-1}	60 min	[50]
	Silver	DNA	1.27 nM	N/A	[58]
Surface plasmon resonance	Gold	DNA	10^4 CFU mL^{-1}	40 min	[51]
	Gold	rpoB gene	N/A	N/A	[52]
Quartz crystal microbalance	Gold	IS6110 insertion element	5 pg	N/A	[53]
	Gold	16S rDNA fragment	20 CFU mL^{-1}	<3 h	[54]
Electrochemical response	Gold	105 bp of IS6110 insertion element	0.01 ng μL^{-1}	6 h	[55]
	Gold	H37rv strain	1.25 ng mL^{-1}	N/A	[56]
	Iron oxide	DNA	0.1 fM	N/A	[60]
	Iron oxide	DNA	6 ng mL^{-1}	N/A	[61]
	Iron oxide	rpoB gene	1 fM	30 min	[62]
	Zirconium dioxide	DNA	0.065 ng mL^{-1}	60 s	[64]
	Zirconium dioxide	DNA	0.00078 μM	N/A	[65]
	Selenium	DNA	8.7×10^{-15} M	50 min	[68]
Electrochemical surface-enhanced Raman spectroscopy	Silver	IS6110 insertion element	280 μg μL^{-1}	N/A	[57]
Fluorescence resonance energy	Selenium	DNA	12.5 ng	N/A	[66]
	Tellurium, Gold	ESAT-6 gene	10 fg	N/A	[67]

5.1 *Noble Metal Nanoparticles and Metal Oxide Nanoparticles-Based Diagnostics*

Noble MeNPs have gained substantial attention in molecular diagnostic applications due to their simplicity and adaptability, and have become a vital component in the development of nanotechnology-based pathogen detection [43]. Due to their ease of preparation, inert nature, favorable biocompatibility, high surface area, unique optical properties with their typical bright-red color in colloidal solutions associated with a well-defined surface plasmon resonance (SPR) band in the visible region of the spectrum, and especially their suitability for binding to biomolecules [44].

Metal oxide nanoparticles have a number of advantages that make them ideal for use in biosensor transducers. Some of them are superparamagnetic property (especially for iron), large surface-to-volume ratio, strong chemical activity, and biological compatibility [59], thermal stability, chemical inertness, biocompatibility, and affinity for oxygen-containing groups [63].

6 Metal Nanoparticles for Targeted Treatment

The absence of new drugs for the treatment of bacterial infections, as well as the resistance of M.tb to antitubercular drugs, urges the need for novel antibacterial agents. MeNPs, such as silver and zinc, have been intensively explored as a potential treatment for a variety of medical disorders [69]. MeNPs are recognized as antitubercular agents that disrupt bacterial membrane integrity. MeNPs have the potential to connect and cling to the cell wall of tuberculosis cells and then mechanically kill them [13]. In general, oxidative stress and free radicals known as reactive oxygen species are responsible for MeNP activity [84]. The interaction between the M.tb and the NPs, as well as the bio-reactive characteristics of the dissolved ionic fraction, is thought to represent the antitubercular mechanisms of MeNPs [85–88]. Antibacterial silver nanoparticles (AgNPs) have been used in a variety of medicinal and diagnostic applications, as well as optoelectronics [70] and water disinfection [71]. AgNPs have been shown to have an antibacterial effect on bacteria resistant to antibiotics in a variety of ways [72]. These include disrupting bacterial membranes and cell walls, resulting in cell leakage by increasing membrane permeability [73], initiating lipid peroxidation and lowering glutathione levels, depolarization of mitochondria, and oxidative DNA damage with apoptotic cell death [74], damaging bacterial cell DNA by binding to its sulfur and phosphorus groups [75], and releasing Ag ions, which play an important antibacterial role by interacting with bacteria [76]. Silver nanoparticles in colloidal form, as well as silver ions in suspension with silver nanoparticles in an aqueous media, have a superior antibacterial impact by acting as a catalyst, disrupting critical enzymes required for germs' oxygen metabolism [77].

Zinc nanoparticles, on the other hand, have effective antimicrobial and UV-blocking capabilities. Because of these qualities, they are commonly utilized in personal care goods including cosmetics and sunscreen [78]. Skin diseases such as lichen planus, eczema, seborrheic dermatitis, psoriasis, and increased skin dryness cause redness and irritation, which can be relieved with zinc oxide nanoparticles administered as a spray [79]. Antibacterial sprays containing a mix of silver and zinc oxide nanoparticles are currently used for immediate alleviation of conjunctivitis, skin irritation, sinusitis, and earache [80]. As indicated in Table 2, many recent researches have investigated the use of metal nanoparticles for the treatment of tuberculosis and MDR-TB. Table 2 depicts the antitubercular activity of metal/metal oxide NPs and TB drugs including metal/metal oxide NPs against several TB bacteria strains.

Importantly, when applied to tuberculosis, all metal-containing treatments, such as silver, gold, titanium, zinc oxide, and gallium, had a bactericidal effect [81]. When first-line TB drugs like Rifampin were synthesized within a metal/metal oxide NP system, a synergistic effect was described, resulting in an increase in antitubercular efficacy and a reduction in their minimal inhibitory concentration [82]. When applied against resistant strains of tuberculosis, silver and zinc oxide NPs only had a bacteriostatic impact [83]. When these metal/metal oxide NPs were cultured with eukaryotic THP-1 and Vero cells, they demonstrated little toxicity, suggesting that an enhanced therapy incorporating a cocktail of metal/metal oxide NPs along with TB drugs could increase antitubercular efficacy and reduce treatment duration.

7 Toxicity Aspect of Metal Nanoparticles

The toxicity of various MeNPs/metal oxides has been widely studied over a long period of time through many research studies. Because of their nano size, the particles are able to enter the circulatory/lymphatic systems and then the tissues and organs. Metallic nanomaterials (composed of a single metal element) are relatively stable and do not undergo dissolution easily. Metal oxide and metal alloy-based nanomaterials, on the other hand, have a lower degree of stability and are more susceptible to dissolution and ion release when exposed to a biological milieu, resulting in reactive oxygen species (ROS) production, oxidative stress to cells, inflammation, and cell signaling modulation. There are some strategies that can be implemented to minimize the toxicity of MeNPs/metal oxides. Targeted distribution of MeNPs/metal oxides can be accomplished through effective entrapment, attachment, or encapsulation of MeNPs/metal oxides into polymer matrix or capping with polymers, resulting in reduced toxicity. Surface modification can also aid to minimize the toxicity of MeNPs/metal oxides. The development of metal-organic frameworks (MOFs) is a popular and recent approach to reducing the toxicity of MeNPs/metal oxides [115].

Table 2 List of various metal nanocomposites used in the therapy of TB

Nano metal composite	Research objective	TB drug utilized	Research outcome	References
Gallium	Investigating the efficacy of gallium NPs against human immunodeficiency and tuberculosis coinfection	N/A	Rod-shaped NPs were obtained having no toxicity on cells Gallium was released from NPs within 15 days. Significant inhibition in growth of M.tb bacteria was observed with the gallium NP compared to the free metal.	[89]
Gallium	Investigating the efficacy of gallium NP on *M. tuberculosis*-infected macrophages	N/A	Gallium NPs were able to regulate these levels These NPs inhibited the growth of TB bacteria for 15 days.	[90]
Gallium	Targeting macrophages by gallium NPs	Rifampin	The morphology of gallium was approximately rectangular. Gallium NP prepared by dendrimers showed faster uptake by THP-1 macrophages. All formulations showed TB bacteria growth inhibition for 15 days	[91]
Iron	Designing therapeutic nanoparticles for lung delivery	Isoniazid	Irregularly shaped particles with an average diameter of 11 mm with a near-uniform size distribution. Higher isoniazid (INH) release was observed from the LC NPs compared with PLGA NPs within 10 h. No cytotoxicity effects were shown with the lowest concentration (10–25 mg/ml) and the highest concentration (500 mg/ml) decreased the cell viability by 52% compared with the control cells.	[82]

(continued)

Table 2 (continued)

Nano metal composite	Research objective	TB drug utilized	Research outcome	References
Iron	Designing theranostic nanoparticles encapsulating isoniazid	Isoniazid	12% of INH was loaded in the NPs with a diameter range of 3.37–6.45 μm based on the micronization method used. A sustained release profile of isoniazid (INH) seen. NPs accumulated inside the L929 fibroblasts with no signs of toxicity.	[92]
Silver	Characterizing silver NPs synthesized by a *Streptomyces* sp. NH28 strain	N/A	Spherical NPs with a mean size of 19.9 nm with a negative zeta potential of −13.8 mV. The antibacterial effect was observed against all strains and didn't show any toxic effects when exposed to L929 fibroblasts.	[94]
Silver	Assessing the efficacy of silver NPs on TB bacteria	N/A	NP with size 43.6 ± 10.7 nm. Decrease the count of TB bacteria in the spleen and lungs by 2× in mice.	[95]
Silver	Designing a green synthesis of silver NPs	N/A	Spherical nanocomposites with a size 11–17.5 nm. IC50 against normal lung cells was 357.2 μg/ml. M.tb was inhibited by an MIC of 1.95 μg/ml	[96]
Silver	Preparing silver NP loaded with antibacterial drugs	Vancomycin	Spherical NP with a size 30 ± 3 nm was prepared successfully. The internalization of the drug inside the bacteria was enhanced through formulation with NPs.	[97]
Silver	Investigating the antimycobacterial activity of silver	N/A	Spherical and tetrahedral silver NP with an average size of 59 nm were prepared. The system showed an antibacterial effect on this TB strain with an MIC of 1 μg/ml	[98]

(continued)

Table 2 (continued)

Nano metal composite	Research objective	TB drug utilized	Research outcome	References
Silver	Fabricating a nanoscale multidrug delivery system	Rifampin, Pyrazinamide	FTIR confirmed the successful synthesis of silver, rifampin (RIF), and polymer-containing pyrazinamide (PZA). PZA and RIF were in an amorphous phase with a size of 140 nm. Their full release was performed within 12 h The combination therapy showed better antitubercular effect than their single administration	[99]
Silver	Designing and characterizing biodegradable silver NPs	N/A	50% of silver was released within 2.5–5.5 h. Encapsulating silver within NPs increased the minimum inhibition order by 70%.	[100]
Silver	Green synthesis of silver NPs by yeast	N/A	Spherical NPs with a diameter of 17 nm Sacrificed the bacteria by inducing oxidative stress. These NPs showed a 95% reduction in TB bacteria with an administrated dose of 37 mg/ml	[101]
Silver	Investigating the changes in immune response to TB when silver NP is administered	N/A	Different formulations showed different sizes and zeta potentials Importantly, silver NPs reduced cellular viability. Increased IL8, and decreased IL10 mRNA expression when exposed to MDM For the TB-infected MDM, silver NPs suppressed. M.tb-induced expression of IL1β, IL10, and TNFα mRNA, and TB bacteria was inhibited by silver NPs.	[102]

(continued)

Table 2 (continued)

Nano metal composite	Research objective	TB drug utilized	Research outcome	References
Silver	Evaluating the biological risks associated with exposure to NP	N/A	After 4 h of silver NP exposure to MDM, no toxicity was observed But after 24 h, the cell viability was reduced by 60–70%. Silver NP up-regulated Hsp72 leading to suppress NF-kB induced by M.tb, thus hosting immune responses.	[103]
Titanium dioxide	Exploring the antibacterial effect of TiO2 NPs	N/A	Spherical NP with a diameter of 16 nm showed size Concentration-dependent antitubercular effects with a 3–4 time decrease in TB metabolic activity with very minimal toxicity when the maximum dose was applied.	[110]
Zinc oxide	Green synthesis of NPs	N/A	Monodisperse spherical NPs with a diameter of 12–53 nm were prepared. The NPs inhibited the growth of TB at a concentration of12.5 mg/ml.	[111]
Zinc oxide	Green synthesis of ZnO	N/A	Rod-shaped ZnO particles with an average size of 33 nm. *B. subtilis* inhibited bacteria with a MIC value of 78.12 mg/ml. TB growth was inhibited by 25–100 mg/ml of ZnO nanorods.	[112]
Zinc oxide	Green synthesis of ZnO as antitubercular agent	N/A	Spherical NPs with a diameter of 34 nm. The maximum diameter of inhibition zone was observed in at (100 μg/ml) against *M. tuberculosis* (35 ± 1.86).	[113]

(continued)

Table 2 (continued)

Nano metal composite	Research objective	TB drug utilized	Research outcome	References
Zinc oxide	Explore the synergistic effect of ZnO and RIF NPs	Rifampin	Uniformly distributed ZnO NPs with a diameter of 11 nm and zeta potential of +19.1 mV. The MIC for the ZnO NP was 256 mg/ml but 32 mg/ml of ZnO NPs were able to reduce the MIC of RIF from 64 mg/ml to 16 mg/ml which confirms a synergistic effect between both antitubercular agents	[114]
Magnesium oxide, zinc oxide	Exploring the antitubercular effect of MgO and ZnO NPs	N/A	The NPs didn't show cytotoxicity. The inhibitory effects could be associated with the ZnO NPs. These NPs sacrificed MDR and had a synergetic effect to clear resistant strains	[93]
Silver, gold	Evaluating the antitubercular activity of metal NPs	N/A	A combination of gold and silver NP had the most striking antitubercular effect with an MIC of less than 2.56 mg/ml. No antibacterial effect was observed for gold NPs with concentrations of <100 mg/ml. Metal NPs entered macrophage cells.	[104]
Silver, gold	Investigating the efficacy of phytogenic metal nanoparticles	N/A	Mixed Ag and Au spherical NPs prepared by *Syzygium cumini* with a diameter of 10–20 nm showed the most powerful formulation against TB bacteria. 45% cell viability was observed at a dose of 30 mg/ml of the mixture of Au-Ag NPs Silver NPs showed more potent antibacterial effects than gold NPs.	[104]

(continued)

Table 2 (continued)

Nano metal composite	Research objective	TB drug utilized	Research outcome	References
Silver, gold	Comparing the efficacy of silver and gold NPs	N/A	Ag and AuNPs were spherical and polyhedral in morphology, respectively. AgNPs showed better antitubercular efficacy than the gold NPs.	[105]
Silver, zinc	Synthesizing NPs using environmental bacteria	N/A	Polydisperse spherical silver and zinc NPs with a mean size of 39 and 62 nm were prepared. IC50 values for silver and zinc NPs were 5.54 and 6.24 mg/ml Authors concluded that metal NPs could enhance the antibacterial efficacy of many drugs like gentamicin.	[106]
Silver, zinc oxide	Evaluating the antitubercular effects of mixed metal oxides	N/A	Spherical NPs with a diameter of 30–80 nm MIC ratio of 8ZnO:2Ag NPs against M.tb was detected at ratio of ~1/32 of the initial concentration of Ag NPs. ZnO NPs were estimated at ~20 ppm and ~ 60 ppm. Silver did not show antitubercular effects at any of the applied doses while ZnO NPs showed a potent antibacterial activity at ~1/128 and toxic effects on the cells.	[107]
Silver, zinc oxide	Determining the effective ratio of mixed metal NPs	N/A	Spherical particles with a size of 13 nm for Ag NPs and 4 nm for ZnO NPs. 0.663 ppm of 5Ag:5ZnO showed effective antibacterial results with no toxicity to THP-1 cells. The combination of both metals together showed better results.	[108]

(continued)

Table 2 (continued)

Nano metal composite	Research objective	TB drug utilized	Research outcome	References
Silver, zinc oxide	Designing biodegradable microparticles containing mixed metal NPs to be delivered to the lungs	Rifampin	ZnO and silver NPs were formulated within PLGA microparticles with a diameter of 4 mm. Selective uptake of the MPs by M.tb-infected macrophages and zinc and silver ions were released which disrupted the M.tb cell wall. This formulation increased the potency of rifampin (RIF) by 75%.	[109]
Silver, zinc oxide	Investigating the antibacterial effect of zinc and silver	N/A	Spherical silver and ZnO NPs with a size of 5.4 ± 2.6 nm and 9.3 ± 3.9 nm were produced. The MIC of all of the formulations was 1 mg/ml. Silver and zinc NP showed a bacteriostatic effect against MDR and XDR strains of TB.	[83]

8 Conclusion and Future Prospective

TB remains a major global illness with high morbidity and mortality for which current therapeutic options are frequently ineffective; as a result, it remains a serious public health concern. New tactics are being investigated, widening the field of study with inventive novel approaches to improve TB diagnosis and treatment while lowering mortality. The number of newly developed TB drugs is currently limited. Simply doing nanotechnology research can lead to meaningful diagnostic and therapeutic advancements in the fight against tuberculosis. Metal nanoparticles have a critical role in TB diagnosis and treatment. Metal nanoparticles are less labor-intensive and far more sensitive than current diagnostic techniques in the diagnosis of tuberculosis. Several new metal/metal oxide nanoparticles, alone or in combination with other substances, can be investigated in the near future to develop a far more efficient diagnostic tool for tuberculosis. The use of metal nanoparticles in respiratory infection medicines may aid in the treatment of tuberculosis. Because of their tiny size and relative mobility, metal nanoparticles and metal ions can pass through the bilayer of an infected macrophage membrane. Metal nanoparticles are used to encapsulate single and combinations of standard antibiotics; the metal nanoparticles have an additive/synergistic effect, allowing lower doses of drug(s) to be utilized with fewer adverse effects and great efficacy. Research investigations using metal/metal oxide nanoparticles in combination with TB drug(s) showed promising results in enhancing antitubercular efficacy. Importantly, there is a need to develop trustworthy in vivo models to investigate the impact of metal nanoparticles – medication formulations in animal models, as there are relatively a few in vivo research on these drug formulations. Metal-based formulations carry a risk associated with the metal's potential toxicity, and the dose would need to be carefully monitored and modified for human usage. Metal nanoparticles have a lot of potential for more effective transport of TB drugs to the afflicted location, both alone and in combination, to boost their potency, especially when antibacterial metal nanoparticles are present. The introduction of more potent, novel, modified, and designed metal nanoparticles will boost the effectiveness of such systems and, ideally, will minimize the impact of tuberculosis on human health in near future.

References

1. World Health Organization. Global tuberculosis control: epidemiology, strategy, financing: WHO report 2009. World Health Organization; 2009. https://apps.who.int/iris/handle/10665/44035
2. World Health Organization. Global tuberculosis report 2018. World Health Organization; 2018. https://apps.who.int/iris/handle/10665/274453
3. Vieira AC, Magalhães J, Rocha S, Cardoso MS, Santos SG, Borges M, Pinheiro M, Reis S. Targeted macrophages delivery of rifampicin-loaded lipid nanoparticles to improve tuberculosis treatment. Nanomedicine. 2017;12(24):2721–36.
4. Gupta S, Kakkar V. Recent technological advancements in tuberculosis diagnostics–a review. Biosens Bioelectron. 2018;115:14–29.
5. Wang S, Inci F, De Libero G, Singhal A, Demirci U. Point-of-care assays for tuberculosis: role of nanotechnology/microfluidics. Biotechnol Adv. 2013;31(4):438–49.
6. Golichenari B, Velonia K, Nosrati R, Nezami A, Farokhi-Fard A, Abnous K, Behravan J, Tsatsakis AM. Label-free nano-biosensing on the road to tuberculosis detection. Biosens Bioelectron. 2018;113:124–35.
7. Zhou L, He X, He D, Wang K, Qin D. Biosensing technologies for mycobacterium tuberculosis detection: status and new developments. Clin Dev Immunol. 2011;2011:193963.
8. Martinkova P, Kostelnik A, Válek T, Pohanka M. Main streams in the construction of biosensors and their applications. Int J Electrochem Sci. 2017;12(8):7386–403.
9. Oni T, Youngblood E, Boulle A, McGrath N, Wilkinson RJ, Levitt NS. Patterns of HIV, TB, and non-communicable disease multi-morbidity in peri-urban South Africa-a cross sectional study. BMC Infect Dis. 2015;15(1):1–8.
10. Golub JE, Cohn S, Saraceni V, Cavalcante SC, Pacheco AG, Moulton LH, et al. Long-term protection from isoniazid preventive therapy for tuberculosis in HIVinfected patients in a medium-burden tuberculosis setting: the TB/HIV in Rio (THRio) study. Clin Infect Dis. 2015;60(4):639–45.
11. Kim JH, Nam WS, Kim SJ, Kwon OK, Seung EJ, Jo JJ, Shresha R, Lee TH, Jeon TW, Ki SH, Lee HS. Mechanism investigation of rifampicin-induced liver injury using comparative toxicoproteomics in mice. Int J Mol Sci. 2017;18(7):1417.
12. Ramappa V, Aithal GP. Hepatotoxicity related to antituberculosis drugs: mechanisms and management. J Clin Exp Hepatol. 2013;3(1):37–49.

13. Jafari AR, Mosavi T, Mosavari N, Majid A, Movahedzade F, Tebyaniyan M, et al. Mixed metal oxide nanoparticles inhibit growth of mycobacterium tuberculosis into THP-1 cells. Int J Mycobacteriol. 2016;5(Suppl 1):S181–3.
14. Pati R, Sahu R, Panda J, Sonawane A. Encapsulation of zinc-rifampicin complex into transferrin-conjugated silver quantum-dots improves its antimycobacterial activity and stability and facilitates drug delivery into macrophages. Sci Rep. 2016;6:24184.
15. Kalluru R, Fenaroli F, Westmoreland D, Ulanova L, Maleki A, Roos N, et al. Polylactide-co-glycolide-rifampicinnanoparticles efficiently clear Mycobacterium bovis BCG infection in macrophages and remain membrane-bound in phago- lysosomes. J Cell Sci. 2013;126(14):3043–54.
16. Jafari A, Mosavari N, Movahedzadeh F, Nodooshan SJ, Safarkar R, Moro R, et al. Bactericidal impact of ag, ZnO and mixed AgZnO colloidal nanoparticles on H37Rv mycobacterium tuberculosis phagocytized by THP-1 cell lines. Microb Pathog. 2017;110:335–44.
17. Jafari A, Jafari Nodooshan S, Safarkar R, Movahedzadeh F, Mosavari N, Novin Kashani A, et al. Toxicity effects of AgZnO nanoparticles and rifampicin on mycobacterium tuberculosis into the macrophage. J Basic Microbiol. 2018;58(1):41–51.
18. Bhatt K, Salgame P. Host innate immune response to mycobacterium tuberculosis. J Clin Immunol. 2007;27(4):347–62.
19. Vergne I, Chua J, Singh SB, Deretic V. Cell biology of mycobacterium tuberculosis phagosome. Annu Rev Cell Dev Biol. 2004;20:367–94.
20. Mueller P, Pieters J. Modulation of macrophage antimicrobial mechanisms by pathogenic mycobacteria. Immunobiology. 2006;211(6–8):549–56.
21. Sherman S, Rohwedder JJ, Ravikrishnan KP, Weg JG. Tuberculous enteritis and peritonitis. Report of 36 general hospital cases. Arch Intern Med. 1980;140(4):506–8.
22. Cooper AM. Cell-mediated immune responses in tuberculosis. Annu Rev Immunol. 2009;27:393–422.
23. Horvath CN, Shaler CR, Jeyanathan M, Zganiacz A, Xing Z. Mechanisms of delayed anti-tuberculosis protection in the lung of parenteral BCG-vaccinated hosts: a critical role of airway luminal T cells. Mucosal Immunol. 2012;5(4):420–31.
24. Vergne I, Chua J, Lee HH, Lucas M, Belisle J, Deretic V. Mechanism of phagolysosome biogenesis block by viable mycobacterium tuberculosis. Proc Natl Acad Sci U S A. 2005;102(11):4033–8.
25. Sturgill-Koszycki S, Schlesinger PH, Chakraborty P, Haddix PL, Collins HL, Fok AK, et al. Lack of acidification in mycobacterium phagosomes produced by exclusion of the vesicular proton-ATPase. Science. 1994;263(5147):678–81.
26. Miller BH, Fratti RA, Poschet JF, Timmins GS, Master SS, Burgos M, et al. Mycobacteria inhibit nitric oxide synthase recruitment to phagosomes during macrophage infection. Infect Immun. 2004;72(5):2872–8.
27. Vergne I, Fratti RA, Hill PJ, Chua J, Belisle J, Deretic V. Mycobacterium tuberculosis phagosome maturation arrest: mycobacterial phosphatidylinositol analog phosphatidylinositol mannoside stimulates early endosomal fusion. Mol Biol Cell. 2004;15(2):751–60.
28. Jamwal SV, Mehrotra P, Singh A, Siddiqui Z, Basu A, Rao KV. Mycobacterial escape from macrophage phagosomes to the cytoplasm represents an alternate adaptation mechanism. Sci Rep. 2016;6:23089.
29. Stamm LM, Morisaki JH, Gao LY, Jeng RL, McDonald KL, Roth R, et al. Mycobacterium marinum escapes from phagosomes and is propelled by actin-based motility. J Exp Med. 2003;198(9):1361–8.
30. Stamm LM, Pak MA, Morisaki JH, Snapper SB, Rottner K, Lommel S, et al. Role of the WASP family proteins for Mycobacterium marinum actin tail formation. Proc Natl Acad Sci U S A. 2005;102(41):14837–42.
31. Sreejit G, Ahmed A, Parveen N, Jha V, Valluri VL, Ghosh S, et al. The ESAT-6 protein of mycobacterium tuberculosis interacts with beta-2-microglobulin (β2M) affecting antigen presentation function of macrophage. PLoS Pathog. 2014;10(10):e1004446.

32. van der Wel N, Hava D, Houben D, Fluitsma D, van Zon M, Pierson J, et al. M. Tuberculosis and M. leprae translocate from the phagolysosome to the cytosol in myeloid cells. Cell. 2007;129(7):1287–98.
33. Zuber B, Chami M, Houssin C, Dubochet J, Griffiths G, Daffé M. Direct visualization of the outer membrane of mycobacteria and corynebacteria in their native state. J Bacteriol. 2008;190(16):5672–80.
34. de Chastellier C. The many niches and strategies used by pathogenic mycobacteria for survival within host macrophages. Immunobiology. 2009;214(7):526–42.
35. Sibley DL, Charron AJ, Hakansson S, Mordue DG. Invasion and intracellular survival by toxoplasma. Immunol Rev. 2013;240(1):72–91.
36. Yavvari PS, Gupta S, Arora D, Nandicoori VK, Srivastava A, Bajaj A. Clathrin-independent killing of intracellular mycobacteria and biofilm disruptions using synthetic antimicrobial polymers. Biomacromolecules. 2017;18(7):2024–33.
37. Ríos-Barrera VA, Campos-Peña V, Aguilar-León D, Lascurain LR, Meraz-Ríos MA, Moreno J, et al. Macrophage and T lymphocyte apoptosis during experimental pulmonary tuberculosis: their relationship to mycobacterial virulence. Eur J Immunol. 2006;36(2):345–53.
38. Aderem A, Underhill DM. Mechanisms of phagocytosis in macrophages. Annu Rev Immunol. 1999;17(1):593–623.
39. Hamilton RF, Buckingham S, Holian A. The effect of size on ag nanosphere toxicity in macrophage cell models and lung epithelial cell lines is dependent on particle dissolution. Int J Mol Sci. 2014;15(4):6815–30.
40. Kruth HS. Receptor-independent fluid-phase pinocytosis mechanisms for induction of foam cell formation with native LDLparticles. Curr Opin Lipidol. 2011;22(5):386–93.
41. Jafari A, Nagheli A, Foumani AA, Soltani B, Goswami R. The role of metallic nanoparticles in inhibition of mycobacterium tuberculosis and enhances phagosome maturation into the infected macrophage. Oman Med J. 2020;35(6):e194.
42. Holzinger M, Le Goff A, Cosnier S. Nanomaterials for biosensing applications: a review. Front Chem. 2014;2:63.
43. Doria G, Conde J, Veigas B, Giestas L, Almeida C, Assunção M, et al. Noble metal nanoparticles for biosensing applications. Sensors. 2012;12(2):1657–87.
44. Daniel MC, Astruc D. Gold nanoparticles: assembly, supramolecular chemistry, quantum-size-related properties, and applications toward biology, catalysis, and nanotechnology. Chem Rev. 2004;104(1):293–346.
45. Baptista PV, Koziol-Montewka M, Paluch-Oles J, Doria G, Franco R. Gold-nanoparticle-probe-based assay for rapid and direct detection of *mycobacterium tuberculosis* DNA in clinical samples. Clin Chem. 2006;52(7):1433–4.
46. Silva LB, Veigas B, Doria G, Costa P, Inácio J, Martins R, et al. Portable optoelectronic biosensing platform for identification of mycobacteria from the *mycobacterium tuberculosis* complex. Biosens Bioelectron. 2011;26:2012–7.
47. Kaewphinit T, Santiwatanakul S, Chansiri K. Colorimetric DNA based biosensor combined with loopmediated isothermal amplification for detection of *mycobacterium tuberculosis* by using gold nanoprobe aggregation. Sens Transducer. 2013;149(2):123–8.
48. Bernacka-Wojcik I, Lopes P, Vaz AC, Veigas B, Wojcik PJ, Simões P, et al. Bio-microfluidic platform for gold nanoprobe based DNA detection—application to *mycobacterium tuberculosis*. Biosens Bioelectron. 2013;48:87–93.
49. Veigas B, Pedrosa P, Carlos FF, Mancio-Silva L, Grosso AR, Fortunato E, et al. One nanoprobe, two pathogens: gold nanoprobes multiplexing for point-of-care. J Nanobiotechnol. 2015;13:48.
50. Tsai T-T, Huang C-Y, Chen C-A, Shen S-W, Wang M-C, Cheng C-M, et al. Diagnosis of tuberculosis using colorimetric gold nanoparticles on a paper-based analytical device. ACS Sens. 2017;2:1345–54.
51. Xiang Y, Zhu X, Huang Q, Zheng J, Fu W. Realtime monitoring of mycobacterium genomic DNA with target-primed rolling circle amplification by a Au nanoparticle-embedded SPR biosensor. Biosens Bioelectron. 2015;66:512–9.

52. Matsishin M, Rachkov A, Lopatynskyi A, Chegel V, Soldatkin A, El'skaya A. Selective amplification of SPR biosensor signal for recognition of rpoB gene fragments by use of gold nanoparticles modified by thiolated DNA. Nanoscale Res Lett. 2017;12:252.
53. Kaewphinit T, Santiwatanakul S, Chansiri K. Gold nanoparticle amplification combined with quartz crystal microbalance DNA based biosensor for detection of *mycobacterium tuberculosis*. Sens Transducer J. 2012;146(11):156–63.
54. Zhang J, Huang J, He F. The construction of *mycobacterium tuberculosis* 16S rDNA MSPQC sensor based on exonuclease III-assisted cyclic signal amplification. Biosens Bioelectron. 2019;138:111322.
55. Torres-Chavolla E, Alocilja EC. Nanoparticle based DNA biosensor for tuberculosis detection using thermophilic helicase-dependent isothermal amplification. Biosens Bioelectron. 2011;26:4614–8.
56. Thiruppathiraja C, Kamatchiammal S, Adaikkappan P, Santhosh DJ, Alagar M. Specific detection of *mycobacterium* sp. genomic DNA using dual labeled gold nanoparticle based electrochemical biosensor. Anal Biochem. 2011;417:73–9.
57. Karaballi RA, Nel A, Krishnan S, Blackburn J, Brosseau CL. Development of an electrochemical surface enhanced Raman spectroscopy (EC-SERS) aptasensor for direct detection of DNA hybridization. Phys Chem Chem Phys. 2015;17:21356–63.
58. Teengam P, Siangproh W, Tuantranont A, Vilaivan T, Chailapakul O, Henry CS. Multiplex paper-based colorimetric DNA sensor using pyrrolidinyl peptide nucleic acid-induced AgNPs aggregation for detecting MERS-CoV, MTB and HPV oligonucleotide. Anal Chem. 2017;89(10):5428–35.
59. Hasanzadeh M, Shadjou N, de la Guardia M. Iron and iron-oxide magnetic nanoparticles as signal-amplification elements in electrochemical biosensing. TrAC Trends Anal Chem. 2015;72:1–9.
60. Prabhakar N, Solanki PR, Kaushik A, Pandey MK, Malhotra BD. Peptide nucleic acid immobilized biocompatible silane nanocomposite platform for mycobacterium tuberculosis detection. Electroanalysis. 2010;22(22):2672–82.
61. Costa MP, Andrade CAS, Montenegro RA, Melo FL, Oliveira MDL. Self-assembled monolayers of mercaptobenzoic acid and magnetite nanoparticles as an efficient support for development of tuberculosis genosensor. J Coll Interf Sci. 2014;433:141–8.
62. Haddaoui M, Sola C, Raouafi N, Korri-Youssoufi H. E-DNA detection of rpoB gene resistance in *mycobacterium tuberculosis* in real samples using Fe3O4/polypyrrole nanocomposite. Biosens Bioelectron. 2019;128:76–82.
63. Bellezza F, Cipiciani A, Quotadamo MA. Immobilization of myoglobin on phosphate and phosphonate grafted-zirconia nanoparticles. Langmuir. 2005;21(24):11099–104.
64. Das M, Sumana G, Nagarajan R, Malhotra BD. Zirconia based nucleic acid sensor for *mycobacterium tuberculosis* detection. Appl Phys Lett. 2010;96:133703.
65. Das M, Dhand C, Sumana G, Srivastava AK, Nagarajan R, Nain L, et al. Electrophoretic fabrication of chitosan-zirconium-oxide nanobiocomposite platform for nucleic acid detection. Biomacromolecules. 2011;12:540–7.
66. Gazouli M, Liandris E, Andreadou M, Sechi LA, Masala S, Paccagnini D, et al. Specific detection of unamplified mycobacterial DNA by use of fluorescent semiconductor quantum dots and magnetic beads. J Clin Microbiol. 2010;48(8):2830–5.
67. Shojaei TR, Salleh MAM, Tabatabaei M, Ekrami A, Motallebi R, Rahmani-Cherati T, et al. Development of sandwich-form biosensor to detect *mycobacterium tuberculosis* complex in clinical sputum specimens. Braz J Infect Dis. 2014;18(6):600–8.
68. Zhang C, Lou J, Tu W, Bao J, Dai Z. Ultrasensitive electrochemical biosensing for DNA using quantum dots combined with restriction endonuclease. Analyst. 2015;140:506–11.
69. Anderson SD, Gwenin VV, Gwenin CD. Magnetic functionalized nanoparticles for biomedical, drug delivery and imaging applications. Nanoscale Res Lett. 2019;14(1):1–16.
70. Chen D, Qiao X, Qiu X, Chen J. Synthesis and electrical properties of uniform silver nanoparticles for electronic applications. J Mater Sci. 2009;44(4):1076–81.

71. Syafiuddin A, Salmiati S, Hadibarata T, Kueh AB, Salim MR, Zaini MA. Silver nanoparticles in the water environment in Malaysia: inspection, characterization, removal, modeling, and future perspective. Sci Rep. 2018;8:986.
72. Durán N, Durán M, De Jesus MB, Seabra AB, Fávaro WJ, Nakazato G. Silver nanoparticles: a new view on mechanistic aspects on antimicrobial activity. Nanomedicine. 2016;12(3):789–99.
73. Shrivastava S, Bera T, Roy A, Singh G, Ramachandrarao P, Dash D. Characterization of enhanced antibacterial effects of novel silver nanoparticles. Nanotechnology. 2007;18(22):225103.
74. Dasgupta N, Ranjan S, Mishra D, Ramalingam C. Thermal co-reduction engineered silver nanoparticles induce oxidative cell damage in human colon cancer cells through inhibition of reduced glutathione and induction of mitochondria-involved apoptosis. Chem Biol Interact. 2018;295:109–18.
75. Durán N, Marcato PD, Conti RD, Alves OL, Costa F, Brocchi M. Potential use of silver nanoparticles on pathogenic bacteria, their toxicity and possible mechanisms of action. J Braz Chem Soc. 2010;21(6):949–59.
76. Zhong X, Song Y, Yang P, Wang Y, Jiang S, Zhang X, Li C. Titanium surface priming with phase-transited lysozyme to establish a silver nanoparticle-loaded chitosan/hyaluronic acid antibacterial multilayer via layer-by-layer self-assembly. PLoS One. 2016;11(1):e0146957.
77. Suganya KU, Govindaraju K, Kumar VG, Dhas TS, Karthick V, Singaravelu G, Elanchezhiyan M. Size controlled biogenic silver nanoparticles as antibacterial agent against isolates from HIV infected patients. Spectrochim Acta A Mol Biomol Spectrosc. 2015;144:266–72.
78. Jiang J, Pi J, Cai J. The advancing of zinc oxide nanoparticles for biomedical applications. Bioinorg Chem Appl. 2018;2018:1062562.
79. Gupta M, Mahajan VK, Mehta KS, Chauhan PS. Zinc therapy in dermatology: a review. Dermatol Res Pract. 2014;2014:709152.
80. Salem W, Leitner DR, Zingl FG, Schratter G, Prassl R, Goessler W, Reidl J, Schild S. Antibacterial activity of silver and zinc nanoparticles against vibrio cholerae and enterotoxic Escherichia coli. Int J Med Microbiol. 2015;305(1):85–95.
81. Simões MF, Ottoni CA, Antunes A. Mycogenic metal nanoparticles for the treatment of mycobacterioses. Antibiotics. 2020;9(9):569.
82. Wyszogrodzka-Gaweł G, Dorożyński P, Giovagnoli S, Strzempek W, Pesta E, Węglarz WP, Gil B, Menaszek E, Kulinowski P. An inhalable theranostic system for local tuberculosis treatment containing an isoniazid loaded metal organic framework Fe-MIL-101-NH2—from raw MOF to drug delivery system. Pharmaceutics. 2019;11(12):687.
83. Heidary M, Bostanabad SZ, Amini SM, Jafari A, Nobar MG, Ghodousi A, Kamalzadeh M, Darban-Sarokhalil D. The anti-mycobacterial activity of ag, ZnO, and ag-ZnO nanoparticles against MDR-and XDR-mycobacterium tuberculosis. Infect Drug Res. 2019;12:3425–35.
84. Nel AE, Mädler L, Velegol D, Xia T, Hoek EM, Somasundaran P, et al. Understanding biophysicochemical interactions at the nano-bio interface. Nat Mater. 2009;8(7):543–57.
85. Ivask A, Elbadawy A, Kaweeteerawat C, Boren D, Fischer H, Ji Z, et al. Toxicity mechanisms in Escherichia coli vary for silver nanoparticles and differ from ionic silver. ACS Nano. 2014;8(1):374–86.
86. Bondarenko O, Ivask A, Käkinen A, Kurvet I, Kahru A. Particle-cell contact enhances antibacterial activity of silver nanoparticles. PLoS One. 2013;8(5):e64060.
87. Morones JR, Elechiguerra JL, Camacho A, Holt K, Kouri JB, Ramírez JT, et al. The bactericidal effect of silver nanoparticles. Nanotechnology. 2005;16(10):2346–53.
88. Martínez-Gutierrez F, Thi EP, Silverman JM, de Oliveira CC, Svensson SL, Vanden Hoek A, et al. Antibacterial activity, inflammatory response, coagulation and cytotoxicity effects of silver nanoparticles. Nanomedicine. 2012;8(3):328–36.
89. Choi SR, Britigan BE, Narayanasamy P. Ga (III) nanoparticles inhibit growth of both mycobacterium tuberculosis and HIV and release of interleukin-6 (IL-6) and IL-8 in coinfected macrophages. Antimicrob Agents Chemother. 2017;61(4):e02505–16.

90. Choi SR, Britigan BE, Narayanasamy P. Treatment of virulent mycobacterium tuberculosis and HIV coinfected macrophages with gallium nanoparticles inhibits pathogen growth and modulates macrophage cytokine production. Msphere. 2019;4(4):e00443–19.
91. Choi SR, Britigan BE, Moran DM, Narayanasamy P. Gallium nanoparticles facilitate phagosome maturation and inhibit growth of virulent mycobacterium tuberculosis in macrophages. PLoS One. 2017;12(5):e0177987.
92. Wyszogrodzka G, Dorożyński P, Gil B, Roth WJ, Strzempek M, Marszałek B, Węglarz WP, Menaszek E, Strzempek W, Kulinowski P. Iron-based metal-organic frameworks as a theranostic carrier for local tuberculosis therapy. Pharm Res. 2018;35:144.
93. Jafari A. Evaluation of magnesium oxide and zinc oxide nanoparticles against multi-drug-resistance mycobacterium tuberculosis. Ind J Tuberc. 2021;68(2):195–200.
94. Składanowski M, Golinska P, Rudnicka K, Dahm H, Rai M. Evaluation of cytotoxicity, immune compatibility and antibacterial activity of biogenic silver nanoparticles. Med Microbiol Immunol. 2016;205(6):603–13.
95. Kalmantaeva OV, Firstova VV, Grishchenko NS, Rudnitskaya TI, Potapov VD, Ignatov SG. Antibacterial and immunomodulating activity of silver nanoparticles on mice experimental tuberculosis model. Appl Biochem Microbiol. 2020;56(2):226–32.
96. Abdel-Aziz MM, Elella MH, Mohamed RR. Green synthesis of quaternized chitosan/silver nanocomposites for targeting mycobacterium tuberculosis and lung carcinoma cells (A-549). Int J Biol Macromol. 2020;142:244–53.
97. Sun F, Oh S, Kim J, Kato T, Kim HJ, Lee J, Park EY. Enhanced internalization of macromolecular drugs into mycobacterium smegmatis with the assistance of silver nanoparticles. J Microbiol Biotechnol. 2017;27(8):1483–90.
98. Selim A, Elhaig MM, Taha SA, Nasr EA. Antibacterial activity of silver nanoparticles against field and reference strains of mycobacterium tuberculosis, Mycobacterium bovis and multiple-drug-resistant tuberculosis strains. Rev Off Int Epizoot. 2018;37:823–30.
99. Praphakar RA, Jeyaraj M, Ahmed M, Kumar SS, Rajan M. Silver nanoparticle functionalized CS-g-(CA-MA-PZA) carrier for sustainable anti-tuberculosis drug delivery. Int J Biol Macromol. 2018;118:1627–38.
100. Lim YH, Tiemann KM, Heo GS, Wagers PO, Rezenom YH, Zhang S, Zhang F, Youngs WJ, Hunstad DA, Wooley KL. Preparation and in vitro antimicrobial activity of silver-bearing degradable polymeric nanoparticles of polyphosphoester-block-poly (L-lactide). ACS Nano. 2015;9(2):1995–2008.
101. Sivaraj A, Kumar V, Sunder R, Parthasarathy K, Kasivelu G. Commercial yeast extracts mediated green synthesis of silver chloride nanoparticles and their anti-mycobacterial activity. J Clust Sci. 2020;31(1):287–91.
102. Sarkar S, Leo BF, Carranza C, Chen S, Rivas-Santiago C, Porter AE, Ryan MP, Gow A, Chung KF, Tetley TD, Zhang J. Modulation of human macrophage responses to mycobacterium tuberculosis by silver nanoparticles of different size and surface modification. PLoS One. 2015;10(11):e0143077.
103. Sarkar S, Carranza C, Theodorou I, Fen LB, Ellis T, Porter A, Ryan M, Zhang J, Tetley T, Schwander S. Impact of silver and carbon nanoparticle exposures on macrophage responses to mycobacterium tuberculosis (M. tb). J. Immunol. 2016;196(1 Supplement):200.21.
104. Singh R, Nawale L, Arkile M, Wadhwani S, Shedbalkar U, Chopade S, Sarkar D, Chopade BA. Phytogenic silver, gold, and bimetallic nanoparticles as novel antitubercular agents. Int J Nanomedicine. 2016;11:1889–97.
105. Singh R, Nawale LU, Arkile M, Shedbalkar UU, Wadhwani SA, Sarkar D, Chopade BA. Chemical and biological metal nanoparticles as antimycobacterial agents: a comparative study. Int J Antimicrob Agents. 2015;46(2):183–8.
106. Punjabi K, Mehta S, Chavan R, Chitalia V, Deogharkar D, Deshpande S. Efficiency of biosynthesized silver and zinc nanoparticles against multi-drug resistant pathogens. Front Microbiol. 2018;9:2207.

107. Jafari AR, Mosavi T, Mosavari N, Majid A, Movahedzade F, Tebyaniyan M, Kamalzadeh M, Dehgan M, Jafari S, Arastoo S. Mixed metal oxide nanoparticles inhibit growth of mycobacterium tuberculosis into THP-1 cells. Int J Mycobacteriol. 2016;5:S181–3.
108. Jafari A, Mosavari N, Movahedzadeh F, Nodooshan SJ, Safarkar R, Moro R, Kamalzadeh M, Majidpour A, Boustanshenas M, Mosavi T. Bactericidal impact of ag, ZnO and mixed AgZnO colloidal nanoparticles on H37Rv mycobacterium tuberculosis phagocytized by THP-1 cell lines. Microb Pathog. 2017;110:335–44.
109. Ellis T, Chiappi M, García-Trenco A, Al-Ejji M, Sarkar S, Georgiou TK, Shaffer MS, Tetley TD, Schwander S, Ryan MP, Porter AE. Multimetallic microparticles increase the potency of rifampicin against intracellular mycobacterium tuberculosis. ACS Nano. 2018;12(6):5228–40.
110. Ramalingam V, Sundaramahalingam S, Rajaram R. Size-dependent antimycobacterial activity of titanium oxide nanoparticles against mycobacterium tuberculosis. J Mater Chem B. 2019;7(27):4338–46.
111. Patil BN, Taranath TC. Limonia acidissima L. leaf mediated synthesis of zinc oxide nanoparticles: a potent tool against mycobacterium tuberculosis. Int J Mycobacteriol. 2016;5(2):197–204.
112. Mahendra C, Chandra MN, Murali M, Abhilash MR, Singh SB, Satish S, Sudarshana MS. Phyto-fabricated ZnO nanoparticles from Canthium dicoccum (L.) for antimicrobial, anti-tuberculosis and antioxidant activity. Process Biochem. 2020;89:220–6.
113. Vijayakumar S, Nilavukkarasi M, Sakthivel B. Bio-synthesized zinc oxide nanoparticles for anti-tuberculosis agent: scientifically unexplored. Gene Rep. 2020;20:100764.
114. Mistry N, Bandyopadhyaya R, Mehra S. Zno nanoparticles and rifampicin synergistically damage the membrane of mycobacteria. ACS Appl Nano Mater. 2020;3(4):3174–84.
115. Manuja A, Kumar B, Kumar R, Chhabra D, Ghosh M, Manuja M, Brar B, Pal Y, Tripathi BN, Prasad M. Metal/metal oxide nanoparticles: toxicity concerns associated with their physical state and remediation for biomedical applications. Toxicol Rep. 2021;8:1970–8.

Tuberculosis: Current Treatment Options and Future Scope

Yesha B. Shah, Priya S. Mistry, Tejas M. Dhameliya, Ketan M. Ranch, Sai H. S. Boddu, Shery Jacob, B. Mahalakshmi, and Jwala Renukuntla

Abstract Tuberculosis (TB) is a communicable disease that mainly affects the lungs. TB is the major cause of ill health and the leading cause of death from a single infectious agent. The emergence of multidrug-resistant tuberculosis (MDR-TB) has led to the failure of first-line antituberculosis therapy. An appropriate combination of anti-TB drugs or substitution with second-line agents are required for improving the treatment success rates of MDR and extensively drug-resistant (XDR) TB. Only a few drugs such as bedaquiline, delamanid, and pretomanid have been approved for treating MDR-TB since the last four decades. There is a dire need for the development of more effective TB drugs, adjunct therapies, and vaccines in order to improve the treatment outcomes. In this chapter, we made efforts to provide an overview of the current treatment options and challenges of TB therapy. In addition, we discussed the latest treatment strategies, new chemical entities, herbal drugs, new drug regimens, and vaccines being developed to treat both drug-susceptible and drug-resistant TB disease.

Keywords Tuberculosis · Treatment · Drug delivery · Formulations · Anti-TB drugs

Y. B. Shah · P. S. Mistry · T. M. Dhameliya (✉) · K. M. Ranch
L. M. College of Pharmacy, Ahmedabad, Gujarat, India
e-mail: tejas.dhameliya@lmcp.ac.in; tmdhameliya@lmcp.ac.in

S. H. S. Boddu (✉)
College of Pharmacy and Health Sciences, Ajman University, Ajman, UAE
Center of Medical and Bio-allied Health Sciences Research, Ajman University, Ajman, UAE
e-mail: s.boddu@ajman.ac.ae

S. Jacob
Department of Pharmaceutical Sciences, College of Pharmacy, Gulf Medical University, Ajman, United Arab Emirates

B. Mahalakshmi
Dept. of Microbiology, S.V. Medical College, Tirupati, AP, India

J. Renukuntla
Department of Basic Pharmaceutical Sciences, Fred Wilson School of Pharmacy, High Point University, High Point, NC, USA

R. Shegokar, Y. Pathak (eds.), *Tubercular Drug Delivery Systems*,
https://doi.org/10.1007/978-3-031-14100-3_4

Abbreviations

ATP	Adenine triphosphate
BCG	Bacillus Calmette–Guérin
CNS	Centeral nervous system
CSF	Cerebro spinal fluid
DDS	Drug delivery system
DNA	Deoxy ribonucleic acid
DOTS	Directly observed therapy shourtcourse
DPI	Dry powder injection
GI	Gastro intestinal
HIV	Human immunodeficiency virus
IG	Immunoglobulines
IL-12	Interleukin 12
INFγ	Interferon gamma
LBG	Locust bean gum
LTBI	Latent tuberculosis infection
MDR-TB	Multidrug-resistant tuberculosis
PDE	Phosphodiesterase
pMDI	Pressurized meter dose inhaller
PPARγ	Paroxisome proliferator activated response
RNA	Ribonucleic acid
TB	Tuberculosis
TBM	Tubercular meningitis
TNFα	Tumour necrosis factor alpha
WHO	Worlds Health Organization
XDR-TB	Extensive drug-resistant tuberculosis

1 Introduction

Tuberculosis (TB) continues to pose a threat to humans due to the lack of effective treatment options and increasing occurrence of multidrug-resistant strains of *Mycobacterium tuberculosis* (*Mtb*). TB primarily affects the lungs and later spreads to a number of human organs, leading to fatal symptoms like fatigue, fever, persistent cough, weakness, night sweats, gastrointestinal symptoms, and weight loss. TB is a curable disease; however, the delay in the treatment could result in a devastating impact on the patient's health. *Mtb* uses a variety of ways for survival in the host lesions and later escapes immunosurveillance of patients. Several factors such as poverty, malnutrition, over population growth, poor quality of detection, health status, old age, and medical conditions result in rapid disease progression. As per the World Health Organization (WHO) report, nearly ten million people were diagnosed with TB in 2020, wherein 1.5 million succumbed to death including 208,000 deaths among human immunodeficiency virus (HIV)-positive populations [1]. One of the primary reasons for such high mortality is that TB gradually develops resistance to drugs, leading to multidrug-resistant tuberculosis (MDR-TB).

In MDR-TB, the bacteria are resistant to at least two first-line anti-TB drugs such as isoniazid and rifampicin. MDR-TB emergence could be attributed to poor patient compliance resulting from longer treatment duration, potential adverse effects, and poor pharmacokinetics of the directly observed therapy short-course (DOTS). Extensively drug-resistant TB (XDR-TB) is an uncommon type of MDR-TB, which are resistant to at least isoniazid and rifampicin along with a second-line anti-TB drug. The emergence of XDR-TB has complicated treatment process by reducing the effectiveness of current drugs. There is a strong desire to identify new treatment strategies and adjunct therapies such as immunoregulatory/immunosuppressive therapy and supplementary cytokines for XDR- and MDR-TB treatment [2]. This chapter aims to summarize the current treatment regimens and challenges in the management of TB. Further a narrative is included on the latest advances in improving the treatment outcomes of TB.

2 Current Treatment and Drug Regimen

Anti-TB treatment guidelines consist of first-line anti-TB agents, which are reserved for patients having drug susceptible TB, and second-line anti-TB agents are used to treat TB with resistance to first line drugs. Currently marketed first-line anti-TB drugs such as isoniazid, rifampicin, pyrazinamide, and ethambutol are given as two-month regimen followed by four-month treatment with isoniazid and rifampicin. The treatment regimen for TB during intensive and continuation phase along with the dosage time [3] and therapeutics for children [4] have been presented in Tables 1 and 2, respectively.

Streptomycin may be used as a first-line anti-TB drug during this two-month regimen. Other injectables such as kanamycin, amikacin, capreomycin or viomycin/tuberactinomycin B, fluoroquinolones and bacteriostatic drugs like para-amino

Table 1 Treatment regimens for tuberculosis as per CDC [3]

PHASE	DRUG	DOSAGE
Intensive phase	INH, RIF, PZA, EMB	7 days/week for 8 weeks or 5 days/week for 8 weeks.
	INH, RIF, PZA, EMB	3 times weekly for 8 weeks
	INH, RIF, PZA, EMB	7 days/week for 14 doses then twice weekly for 12 doses
Continuation phase	INH, RIF	7 days/week for 18 weeks or 5 days/week for 18 weeks
	INH, RIF	3 times weekly for 18 weeks
	INH, RIF	Twice weekly for 18 weeks

INH isoniazid, *RIF* rifampin, *PZA* pyrazinamide, *EMB* ethambutol

Table 2 Anti-tubercular treatment regimen in children [4]

Drugs	Daily dose	Adverse reaction
INH	10–15 mg/kg	Mild hepatic enzyme elevation, hepatitis, peripheral neuritis, hypersensitivity
RIF	10–20 mg/kg	Orange discoloration of secretions, vomiting, hepatitis, influenza like symptoms, thrombocytopenia, pruritis
PZA	30–40 mg/kg	Hepatotoxic effects, hyperuricemia, arthralgias, GI upset
EMB	15–25 mg/kg	Optic neuritis, decreased red-green decertation, GI disturbances, hypersensitivity
Streptomycin	12–18 mg/kg	Irreversible auditory nerve damage

INH isoniazid, *RIF* rifampin, *PZA* pyrazinamide, *EMB* ethambutol

salicylic acid, ethionamide, prothionamide, and D-cycloserine are used as adjuvant therapeutic second-line anti-TB drugs in MDR-TB. Moxifloxacin as a part of daily regimen has been used to reduce the duration of treatment of drug-susceptible TB. However, its use has been found to be inferior in comparison with the standard drug regimen. Fluoroquinolones have been used in pediatric patients for the treatment of MDR-TB, but they have shown lower serum concentrations due to faster elimination in children of age less than 5 years [5]. Some other repurposed drugs such as clofazimine, amoxicillin-clavulanate, and clarithromycin have also shown in vitro anti-mycobacterial activity [6].

The treatment of latent TB infection (LTBI, having neither symptomatic nor contagious effects) needs to be monitored in those patients with high potential for development of active TB. Till date, isoniazid has been used in the treatment of LTBI for 9 months and 6 months in immunocompetent patients. Alternatively, rifampicin for duration of 4 months with zero resistance to the drug has also been used. Further, a combination of rifampicin with pyrazinamide or isoniazid has shown promising efficacy while reducing the hepatic side effects [7]. The duration of TB treatment is generally long due to sustainability of mycobacteria in the host cells for months, even after anti-tuberculotic therapy with drugs. This can be attributed to the bacterial persistence during their slow-replicating phase having a scope of induction by the host immune system [8]. Hence, researchers have embarked to develop the immunomodulators for treatment of TB [9, 10].

Anti-TB drugs exhibit their anti-tubercular action through various mechanisms such as (a) alteration of proteinous or cellular targets, (b) interference in biosynthesis of mycolic acid for the mycobacterial cell wall (e.g., isoniazid, pyrazinamide, ethionamide, prothionamide, and thioacetazone), (c) interference in the synthesis of arabinogalactam and peptidoglycan (e.g., ethambutol, D-cycloserine, amoxicillin, and clofazimine), (d) obstruction in the synthesis of key proteins and thus increasing cell membrane permeability (e.g., streptomycin, kanamycin, amikacin, capreomycin, viomycin, clarithromycin, and linezolid), (e) inhibition of DNA-dependent RNA polymerase in bacterial cells (e.g., rifampicin, rifapentine, and fluoroquinolones), and (f) antagonization of siderophore production in *M. tuberculosis* (e.g., *p*-amino salicylic acid). Recently approved novel anti-TB drugs and new chemical entities in the pipeline work through the inhibition of proton pumps $F_0F_1H^+$ ATPase (bedaquiline), *Mtb* cytochrome P450 monooxygenases, FtsZ-targeting compounds, inhibitors of branched-chain amino acid biosynthesis, nucleoside monophosphate kinase inhibitors, pyrimidine or purine nucleoside analogues, and signaling kinase inhibitors [11]. Apart from repurposing of existing antibiotics against TB, a logical strategy of enhancing immune response in TB patients is to develop the concept of host-directed therapies. Some of the nutrients, including vitamin A (alveolar regeneration), vitamin D (phototherapy to boost host protective activities), vitamin A, B, C, and E (anti-oxidants) have shown promising results as supportive agents to cure TB [12].

In 1990, the concept of DOTS therapy was introduced by WHO, where antibiotics are administered under the direct supervision of healthcare professionals to maintain strict drug adherence resulting in higher than 80% of cure rates [13]. Novel drugs with in vitro inhibition of bacilli such as bedaquiline (belongs to diarylquinoline class) [14], pretomanid (belongs to nitorimidazo-oxazine class) [15], and delamanid (belongs to nitrodihydro imidazooxazole class) have been reported (Fig. 1) [16]. The approval of newer drug classes with novel mechanisms have opened up the avenue of innovative and aggressive therapy against MDR- and XDR-TB. Among the newly emerging drug targets, serine production pathway consisting of three enzymes serA1, serC, serB2 has been recognized as the key pathway for the growth of *Mtb*. Among these enzymes, serB2 was reported to be involved in virulence mechanism of the mycobacteria [17].

The drugs used for treatment of TB have undoubtedly cured a huge number of TB cases, but these drugs still carry the certain drawbacks on long treatment such as development of drug resistant strains and treatment-related side effects. Attempts are being made to deliver the drugs by alternate routes such as inhalation route. Experimental studies in animals have shown that rifapentine could reduce the treatment duration for latent as well as active TB infections, but this does not align with the clinical trials. A few reasons for failure of orally administered rifapentine are

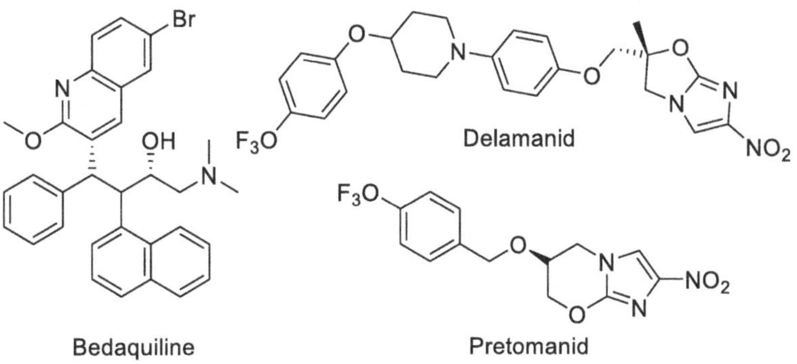

Fig. 1 Recently approved anti-tubercular drugs

low oral bioavailability, first-pass metabolism, and expression of p-glycoprotein in the gut which limits rifapentine absorption and high plasma protein binding. Thus, novel route like inhalation is considered to improve its efficacy which can shorten the duration of the TB treatment [18].

3 Challenges in Treatment of Tuberculosis

The major challenges with the treatment of TB include insufficient diagnosis or lack of effective medicines, existence of MDR or XDR-TB, drug-drug interactions between anti-TB drugs and anti-retroviral drugs in HIV co-infected tuberculotic patients, need of child-friendly and efficacious treatment for pediatric patients, and poor patient adherence because of the long course of the treatment which leads to relapse or resistant TB. An output-based novel, short-term treatment regimens have been identified by Lee et al. involving (i) ethambutol, clofazimine, and prothionamide, and secondly (ii) clofazimine, ethambutol, and bedaquiline given for 4 weeks rather than six-month regimen of first-line drugs like isoniazid, rifampicin, pyrazinamide, and ethambutol. These newly reported treatment regimens were found more efficacious in reducing time for the treatment by 75% in comparison with standard anti-TB drug regimen through a possibility of relapse-free treatment [19].

The major hurdle for anti-TB treatment is poor cellular permeability of drugs and their inferior concentrations at the site of action. Novel micro and nanometric drug delivery systems (DDS) have been reported to offer benefits including uniform distribution of drugs, increased bioavailability, sustained release, fewer adverse effects, superior patient compliance, and targeted delivery of anti-TB medicines. The emergence of multidrug-resistant bacterial strains with nanoparticulate systems does not arise as it acts via direct contact with the bacterial cell wall rather than penetration through the cell wall. This indicates that nanocarrier systems would be less prone to promoting resistance in bacteria than antibiotics. Different types of colloidal antimycobacterial drug delivery systems (DDS) include vesicular DDS, particulate DDS, supramolecular DDS, specialized DDS, and complex conjugate DDS (Fig. 2). The use of nanotechnology in the design and development of formulations provided a cost-effective and low dosage outcome for the treatment of TB [20]. An approach to improve bioavailability is microencapsulation of drug in liposomes or microspheres for sustained release and longer duration of effect, where drugs are entrapped in the biodegradable polymer [21]. Delivery of anti-TB drugs to alveolar macrophages via inhalation therapy using nanotechnology has shown promising results in treating TB with advantages such as on-site drug delivery to avoid side effects such as hepatotoxicity as caused by conventional anti-TB therapy. After the successful entry into the cell, nanocarrier systems can further penetrate into the niche environment of *Mtb* and show better activity [22].

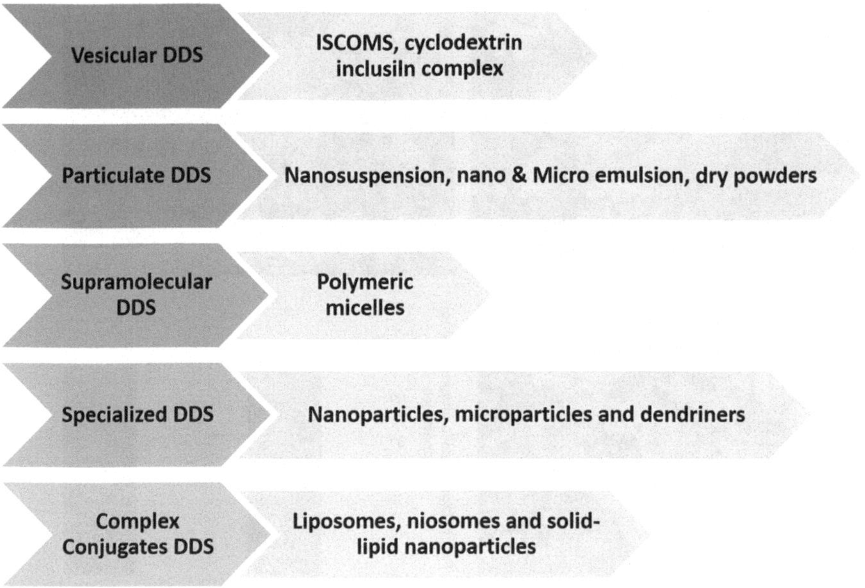

Fig. 2 Colloidal drug delivery system for anti-tubercular drugs or agents

4 Recent Advancements in Treatment of Tuberculosis

4.1 Pharmaceutical Formulations

Though conventional treatment options delivered via oral or parenteral routes are quite effective in most patients, they require long-term administration and thereby results in serious systemic side-effects. The conventional treatment options often result in patient adherence issues, which further contributes to the emergence of serious drug-resistant TB. The exploration of various inhalational devices such as nebulizers, pressurized metered dose inhalers (pMDI), and dry powder inhalers (DPI) should be undertaken for tuberculotic patients for effective delivery of anti-TB drugs [23, 24]. Pulmonary route of drug administration has various advantages in TB patients as the primary target of aerobic bacilli is lungs. Nebulizers have been used for transformation of liquid droplets into aerosol particle for deeper delivery of drugs in the lung tissue. The inhalational devices have been developed for nebulization via jet nebulization, ultrasonic nebulization, and mesh nebulization [25]. pMDIs are preferred compared to normal MDIs as they are cost effective and provide accurate, reliable, and reduced dosing.

DPIs have gained popularity due to improved formulation stability and ability to show sustained drug release and result in better uptake by the alveolar macrophages (AMs). DPIs can be prepared with novel polymers and can be loaded with either a single drug or a combination of multiple drugs to effectively manage TB. For example, to improve the delivery of isoniazid or rifabutin to the alveoli of tuberculotic

Fig. 3 Microphotographs of LBG-based microparticles viewed by scanning electron microscopy. (**a**) Unloaded LBG microparticles; (**b**) LBG:INH = 10:1 (w/w) microparticles; (**c**) LBG:RFB = 10:0.5 (w/w) microparticles, the latter representative of formulations containing different amounts of RFB; (**d**) uptake of fluorescently labelled LBG microparticles by macrophage-differentiated THP-1 cells and NR8383 cells upon exposure to 50 µg/cm^2, for a period of 2 h. Results are expressed as mean ± SEM (n > 3). *INH* isoniazid, *LBG* locust bean gum, *RFB* rifabutin. (Reproduced with permission from Alves et al. [26])

patients, a novel approach for inhalation delivery with locust bean gum (polysaccharide composed of galactose and mannose) as a carrier system was explored by Grenha and co-workers. Microparticles exhibited a good aerodynamic diameter value ranging between 1.15 and 1.67 µm, which enables deep lung delivery (Fig. 3). Cytotoxic evaluation of rifabutin-loaded microparticles in A549-lung epithelial cells and macrophages (THP-1 cells) revealed a toxic effect at the highest concentrations. Locust bean gum particles were easily captured by macrophages with more than 94% of macrophage population exhibiting fluorescent signal [26]. However, in vivo studies on the long-term safety profile of DPIs are required to identify the actual potential of this approach.

The conventional TB treatment failure due to patient non-compliance has strengthened the need for advancement of not only new drugs, but also new drug delivery systems, including pulmonary drug delivery system. The pulmonary delivery has advantages of protection of drugs from degradation and reduced metabolic activity through avoidance of first-pass hepatic metabolism. To overcome the poor blood-flow caused by lung lesions or granulomas leading to the subtherapeutic level of anti-TB drugs, direct inhalation of drugs to the lungs should be adopted to increase the drug concentration in cells surrounding the granulomas.

The effectivity of drugs against *Mtb* is reduced significantly due to drug resistance. As a result, researchers are coming up with novel techniques such as incorporation of drugs into nanoformulations. Several types of nanocarrier systems have been reported in the literature for TB therapy. These include (a) vesicular systems including liposomes, niosomes, and solid lipid nanoparticles, (b) particulate systems including microparticles, nanoparticles, and dendrimers, (c) supramolecular systems like polymeric micelles, (d) specialized systems including nanosuspensions, microemulsions, nanoemulsions, and dry powders, (e) complex conjugate systems like immunostimulating complex and cyclodextrin inclusion complexes, and (f) carrier-based systems like nanofibers, nanotubes, nanobeads, and quantum dots [27]. Nanocarrier systems are considered to offer several advantages over conventional systems such as enhanced therapeutic efficacy, increased bioavailability, higher drug concentration at the target site, and fewer side effects [28].

The nano-antimicrobials, being one of the efficient targeted drug delivery system, hold a huge potential importance with specific bactericidal characteristics as an effective alternative means for superbug and *Mtb* infections [29]. An elaborate study of metal nanoparticles (MNPs) comprised of salts of metals such as Ag, Au, Ti, Zn, and Ga and their encapsulation with anti-TB drugs revealed their antimicrobial property against resistant mycobacteria. Especially, zinc metal showed antibacterial activity via UV-blockade leading to extra-ordinary bactericidal effect with improvised and shortened duration of treatment against *Mtb* infections when combined with other anti-TB drugs [30].

Liposomal formulations of various conventional anti-TB drugs have shown reduced toxicity, improved pharmacokinetic profile, reduced dosages, increased patient compliance, and target specific action of the drug. Av-Gay et al. have reported the liposomal formulation of ethambutol with an encapsulation of 76–92% in order to reduce the duration of anti-tubercular drug regimen [31]. The in vitro release study through disk-diffusion assay against *Mycobacterium bovis* BCG showed the efficacy and bioavailability of the liposomal formulation similar to the free ethambutol. Further, in vivo studies should be conducted to evaluate the liposomal formulation in an animal model. Surface-modification of liposomes with hydrophilic polymers such as polyethylene glycol prevents the liposomes from getting recognized by reticuloendothelial system and hence increases the blood circulation time. In a study by Deol et al., stealth liposomes were prepared through surface modification with O-stearylamylopectin [32]. Surface modification of liposomes resulted in higher affinity toward lung tissue of mice. Likewise, PEGylated liposomes containing anti-TB drugs were linked to interfering RNA (siRNA) to target transforming growth factor-β1 (TGF-β1). As visualized by transmission electron microscopy and scanning electron microscopy, the surface functionalized liposomes were efficiently endocytosed by human macrophages exhibiting good selectivity and minimal cytotoxicity (Fig. 4) [27].

Lipid nanoparticles offer several advantages in TB treatment due to their ability to improve biopharmaceutical and pharmacokinetic attributes of drugs. The entrapment of drugs into lipid nanoparticles might result in a dose reduction and sustained delivery. Further, targeted delivery to the lungs could be obtained through surface

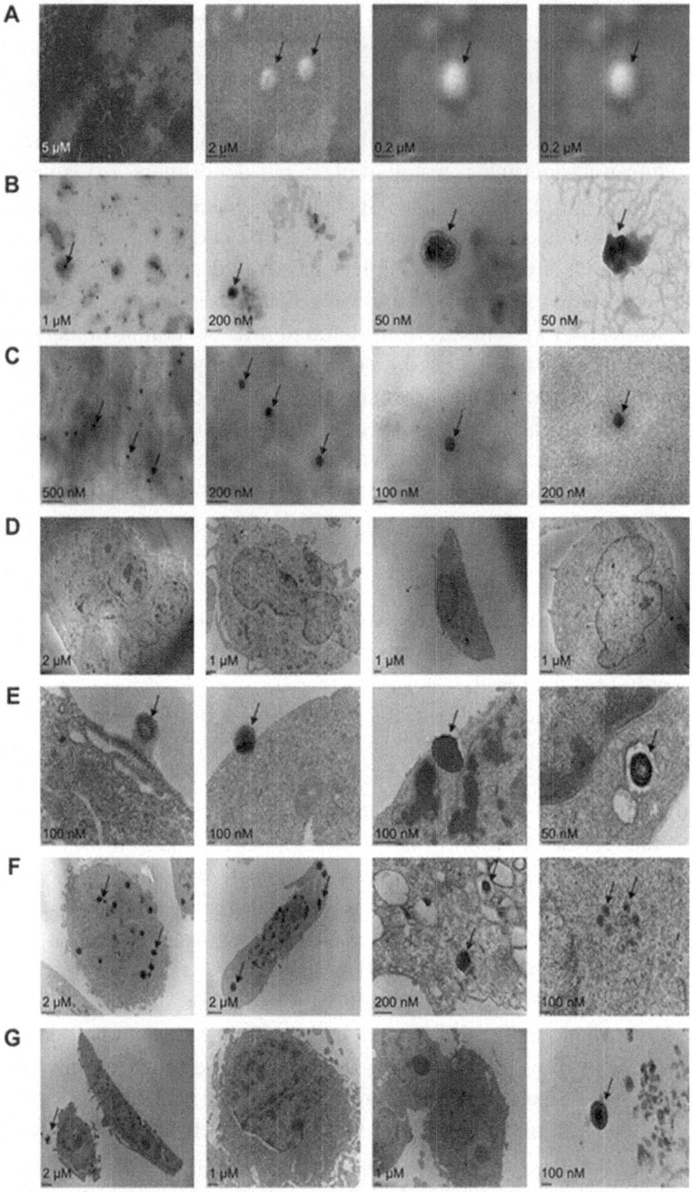

Fig. 4 The morphology of newly synthesized NP-siRNA liposomes under SEM and TEM. (**a**) SEM images showing the round shapes of newly synthesized NP-siRNA liposomes. (**b**) TEM images showing the spherical shapes of the newly synthesized NP-siRNA liposomes. (**c**) TEM images showing the irregular shapes of PEG. (**d**) TEM images showing the macrophages without exposure to the NP-siRNA liposomes. (**e**) TEM images showing the endocytosis of NP-siRNA liposomes by macrophages at 37 °C. (**f**) TEM images showing the endocytosed NP-siRNA liposomes in macrophages at 37 °C. (**g**) TEM images showing lack of endocytosis of NP-siRNA liposomes by macrophages at 4 °C. Arrows point to NP-siRNA liposomal nanoparticles. (Reproduced with permission from Niu et al. [27])

modification of lipid nanoparticles [33]. In a recent study, a mannose-modified macrophage-targeting solid-lipid nanoparticle containing a pH-sensitive prodrug of isoniazid (isonicotinic acid octylidene-hydrazide) was used in the treatment of latent TB infection. The surface-modified SLNs exhibited a fourfold increase of intracellular antibiotic efficacy and macrophage uptake because of the pH-sensitive degradation of isonicotinic acid octylidene-hydrazide and macrophage-targeting ability of solid lipid nanoparticles [34].

Nanoemulsions have become popular due to their ability to incorporate a wide range of therapeutic agents, increased drug loading, and improved stability. When administered via the oral route, nanoemulsions prolong the residence time of drugs in the gastrointestinal tract and avoid the first-pass metabolism of drugs through lymphatic uptake [35]. A recent review by Rajput et al. summarized the usefulness of various nanocarrier-based systems such as solid-lipid nanoparticles, polymeric nanoparticles, liposomes, nanomicelles, niosomes, phytosomes, microemulsions, nanoemulsions, and dendrimers in the management of TB [36]. Despite the encouraging results with the use of nanocarrier systems in TB treatment, it is a little early to jump into conclusion. Further studies are required in this area to achieve sufficient drug-loading into nanocarrier systems along with scalability of optimized formulations.

4.2 New Drugs or New Chemical Entities

Given the widespread implications of TB, several research groups are involved in the design and evaluation of new anti-TB drug leads. Several strategies have been attempted such as scaffold hopping [37], screening libraries of chemical and natural products to identify the potential targets critical to the microorganism using bioinformatics and genetic tools, transcription studies, and whole cell methods such as low oxygen recovery assay (LORA) and resazurin microtiters assay (REMA) [38]. Due to better understanding of physiology of TB infection considering the advancements in genetics, immunological methods, imaging technologies, the host response has been improved to deadly disease. The new strategies for an effective therapy include identification of new target through knowledge of bacterial physiology, the application of genetic synergy to combination therapy, and approaches to change host response [39]. New drug candidates for TB in clinical trial with novel mechanism of action showing action against resistant *Mtb* strains have been considered to replace the 40-year-old therapeutic agents for TB [40].

Extensive literature review on anti-tubercular motifs and their timely updates provide intuition to the medicinal chemists to derive innovative chemical structures for designing the new scaffolds [41–48]. Recently, we have reported the quinquennial coverage of anti-TB scaffolds identified during 2015–2020, [49] and synthetic strategies of indole containing heterocyclic scaffolds as anti-TB agents [50]. Dupont and co-workers have identified a piperidinol containing lead through in vitro studies against drug-sensitive *Mtb*, MDR, and XDR-TB acting through the inhibition of mmpL3 [51]. Sriram and co-workers have reported keto-acid reductoisomerase inhibitors with inhibitory constant (K_i) of 3.06 μM as novel anti-mycobacterial agents [52]. Brown and his co-workers revealed non-cytotoxic benzoxa [2,1,3]diazole substituted acid hydrazide against *Mtb* using REMA assay with MIC up to 12 μM [53]. Further research into finding of new drugs is required through the drug discovery approaches to fight against resistant TB.

4.3 Herbal Drugs

For many centuries, polyherbal remedies are being used for treating various diseases. Studies show that polyherbal medicines contain several active constituents that possess pharmacological functions and act synergistically against infections. Medicinal plants and their phytochemical constituents have gained popularity in treating TB due to their ability to show therapeutic activity without serious side-effects [54–56]. In 2017, Payne and co-workers have reported uridylpeptide sansanmycin analogues as *Mtb* phosphor-MurNAc-pentapeptide translocase enzyme inhibitors affecting the synthesis of lipid I through alterations in peptidoglycan synthesis [57]. Upon in vitro testing of three leads against infected THP-1 macrophages, the compounds were found active with IC_{50} in the range of 0.11–4.33 μM. Along with these natural product derivatives, dihydrosansamycin analogues have been evaluated against *Mtb*. However, none of these agents were found effective in comparison with isoniazid and rifampin. Recently, plants namely *Zanthoxylum leprieurii* (stem bark), *Lantana camara* (constating of verbascoside), *Cryptolepis sanguinolenta*, and their extracts have been reported to possess anti-tuberculotic potential against drug-resistant strains of mycobacteria [58]. Nevertheless, further in vivo studies are warranted to investigate the safety profile and mechanism of actions. So far, no plant-derived molecule made it to the market or undergoing clinical trials for the treatment of mycobacterial infections. This could be attributed to a number of challenges such as the low yield/activity of purified compounds and the existence of natural products in multiple stereoisomers. For instance, triterpenes contain ten or more chiral centers. An integrated approach is required for identification of plants and bioactive molecules with anti-mycobacterial activity.

4.4 Recent Trials of New TB Drugs and Regimens

Over the last decade, a significant progress has been made in research and development of new drugs, with approximately 8 new/repurposed drugs in phase I, phase II, or phase III trials for treating drug-susceptible, MDR-TB, XDR-TB, or LTB infections (Table 3) [59, 60]. In general, treatment options for XDR-TB are limited. Therefore, further developments on the safety, tolerability, and efficacy of new treatment regimens are of paramount importance. Many new chemical entities

Table 3 Anti-TB drug candidates currently in clinical development

Phase of clinical trials	Anti-tubercular dug candidates
Phase I	TBA-354
Phase II	Sutezolid (PNU-100480) SQI09 Rifapentine for DS-TB AZD5847 Bedaquiline-Pretomanid- Pyrazinamide regimen
Phase III	Bedaquiline (TMC-207) with OBR for MDR-TB Delamanid (OPC-67683) with OBR for MDR-TB Rifapentine for LTBI Pretomanid-Moxifloxacin- Pyrazinamide regimen

identified through academic and industrial collaborations have been under evaluation in clinical trials either individually or in a combination with other drugs [59, 60]. Recently, Tarning and co-workers have evaluated pharmacokinetics parameters of anti-TB drugs in pediatric patients with tubercular meningitis with a view of optimizing six different prescribed dosages of drugs through non-linear mixed effect modeling using Monte Carlo simulations on 5000 virtual children. Hundred Vietnamese pediatric patients with tuberculosis meningitis were subjected to treatment of isoniazid, rifampin, pyrazinamide, and ethambutol for 8 months. The pharmacokinetics of rifampin and pyrazinamide was studied using one compartment disposition model and that of ethambutol and isoniazid was accessed using two compartment disposition model. The first-line anti-TB drug, rifampicin, showed better cerebrospinal fluid penetration in inflamed tissues owing to the abundant presence of proteins in comparison with of that of isoniazid and pyrazinamide. Further, the study has revealed the necessity of the reduction in usual dosage of rifampin (50 mg/kg body weight) to avoid disability of neurons [61]. In contrary, previous study findings in murine model have shown that increasing the dose of rifampicin to 80 mg/kg/day for 3 weeks would provide better treatment avoiding serious side effects [62].

Several research groups have been working on the combination of more than one anti-TB drugs to evaluate their efficacy. A retrospective study to evaluate the efficacy of linezolid, moxifloxacin, and thioridazine was conducted on 17 HIV-negative patients with pulmonary XDR-TB in Argentina population. Linezolid was given to all patients, while moxifloxacin and thioridazine was only administered in 14 patients. This study concluded that the combination of linezolid, moxifloxacin, and thioridazine is recommended for use in XDR-TB patients [63].

4.5 *Vaccines*

Currently, the only available vaccine for TB is BCG vaccine, which is a live attenuated strain of *M. bovis.* BCG vaccine is shown to have efficacy in protecting infants from some forms of TB; however, it is inefficient in adults. Thus, there is a need for developing newer vaccine strategies along with the strategies to improve the efficacy of BCG vaccine [64]. Three generations of TB vaccines are available: first-generation or conventional vaccines (live or attenuated microorganisms), second-generation or subunit vaccines (recombinant DNA technology), and third generation vaccines that utilize genetically engineered DNA [65]. Conventional vaccines tend to activate both humoral/cell-mediated immunity. Most extensively reviewed subunit vaccines for TB are composed of proteins, polyproteins, and other subunits. Despite the increased safety, stability, and ability to boost prior BCG immunization, further research on subunit vaccines is needed to identify adjuvants that induce a strong cell-mediated immune (CMI) response. The limitations of first- and second-generation vaccines have prompted the development of new platforms such as DNA vaccines. The DNA vaccine enables the injection of DNA into host muscles resulting in the expression of the encoded protein [66]. DNA vaccines protect against TB through TH1 CD4 response leading to the formation of interferon-γ (INF-γ) and interleukins (IL-4) or cytokines. DNA vaccines do not induce antivector immunity and hence can be applied in "prime/boost" immunization regimens. In addition, DNA vaccines require less manufacturing time and cost as it involves only single-step cloning into plasmid vectors.

Currently tested vaccines aim to decrease the initial bacterial burden by converting immunologically active mycobacterium to a dormant state. However, the post-exposure vaccines fail to reduce the rate of reinfection in patients. The live mycobacterial vaccines, subunit and live vector-based vaccines that boost BCG prime, killed whole bacterial vaccines as adjuvants have been the prime candidates for vaccine development [67]. Several vaccines are currently under phase I, phase II, and phase III clinical trials and these are summarized in Table 4.

Novel strategies are being designed to improve BCG and to develop attenuated *M. tuberculosis* strains for use as an alternative to BCG. These include endosome BCG, overexpressing antigens, combined endosome escape and antigen overexpression, introduction of deleted genes, and attenuated TB vaccines [68]. COVID-19 pandemic has proved that political commitment along with big investments in research and development can result in faster access to life-saving vaccines. This clearly indicates that increased funding in developing new TB vaccines can be a game-changer in alleviating the suffering and deaths in TB patients.

4.6 Others

Studies have been reported wherein granulomatous TB was treated using light radiation or photodynamic therapy in a mice model resulting in the inactivation of tubercular viable bacilli [69]. A new approach has been explored to deliver rifampicin to the site of action using a safe, efficacious, and non-invasive transdermal ultrasound techniques via transdermal patches. This treatment approach showed cure rates of 87.1% and efficacy rates of 93.55% (Fig. 5) [70].

Table 4 TB vaccines currently under trials [64]

Candidate	Strategy	Status
Viral vectors boost		
MVA85A	Boost response to BCG, also considered post-exposure vaccine	Phase IIb
Crucell AD35	Boost response to BCG	Phase IIb
AdAg85A	A replication-deficient adenovirus 5 vector expressing Antigen 85A. Boost response to BCG, also considered post-exposure vaccine and a BCG replacement	Phase I
ChAdOx1.85A + MVA85A	A prime/boost regimen containing prime with a chimpanzee Adenovirus (ChAd) expressing Antigen 85A (ChAdOx1.85A) followed by a boost with modified Vaccinia Ankara virus (MVA) expressing Antigen 85A	Phase I
TB-FLU-04 L	Influenza virus strain A/Puerto Rico/8/34 H1N1 and MTb antigens Ag85A and early secretory antigenic target 6 (ESAT6)	Phase I
Protein + adjuvant		
M72 + AS01	AS01E is used an adjuvant along with a fusion protein of 2 antigens, Rv1196, and Rv0125). Boost response to BCG, also considered post-exposure vaccine	Phase IIb
H-1 + IC31 H-1 + CAF01	Boost response to BCG, also considered post-exposure vaccine and a BCG replacement	Phase IIa
HyVac4 + IC31	Boost response to BCG	Phase I
H56 + IC31	Fusion protein Ag85B-ESAT-6-Rv2660c in IC31 adjuvant	Phase I
ID83 + GLA-SE	Boost response to BCG, also considered post-exposure vaccine	Phase I
Modified BCG		
rBCG30	Enhances immunogenicity of BCG and is a post-exposure vaccine	Phase I
rBCGureC:Hly	BCG replacement or immunogenicity enhancer	Phase IIa
Killed whole cell preps		
M. indicus pranii (MIP)	A purified killed vaccine of Mtb fragments based on killed *M. indicus pranii* organisms. Adjunct for chemotherapy in HIV-infected patients	Phase III
M. vaccae	Adjunct for chemotherapy in HIV-infected patients	Phase III
RUTI	A purified killed vaccine of Mtb fragments	Phase IIa
DAR-901	Inactivated, a whole cell mycobacterial vaccine prepared from the Master Cell Bank for SRL172	Phase I
Viable vaccines		
MTBVAC	A live attenuated *M. tuberculosis* vaccine developed being developed for newborns and secondary adolescences. Studies are initiated in newborns in 2019	Phase IIa
VPM1002	A live-attenuated, recombinant BCG, altered to include a gene from Listeria monocytogenes that codes for listeriolysin O protein	Phase III

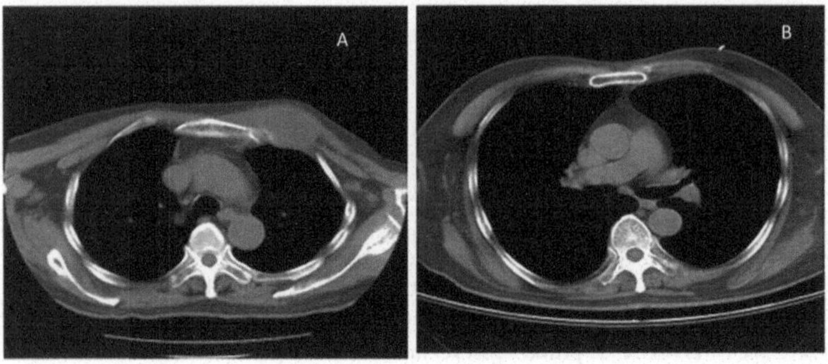

Fig. 5 (**a**) Chest wall tuberculosis prior to transdermal ultrasound treatment. (**b**) Cured tuberculosis following transdermal ultrasound treatment. (Reproduced with permission from Han et al. [70])

5 Conclusions

TB is an infectious disease that mainly affects lungs and can result in fatal effects. To some extent, the infection is controllable and treatable by using the anti-TB regimen. However, the gradual progression of TB into MDR-TB and XDR-TB has posed challenges to TB therapy. To further complicate the issue, spreading of the HIV infection increased the resistance of *Mtb* strains to the highly effective first-line anti-TB drugs. These strains pose a significant threat, especially in immune impaired patients, and clearly justify the need of some innovative drugs and drug delivery systems as outlined in the present work. During the last decade, multiple vaccine candidates have entered into the clinical trials, and these vaccines are expected to provide safety as well as immunogenicity from TB disease in adolescents and adults. To conclude, addressing the challenges posed in the treatment of TB will require a collective approach that includes: the design of effective drug regimens that are administered over a shorter duration, providing free essential healthcare to everyone, including those in greatest need, reducing poverty, and increased commitment from governments to facilitate rapid detection of MDR-TB and providing access to required therapy.

Conflict of Interest The authors declare no competing financial interest.

References

1. Global Tuberculosis Report 2021. https://www.who.int/teams/global-tuberculosis-programme/tb-reports/global-tuberculosis-report-2021. Accessed 27 Oct 2021.
2. Zumla A, Raviglione M, Hafner R, et al. Tuberculosis. New Eng J Med. 2013;368:745–55.
3. Treatment for TB disease. https://www.cdc.gov/tb/topic/treatment/tbdisease.htm. Accessed 25 Dec 2021.
4. PMDT guideline: Ministry of Health and Family Welfare. https://tbcindia.gov.in/showfile.php?lid=3155. Accessed 14 Jan 2022.

5. Thee S, Garcia-Prats AJ, Donald PR, et al. Fluoroquinolones for the treatment of tuberculosis in children. Tuberculosis. 2015;95:229–45.
6. Jadhavar PS, Vaja MD, Dhameliya TM, Chakraborti AK. Oxazolidinones as anti-tubercular agents: discovery, development and future perspectives. Curr Med Chem. 2015;22:4379–97.
7. Bocchino M, Matarese A, Sanduzzi A. Current treatment options for latent tuberculosis infection. J Rheumatol. 2014;41:71–7.
8. Ginsberg AM. Tuberculosis drug development: Progress, challenges, and the road ahead. Tuberculosis. 2010;90:162–7.
9. Churchyard GJ, Kaplan G, Fallows D, et al. Advances in immunotherapy for tuberculosis treatment. Clin Chest Med. 2009;30:769–82.
10. Dawson R, Condos R, Tse D, et al. Immunomodulation with recombinant interferon-γ1b in pulmonary tuberculosis. PLoS One. 2009;4:e6984.
11. Janin YL. Antituberculosis drugs: ten years of research. Bioorg Med Chem. 2007;15:2479–513.
12. O'Connor G, Gleeson LE, Fagan-Murphy A, et al. Sharpening nature's tools for efficient tuberculosis control: a review of the potential role and development of host-directed therapies and strategies for targeted respiratory delivery. Adv Drug Deliv Rev. 2016;102:33–54.
13. Haydel SE. Extensively drug-resistant tuberculosis: a sign of the times and an impetus for antimicrobial discovery. Pharmaceuticals. 2010;3:2268–90.
14. SIRTURO™ (bedaquiline). http://www.sirturo.com/. Accessed 15 Apr 2015.
15. Rakesh BDF, Scherman MS, et al. Synthesis and evaluation of pretomanid (PA-824) oxazolidinone hybrids. Bioorg Med Chem Lett. 2016;26:388–91.
16. (2015) Otsuka pharmaceutical media release. Otsuka wins European marketing authorization for Deltyba(TM) (delamanid). http://www.otsuka.co.jp/en/company/release/2014/0430_01.html. Accessed 18 May 2016.
17. Haufroid M, Wouters J. Targeting the serine pathway: a promising approach against tuberculosis? Pharmaceuticals. 2019;12:66.
18. Chan JGY, Bai X, Traini D. An update on the use of rifapentine for tuberculosis therapy. Expert Opin Drug Deliv. 2014;11:421–31.
19. Lee BY, Clemens DL, Silva A, et al. Drug regimens identified and optimized by output-driven platform markedly reduce tuberculosis treatment time. Nat Commun. 2017;8:14183.
20. Kaur M, Garg T, Rath G, Goyal AK. Current nanotechnological approaches for an effective delivery of bio-active drug molecules in the treatment of acne. Crit Rev Ther Drug Carr Syst. 2014;31:49–88.
21. du Toit LC, Pillay V, Danckwerts MP. Tuberculosis chemotherapy: current drug delivery approaches. Respir Res. 2006;7:118.
22. Lawlor C, Kelly C, O'Leary S, et al. Cellular targeting and trafficking of drug delivery systems for the prevention and treatment of MTb. Tuberculosis. 2011;91:93–7.
23. Kaur R, Kaur R, Singh C, et al. Inhalational drug delivery in pulmonary aspergillosis. Crit Rev Ther Drug Carrier Syst. 2019;36:183–217.
24. Leite JM da S, Patriota YBG, Roca MF de La, Soares-Sobrinho JL. New perspectives in drug delivery systems for the treatment of tuberculosis. Curr Med Chem. 2021;28 https://doi.org/10.2174/0929867328666210629154908.
25. Chandel A, Goyal AK, Ghosh G, Rath G. Recent advances in aerosolised drug delivery. Biomed Pharmacother. 2019;112:108601.
26. Alves AD, Cavaco JS, Guerreiro F, et al. Inhalable antitubercular therapy mediated by locust bean gum microparticles. Molecules. 2016;21:702.
27. Niu N-K, Yin J-J, Yang Y-X, et al. Novel targeting of PEGylated liposomes for codelivery of TGF-β1 siRNA and four antitubercular drugs to human macrophages for the treatment of mycobacterial infection: a quantitative proteomic study. Drug Des Devel Ther. 2015;9:4441–70.
28. Hadrich G, Boschero RA, Appel AS, et al. Tuberculosis treatment facilitated by lipid nanocarriers: can inhalation improve the regimen? Assay Drug Dev Technol. 2020;18:298–307.
29. Minakshi P, Ghosh M, Brar B, et al. Nano-antimicrobials: a new paradigm for combating Mycobacterial resistance. Curr Pharm Des. 2019;25:1554–79.

30. Alzahabi KH, Usmani O, Georgiou TK, et al. Approaches to treating tuberculosis by encapsulating metal ions and anti-mycobacterial drugs utilizing nano- and microparticle technologies. Emerg Top Life Sci. 2021;4:581–600.
31. Wiens T, Redelmeier T, Av-Gay Y. Development of a liposome formulation of ethambutol. Antimicrob Agents Chemother. 2004;48:1887–8.
32. Deol P, Khuller GK. Lung specific stealth liposomes: stability, biodistribution and toxicity of liposomal antitubercular drugs in mice. Biochim Biophys Acta – Gen Subj. 1997;1334:161–72.
33. Gordillo-Galeano A, Ospina-Giraldo LF, Mora-Huertas CE. Lipid nanoparticles with improved biopharmaceutical attributes for tuberculosis treatment. Int J Pharm. 2021;596:120321.
34. Ma C, Wu M, Ye W, et al. Inhalable solid lipid nanoparticles for intracellular tuberculosis infection therapy: macrophage-targeting and pH-sensitive properties. Drug Deliv Transl Res. 2021;11:1218–35.
35. Tayeb HH, Sainsbury F. Nanoemulsions in drug delivery: formulation to medical application. Nanomedicine. 2018;13:2507–25.
36. Rajput A, Mandlik S, Pokharkar V. Nanocarrier-based approaches for the efficient delivery of anti-tubercular drugs and vaccines for management of tuberculosis. Front Pharmacol. 2021;12:749945.
37. Hu Y, Stumpfe D, Bajorath J. Recent advances in scaffold hopping. J Med Chem. 2017;60:1238–46.
38. Palomino J, Ramos D, da Silva P. New anti-tuberculosis drugs: strategies, sources and new molecules. Curr Med Chem. 2009;16:1898–904.
39. Baer CE, Rubin EJ, Sassetti CM. New insights into TB physiology suggest untapped therapeutic opportunities. Immunol Rev. 2015;264:327–43.
40. Rivers EC, Mancera RL. New anti-tuberculosis drugs in clinical trials with novel mechanisms of action. Drug Discov Today. 2008;13:1090–8.
41. Nayyar A, Jain R. Recent advances in new structural classes of anti-tuberculosis agents. Curr Med Chem. 2005;12:1873–86.
42. Kumar V, Patel S, Jain R. New structural classes of antituberculosis agents. Med Res Rev. 2018;38:684–740.
43. Sharma A, De Rosa M, Singla N, et al. Tuberculosis: an overview of the immunogenic response, disease progression, and medicinal chemistry efforts in the last decade toward the development of potential Drugs for Extensively Drug-Resistant Tuberculosis Strains. J Med Chem. 2021;64:4359–95.
44. Singh P, Jaiyeola B, Kerru N, et al. A review of recent advancements in anti-tubercular molecular hybrids. Curr Med Chem. 2017;24:4180–212.
45. Angula KT, Legoabe LJ, Beteck RM. Chemical classes presenting novel antituberculosis agents currently in different phases of drug development: a 2010–2020 review. Pharmaceuticals. 2021;14:461.
46. Dhameliya TM, Patel KI, Tiwari R, et al. Design, synthesis, and biological evaluation of benzo[d]imidazole-2-carboxamides as new anti-TB agents. Bioorg Chem. 2021;107:104538.
47. Jadhavar PS, Patel KI, Dhameliya TM, et al. Benzimidazoquinazolines as new potent anti-TB chemotypes: design, synthesis, and biological evaluation. Bioorg Chem. 2020;99:103774.
48. Dhameliya TM, Chudasma SJ, Patel TM, Dave BP. A review on synthetic account of 1,2,4-oxadiazoles as anti-infective agents. Mol Divers. 2022; https://doi.org/10.1007/s11030-021-10375-4.
49. Dhameliya TM, Bhakhar KA, Gajjar ND, et al. Recent advancements and developments in search of anti-tuberculosis agents: a quinquennial update and future directions. J Mol Struct. 2022;1248:131473.
50. Bhakhar KA, Sureja DK, Dhameliya TM. Synthetic account of indoles in search of potential anti-mycobacterial agents: a review and future insights. J Mol Struct. 2022;1248:131522.
51. Dupont C, Chen Y, Xu Z, et al. A piperidinol-containing molecule is active against Mycobacterium tuberculosis by inhibiting the mycolic acid flippase activity of MmpL3. J Biol Chem. 2019;294:17512–23.

52. Krishna VS, Zheng S, Rekha EM, et al. Discovery and evaluation of novel Mycobacterium tuberculosis ketol-acid reductoisomerase inhibitors as therapeutic drug leads. J Comput Aided Mol Des. 2019;33:357–66.
53. Brown AK, Aljohani AKB, Gill JH, et al. Identification of novel benzoxa-[2,1,3]-diazole substituted amino acid hydrazides as potential anti-tubercular agents. Molecules. 2019;24:811.
54. Sharma D, Yadav J. An overview of phytotherapeutic approaches for the treatment of tuberculosis. Mini-Reviews Med Chem. 2016;17:167–83.
55. Sharifi-Rad J, Salehi B, Stojanović-Radić ZZ, et al. Medicinal plants used in the treatment of tuberculosis – ethnobotanical and ethnopharmacological approaches. Biotechnol Adv. 2020;44:107629.
56. Abdella G, Mirutse G, Gobena A, Adane W. In vitro anti-mycobacterial activity of selected medicinal plants against Mycobacterium tuberculosis and Mycobacterium bovis strains. BMC Complement Altern Med. 2013;13:291.
57. Tran AT, Watson EE, Pujari V, et al. Sansanmycin natural product analogues as potent and selective anti-mycobacterials that inhibit lipid i biosynthesis. Nat Commun. 2017;8:14414.
58. Tuyiringire N, Deyno S, Weisheit A, et al. Three promising antimycobacterial medicinal plants reviewed as potential sources of drug hit candidates against multidrug-resistant tuberculosis. Tuberculosis. 2020;124:101987.
59. Clinical Portfolio | TB Alliance. https://www.tballiance.org/portfolio. Accessed 26 Oct 2021.
60. Workcing group for new TB drugs. https://www.newtbdrugs.org/. Accessed 26 Oct 2021.
61. Panjasawatwong N, Wattanakul T, Hoglund RM, et al. Population pharmacokinetic properties of antituberculosis drugs in Vietnamese children with tuberculous meningitis. Antimicrob Agents Chemother. 2021;65:e00487–20.
62. De Steenwinkel JEM, Aarnoutse RE, De Knegt GJ, et al. Optimization of the rifampin dosage to improve the therapeutic efficacy in tuberculosis treatment using a murine model. Am J Respir Crit Care Med. 2013;187:1127–34.
63. Abbate E, Vescovo M, Natiello M, et al. Successful alternative treatment of extensively drug-resistant tuberculosis in Argentina with a combination of linezolid, moxifloxacin and thioridazine. J Antimicrob Chemother. 2012;67:473–7.
64. Pitt JM, Blankley S, McShane H, O'Garra A. Vaccination against tuberculosis: how can we better BCG? Microb Pathog. 2013;58:2–16.
65. Méndez-Samperio P. Global efforts in the development of vaccines for tuberculosis: requirements for improved vaccines against Mycobacterium tuberculosis. Scand J Immunol. 2016;84:204–10.
66. Orme IM. Preclinical testing of new vaccines for tuberculosis: a comprehensive review. Vaccine. 2006;24:2–19.
67. Kaufmann SH, Hussey G, Lambert PH. New vaccines for tuberculosis. Lancet. 2010;375:2110–9.
68. Sander C, McShane H. Translational mini-review series on vaccines: development and evaluation of improved vaccines against tuberculosis. Clin Exp Immunol. 2007;147:401–11.
69. O'Riordan K, Sharlin DS, Gross J, et al. Photoinactivation of mycobacteria in vitro and in a new murine model of localized Mycobacterium bovis BCG-induced granulomatous infection. Antimicrob Agents Chemother. 2006;50:1828–34.
70. Han Y, Zhao Q, Yu D, Liu Z. Treatment of chest wall tuberculosis with transdermal ultrasound-mediated drug delivery. Exp Ther Med. 2015;9:1433–7.

Polymeric Nanoparticles in Tuberculosis

Sanjoy Kumar Das, Soumalya Chakraborty, Sourav Bhowmik, Sudipta Roy, and Yashwant Pathak

Abstract Tuberculosis (TB) is an infectious disease caused by the bacteria *Mycobacterium tuberculosis (Mtb).* TB causes the most human deaths than any other diseases from a single infectious agent. Treatments of TB are long and costly and have many limitations. Intracellular bacilli are slow growing and difficult to target, which is augmenting the emergence of multidrug resistance. Targeting intracellular *Mycobacterium tuberculosis* is very challenging, but nanomedicine may offer a solution. Nanomedicine is a significantly growing research area and offers the potential for specific disease targeting, dosage reduction, and intracellular drug delivery. Different polymers of natural or synthetic origin which are commonly used for the fabrication of nanoparticles with antitubercular drugs are outlined in this chapter. Polymeric nanoparticles have recently attracted increasing attention in tuberculosis treatment due to their unique properties. Polymeric nanoparticles provide an innovative therapeutic alternative to improve the limitations and disadvantages of the conventional available treatments for tuberculosis.

Keywords Polymeric nanoparticles · Tuberculosis · Natural polymers · Synthetic polymers

S. K. Das (✉)
Institute of Pharmacy, Jalpaiguri, West Bengal, India
e-mail: sanjoydasju@gmail.com

S. Chakraborty
National Institute of Pharmaceutical Education and Research (NIPER), Sector 67, Mohali, Punjab, India

S. Bhowmik
SRL Limited, Fortis Hospital, Kolkata, West Bengal, India

S. Roy
Bengal School of Technology (A College of Pharmacy) Sugandha, Hoogly, West Bengal, India

Y. Pathak
Taneja College of Pharmacy, University of South Florida, Tampa, FL, USA

Faculty of Pharmacy, Airlangga University, Surabaya, Indonesia

R. Shegokar, Y. Pathak (eds.), *Tubercular Drug Delivery Systems*,
https://doi.org/10.1007/978-3-031-14100-3_5

1 Introduction

Tuberculosis (TB) is an infectious disease caused by the bacteria *Mycobacterium tuberculosis (Mtb).* TB is the world's second greatest cause of death from an infectious disease. The requirement of daily medicine delivery for long periods of time is a significant obstacle in treating the disease (up to 9 months). This often leads to poor adherence by patients, which is not only a risk to the patient's well-being, but due to its highly infectious nature, it represents a serious risk to public health. The current approach to tackle poor adherence is directly observed treatment, where each day during their treatment patients is observed taking their medication by a healthcare worker, which presents a considerable burden to the healthcare system, both in cost and time [1]. Since this therapy is carried out for a long duration, chances of patients discontinuing the course before cure are high, and this compromises the patient's compliance and adherence to treatment. The failure of treatment could lead to the appearance of multidrug-resistant and extensively drug-resistant TB. The drug-resistant strains pose challenges for treatment and eradication of TB [2]. In addition, the drugs currently used for the treatment of TB suffer from serious adverse side effects, such as hepatotoxicity, as well as short plasma half-life and rapid clearance [1]. Traditional TB treatment also entails precise dosages and frequencies, and lengthy treatment periods that lead to patient noncompliance. So, despite the availability of current antibiotics, there is still a large demand for other treatment options [3]. In tuberculosis, the lung is the main target organ. The pathogen is an intracellular agent, with alveolar macrophages serving as both a reservoir and a therapy target. Various polymeric carriers derived from natural or synthetic sources have been employed to target drugs to the deep lung in recent decades. These carriers have particle sizes that range from micro to nanometers. Despite the advantages of micron-sized carriers, nanoparticles have lately become essential in pharmaceutical science due to their unique properties [4].

The delivery of antituberculosis drugs (ATDs) with nanoparticle (NP)-based controlled-delivery devices is one of the promising approaches. Several reports have been published on the advantages of NP drug delivery systems for infectious diseases. Some of the drug delivery systems have been accepted for clinical treatment of different infectious diseases, while a few others are currently under different phases of clinical and preclinical trials. The polymeric NPs offer unique benefits to achieve a slow and sustained release that has the potential to treat chronic diseases like TB [2].

This chapter highlights the pathogenesis of TB, conventional treatments of TB, and its limitations along with the different natural and synthetic polymers used for the preparation of nanoparticles in TB.

2 Pathogenesis of Tuberculosis

Mtb is one of the most successful human pathogens, due to its ability to carry a primary infection to a state of dormancy, persisting in the body even in immune-competent people. In this regard, it is important to mention that there are two billion people infected worldwide and only nine million develop the disease annually [5]. The presence of hereditary or acquired deficiencies of the immune system markedly increases the risk of progression to active TB.

The first stage of tuberculosis is initiated with inhalation of droplets generated by a person with active tuberculosis. These droplets can remain for a longer time in the air. When inhaled, a single droplet may be enough to cause the disease. Most droplets end up in the upper respiratory tract, where the microbes are killed, but a few penetrate further down. The bacteria reach the alveoli in the lungs, where the alveolar macrophages phagocytose them. Several receptors are involved in the uptake process including mannose receptors, Toll-like receptor 2 (TLR2) and Toll-like receptor 4 (TLR4), surfactant protein A receptors, CD14, scavenger receptors, complement receptors, and immunoglobulin receptors [6]. Sometimes macrophages fail to destroy the bacteria either because compounds produced by the microbe inactivate them or because phagosome-lysosome fusion mechanisms are inhibited by *Mtb*, thereby avoiding low pH exposure and hydrolytic surroundings of phagolysosomes.

In the second stage, mycobacterium multiplies in the macrophage, eventually causing its lysis. This results in the cellular damage which attracts the inflammatory cells and blood monocytes to the area. Monocytes differentiate into macrophages and attempt to attack the microbe which is ingested by the macrophages and grow inside the phagocyte. These macrophages again lyse and die due to bacterial load [7]. Two to three weeks after infection, the third stage begins. T cell immunity develops, and lymphocytes drift to the region of infection. Presentation of mycobacterial antigens to the T cells causes their stimulation, resulting in the release of γ-interferon and other cytokines. The γ-interferon activates macrophages to secrete IL 12, TNF-α, IL-8, and other proinflammatory cytokines. Fast growth of the *Mtb* stops, and, at this stage, the host cell develops cell-mediated immunity. Those that are outside of cells are resistant to antibody-activated complement attack due to the high lipid content of mycobacterial cell wall. Cell-mediated immunity is also responsible for much of the pathology of tuberculosis. Tissue damage can also take place when activated macrophages release lytic enzymes, reactive intermediates, and various cytokines. It is at this stage that the immune system, specifically the macrophages, will enclose the microorganisms inside tubercles. In between these structures, the atmosphere is anoxic and acidic and prevents the growth of mycobacteria. In-between these structures is anoxic and acidic, preventing the growth of mycobacteria. This balance between host and mycobacterium is called latency which is one of the hallmarks of TB. In the fifth and final stage, the tubercles may dissolve by many factors such as malnutrition, immunosuppression, steroid use, or HIV infection. For unknown reasons, the centers of tubercles may liquefy, providing an outstanding growth medium for the microbe which now begins to grow rapidly in the extracellular fluid. The large number of bacteria and the immune response against them eventually cause the lung tissue near the tubercles to become necrotic and form a cavity [8]. Most tuberculosis infections stop at stage three.

3 Conventional Treatment and Its Limitation

In 2014, TB strategy was adopted by the World Health Assembly with a goal to design a blueprint for sustainable strategies to reduce the number of TB deaths by 90% by 2030 [9]. Currently the short-term treatment (6 months duration) for tuberculosis is composed of a combination of different anti-tuberculosis drugs (known as first-line drugs), including rifampicin (RIF), isoniazid (INH), pyrazinamide (PYR), ethambutol (ETB), and streptomycin (STM) [9]. The World Health Organization (WHO) and the International Union against TB have suggested a combination of at least two first-line drugs in one dosage form to cope with the complex nature of disease progression. These are commonly known as Fixed Dose Combinations (FDC) [10]. Thus, prescription errors can be reduced and adherence to the treatment can be improved. Conventional treatment regimen for TB is treated with first-line drugs, second-line drugs, and third-line drugs. First-line drugs are used as an oral combination therapy with isoniazid, rifampin, pyrazinamide, and ethambutol for several months. Second-line anti-TB therapy is used to circumvent with multidrug resistance TB. Second-line drugs are more lethal and are more expensive than first-line drugs and treatment may last longer [11–13]. The clinical efficacy of third-line anti-TB drugs is not established. Third-line drugs are also not listed by WHO [13].

In 2016, an estimated 490,000 people all over the globe developed multidrug-resistant TB (MDR-TB) [9]. The main reasons behind the emergence of MDR-TB include mismanagement of TB treatment and person-to-person transmission. Inappropriate use and premature treatment interruption of anti-TB drugs may lead to drug resistance. In this context, in spite of the availability of anti-TB drugs since a longer time, TB still remains to be one of the main preventable causes of death by an infectious disease. Hence, it is important to develop new drug delivery systems that ensure high treatment adherence and low adverse effects and that are adequate for both adults and children [14].

4 Polymeric Nanoparticles

Several nanosized carriers are available for drug delivery purposes that include liposomes, solid lipid nanoparticles (SLNs), polymeric nanoparticles (PNPs), nanosuspensions, nanoemulsions, and many other excellent multifunctional nanosystems, leading to better pharmacokinetics, biodistribution, and bioavailability. Polymeric nanoparticles possess very good biocompatible and biodegradable features that make them sustainable candidates for use as drug delivery carriers [15]. The choice of polymer to develop the polymeric nanoparticles is dependent on different factors like size of the nanoparticles (NPs) required, inherent properties of the drug, surface characteristics, biodegradability, biocompatibility, toxicity, and desired drug release profile [16].

5 Polymer Used

Natural polymers and synthetic polymers are used for the preparation of nanoparticles in TB drug delivery. Natural polymers are of interest for the preparation of nanoparticles in TB treatment such as: (i) polysaccharide-based polymers and (ii) polypeptide- and protein-based polymers. These natural polymers are mainly used for the fabrication of nanoparticles due to their properties, e.g., biodegradability, biocompatibility, and low toxicity. Apart from the use of natural polymers, synthetic polymers are also used for the preparations of nanoparticles [4].

5.1 Polysaccharide-Based Polymers

Polysaccharides are a significant class of hydrophilic polymers with natural origin and biocompatibility that find frequent use in water-based polymer systems and in nanotechnology in particular, which is mainly due to their favorable properties in biological systems, e.g., biodegradability, biocompatibility, and low toxicity. These properties constitute considerable requirements for the utilization of NPs and thus polysaccharides represent an ideal class of building blocks for NP fabrication [17]. Chitosan is the most widely studied polysaccharide to develop polymeric nanoparticles [16].

5.1.1 Chitosan Nanoparticles

Chitosan has multiple properties that can aid in the treatment of TB. For instance, it is known for its biocompatibility and biodegradability which promotes good adhesion to mucosal surfaces. It also increases the absorption of vaccines and drugs, lengthens the duration of therapeutic effects for drug effectivity, and facilitates drug delivery to specific body sites [3]. Specific qualities such as solubility, stability, and improved mucoadhesion through physical and chemical alterations allow chitosan to better serve its purpose and expand its uses. Furthermore, the US Food and Drug Administration has declared chitosan to be safe for human consumption. Traditional TB treatment also includes exact dosages and frequencies, as well as long treatment periods, which might contribute to patient noncompliance. Therefore, there is always a demand for other treatment options so that the problems associated with traditional treatment can be solved [3]. Therefore, several research and efforts are being made by researchers on chitosan-based nanoparticles in TB treatment.

Pourshahab et al. [18] have prepared the spray-dried inhalable powders containing INH-loaded chitosan/tripolyphosphate (TPP) nanoparticles for sustained delivery of the drug to the lung. For the preparation of nanoparticles, they have used ionic gelation method. From the *in vitro* drug release study, they found that the rate of drug release from nanoparticles was decreased with increasing the amount of chitosan. Nanoparticles were spray-dried using excipients such as lactose, mannitol, and maltodextrin alone or with leucine. The *in vitro* deposition data indicated that spray drying of isoniazid-loaded nanoparticles with lactose in the presence of leucine resulted in the production of inhalable powders with the highest fine particle fraction (FPF) (45%) [18].

Garg et al. [19], in the year 2016, have prepared the chitosan nanoparticles (CNPs) by ionic gelation technique followed by spray drying for sustained delivery of anti-tubercular drugs, INH, and RIF, to the lungs. For the preparation of nanoparticles, they have selected the chitosan with low molecular weight. They further investigated the chemotherapeutic efficacy and toxicity against experimental murine TB. They found that the CNPs had a smooth spherical shape with an average size of 230 ± 4.5 nm, with a polydispersity index of 0.180 ± 0.021. Drug encapsulation efficiency was observed to be 70.8 ± 6.62 for RIF and 68.8 ± 7.02 for INH. The higher drug encapsulation was observed for CNPs due to the higher drug/polymer ratio, as well as an excellent ionic gelation between tripolyphosphate (TPP) and chitosan. Kinetic analysis of drug release from optimized formulations indicates that the drug is released from the nanoparticles by a diffusion mechanism. An initial burst release of RIF and INH from nanoparticles could be due to rapid dissolution of drug crystals located on the surface or present beneath the nanoparticle surface. They found that the smooth surface nanoparticles are easily captured by the alveolar macrophage of the lungs, and the low PDI of nanoparticles suggested a homogenous dispersion. Both drugs were detected in different organs (lungs, liver, spleen, and kidney) until 24 h post nebulization. They found that the optimized formulations showed lower cytotoxicity and a significant reduction in the number of bacilli in the lungs, as compared to free drug. Finally, they have concluded that the CNPs may be exploited as a prospective tool for drug delivery to direct drugs to the lung tissues for the treatment of TB [19].

Rawal et al. [20] have developed RIF-loaded nanoparticles by ionic gelation probe sonication method. They further analyzed the prepared nanoparticle with respect to its direct targeting potential of lungs. The size range and the drug entrapment efficiency of the nanoparticles were assessed from 124.1 ± 0.2 to 402.3 ± 2.8 nm and 72.00 ± 0.1%, respectively. The results of the cumulative in vitro drug release studies exhibited that the drug release from the developed nanoparticle sustained up to 24 h. Additionally, pharmacokinetic and toxicity studies carried out with prepared NPs dry powder inhalation (DPI) formulations and compared with conventional DPI and marketed formulation showed rifampicin release for extended periods. From their findings, they suggested the freeze-dried rifampicin nanoparticles as a better targeted delivery system for developing treatment strategy for tuberculosis [20].

RIF is one of the most efficient anti-TB medications and a key component of DOTS (directly observed treatment, short-course) therapy. However, inadequate bioavailability, increased drug resistance, decreased cell permeability, failure to achieve enough drug concentrations at the infected site, and degradation before reaching the target site have all hampered this medicine's usefulness [21]. Therefore, they have used a novel hydrophobic derivative of chitosan (octanoyl chitosan) for the preparation of nanoparticles containing RIF. They have prepared octanoyl chitosan (OC) nanoparticles by using double emulsion solvent evaporation technique without cross-linking. They did not use cross-linking method because this method involves the use of potential toxic agents such as glutaraldehyde and after formulation the removal of cross-linking agents makes the method less effective and tedious. They further optimized the OC NPs by using 3^2 full factorial design. They found that the optimized batch of OC NPs exhibited a smooth and spherical morphology and had a mean hydrodynamic diameter of 253 ± 19.06 nm (PDI 0.323 ± 0.059) and entrapment efficiency of 64.86 ± 7.73% for rifampicin. They further studied MTT assay for the determination of biodegradability and non-cytotoxicity of the polymer, and their results suggested its likely safety in clinical use. They also concluded that further evaluations in animals are required to evaluate its utility and potential clinical use [21].

The drug loading of aminoglycoside (AG) loaded chitosan nanoparticles is very low. This drug loading is low because of the electrostatic repulsion between the positively charged AG and positively charged chitosan [22]. Therefore, dextran sulfate (Mw 500 000), a polyanion, has been used to shield the positively charged AG in order to improve the drug loading. Lu et al. have prepared the AG (streptomycin, gentamicin, and tobramycin)-loaded chitosan nanoparticles with the aim of high drug loading. They further conducted the test of in vivo oral efficacy of streptomycin (SM)-loaded chitosan nanoparticles in a Mycobacterium tuberculosis chronic infection mouse model. They concluded that the chitosan nanoparticles may provide a promising oral drug delivery formulation for AG which usually, in tuberculosis treatment, is administrated as an injectable preparation [22].

In 2018, Wardani et al. [23] worked on the in vitro antibacterial activity of chitosan nanoparticles against Mycobacterium tuberculosis. They have concluded from their work that chitosan nanoparticles have a relatively rougher surface with an uneven structure which exhibited highly amorphous feature, and it has promising anti-tubercular activity by preliminary in vitro techniques. Therefore, they also concluded it has the definite potential as a source of compounds that may be developed further into antimycobacterial drugs [23]. A new nanomedicine antibacterial agent, based on dihydroartemisinin (DHA) and chitosan (CS), has been developed by Gu et al. [24] to overcome MTB's drug-resistant. To enhance DHA's solubility, they have prepared nanoparticles of DHA-loaded CS by an ionic crosslinking method with sodium tripolyphosphate (STPP) as the crosslinking agent. They found that the DHA-CS NPs exhibited an excellent antibacterial effect on the rifampicin-resistant strain (ATCC 35838) and, at a concentration of 8.0μg/ml, the antibacterial impact reaches up to 61.0 ± 2.13% ($n = 3$). From their findings, they have concluded the DHA-CS NPs combined with rifampicin may have potential use for TB treatment [24]. A few examples of chitosan-based nanoparticles in TB are given in Table 1.

Table 1 Chitosan-based nanoparticles in TB

Carrier material	Specification of chitosan used	Encapsulated drug	Particle size	Drug encapsulation efficiency	Remarks	Refs.
Chitosan nanoparticles	Chitosan (CS) (degree of deacetylation: 93%)	Rifampicin (RIF)	221.9 nm	44.17 ± 1.98%	In vitro study suggests that oral sustained release CNs of rifampicin might be an effective drug delivery system for tuberculosis.	[16]
Chitosan nanoparticles	Chitosan (CIF, Cochin)	Isoniazid	661.8–823.8 nm		Chitosan nanoparticles are physically and chemically stable and retain their pharmaceutical properties at various environmental conditions over a period of 3 months.	[25]
Hyaluronic Acid Coated Chitosan Nanoparticles	Chitosan (Analab Fine Chemicals, Mumbai)	Rifampicin	211.23–301.67 nm	82.56 ± 1.26% to 87.21 ± 0.98%	Nanoparticles could effectively target macrophage, a residual site of *Mycobacterium tuberculosis*.	[26]
Chitosan nanoparticles-based dry powder inhalation formulations	Low molecular weight of chitosan (>75% deacetylation)	Prothionamide	1.76μm (aerodynamic particle size)		Animal study also revealed the reduction of dose in pulmonary administration, which will improve the management of tuberculosis.	[27]
Chitosan nanoparticles		Isoniazid	620 ± 10.97 nm		Nanosystem is thus an efficient approach for antitubercular therapy.	[28]

5.1.2 Alginate Nanoparticles

Sodium alginate is a natural polymer with properties such as an aqueous matrix environment, high gel porosity, and biocompatibility, and is approved by the US Food and Drug Administration (USFDA) for oral use [29]. It is a natural polysaccharide, rich in carboxyl group and is easy to bind with positive charge cations such as Ca^{2+}. It is of low toxicity, good biocompatibility, and relatively low cost; therefore, it can be used for the preparation of nanoparticles [30].

Kumar and Bhatt [30] fabricated and evaluated the isoniazid-loaded sodium alginate nanoparticles. They prepared the nanoparticles using ionotropic gelation technique. The particle size, drug loading, and encapsulation efficiency of the fabricated nanoparticles were studied. The in vitro drug release study of the optimized formulation showed 66.56% drug release in 24 h. They concluded that the isoniazid-loaded sodium alginate nanoformulation has the potential to provide enhanced efficacy of isoniazid [30]. Shaji and Shaikh [31] have prepared D-cycloserine (D-CS)-loaded alginate-chitosan nanoparticles using ionotropic gelation method. They further designed and optimized the biodegradable polymeric nanoparticles of D-CS using 2^3 factorial design to study the influence of formulation variables on particle size and entrapment efficiency of polymeric nanoparticles. They found that the optimized batch exhibited the entrapment efficiency of $98.10 \pm 0.24\%$ with particle size 344 ± 5 nm. Further, in vitro release study of the optimized formulation in phosphate buffer saline (pH 7.4) showed a biphasic release pattern with initial burst release of about 34.49% of drug, followed by controlled release up to 24 h. They further concluded that the delivery system for D-CS could be a potential alternative to the existing conventional therapy in multidrug-resistant tuberculosis (MDR-TB) [31]. Alginate-based nanoparticles in tuberculosis treatment are summarized in Table 2.

5.1.3 Guar Gum Nanoparticles

Guar and its derivatives are widely used in many applications including food, drug delivery, and healthcare products because of their natural abundance and their low cost and other desirable functionalities [34]. Guar gum is a water-soluble polysaccharide, and it is composed of sugars, such as galactose and mannose. Additionally, swelling behavior of guar gum at acidic pH imparts an efficacy to protect the antigen in harsh gastric environment. Guar gum is a natural nontoxic, biodegradable, mucoadhesive, cost-effective polymer which can encapsulate a higher amount of antigen [35]. Kaur et al. [35], in the year 2015, developed an effective carrier system containing Ag85A-loaded guar gum nanoparticles for oral vaccination against tuberculosis. They have used nanoprecipitation for the preparation of nanoparticles. They found that the developed particles with an average diameter of 895.5 ± 14.73 nm and high antigen entrapment seem to be optimum for oral vaccine delivery. In vivo studies data revealed that the developed nanocarriers can induce a strong mucosal as well as systemic immune response. Finally, from the experimental evidence, they

Table 2 Alginate-based nanoparticles in TB

Carrier material	Specification of alginate used	Encapsulated drug	Particle size	Drug encapsulation efficiency	Route of administration	Remarks	Refs.
Alginate nanoparticles	Sodium alginate (medium viscosity, 3500 cps for a 2% w/v solution)	Isoniazid (INH), rifampicin (RIF) and pyrazinamide (PZA)	235.5 ± 0 nm	70–90% for INH and PZA and 80–90% for RIF	Pulmonary	Inhalable alginate nanoparticles can serve as an ideal carrier for the controlled release of antitubercular drugs.	[29]
Alginate nanoparticles	Sodium alginate (medium viscosity, 3500 cps for a 2% w/v solution)	ATDs (rifampicin, isoniazid, pyrazinamide and ethambutol)		70–90%	Orally	Potential for intermittent therapy of TB.	[32]
Sodium alginate coated with chitosan and Tween 80	Low viscosity sodium alginate (ALG, MW ~ 4500 Da, viscosity ~8 cP, 39% content of α-L-guluronate, and % 61 content of β-D-mannuronate)	Rifampicin (RIF) in combination with ascorbic acid (ASC)	324.0 ± 40.7 nm	0–36%	Pulmonary	Reducing the risk of systemic toxicity and hence improving the patient compliance.	[33]

concluded that the guar-gum nanoparticle can be utilized for safe and effective vaccine delivery via oral route [35]. Goyal et al. [36] in the year 2016, worked on chemotherapeutic evaluation of guar gum-coated chitosan nanoparticle against experimental tuberculosis. The major objective of their work was to develop and evaluate the therapeutic potential of ATDs-loaded natural polysaccharide comprising of galactomannan subunit in experimental TB. From their work, they have concluded that guar gum-coated chitosan nanoparticles could be a promising carrier for selective delivery of ATDs to alveolar macrophages for efficient management of TB with the interception of minimal side effects [36].

5.2 Polypeptide and Protein-Based Polymers

5.2.1 Gelatin-Based Polymers

Nanoparticles made of biodegradable polymers like proteins and polysaccharides can act as efficient drug delivery vehicles for controlled and targeted release, aiming to improve the therapeutic effects and also to reduce the side effects of the formulated drugs. Over the past few decades, there has been considerable interest in developing protein-based nanoparticles as GRAS (generally regarded as safe) drug delivery devices [37]. Gelatin is a denatured protein which is obtained either by partial acid or alkaline hydrolysis of animal collagen and has been extensively used for the preparation of nanoparticles [4, 37]. Gelatin is a natural versatile biopolymer, and it can be used in different applications because of its low cost, easy availability, biodegradable and biocompatible nature as well as the presence of abundant active groups. The gelatin nanoparticles (GNPs) can be prepared by several different techniques, including desolvation, coacervation-phase separation, emulsification-solvent evaporation, reverse phase microemulsion, and nanoprecipitation [38]. Saraogi et al. [39] developed and characterized the rifampicin-loaded gelatin nanoparticulate delivery system for the effective management of tuberculosis. Gelatin nanoparticles containing rifampicin were prepared by using two-step desolvation method. The gelatin nanoparticles were characterized for size measurements, drug entrapment, and in vitro drug release study. The size of nanoparticles was found to be 264 ± 11.2 nm with low PDI suggesting the narrow particle size distribution. They have conducted the TEM photomicrograph, and this study revealed the gelatin nanoparticles (GPs) were spherical in shape (Fig. 1) [39].

The drug release showed the biphasic pattern of release, i.e., initial burst followed by a sustained release pattern. The gelatin nanoparticles were evaluated for cytotoxicity study on J-774 macrophage cell lines and *in vivo* biodistribution and antitubercular studies on mice model. The biocompatibility of GPs was tested using MTT assay on J774 cells. They found that the cells incubated with GPs and RIF-GP remained nearly 100% viable when compared to the control group at concentrations as high as 1mg/mL (Fig. 2). They found that the cell viability was 94%, 91%, and 65%, respectively, for GPs, RIF-GPs, and RIF-treated cells, at 1 mg/mL

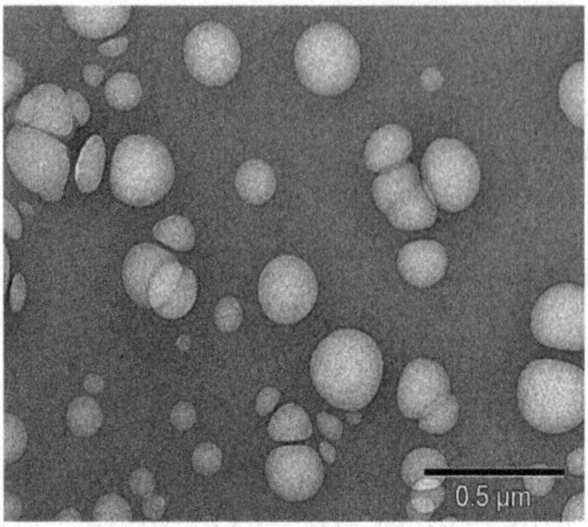

Fig. 1 TEM photomicrograph of gelatin nanoparticles [39]. Reused with permission from Elsevier

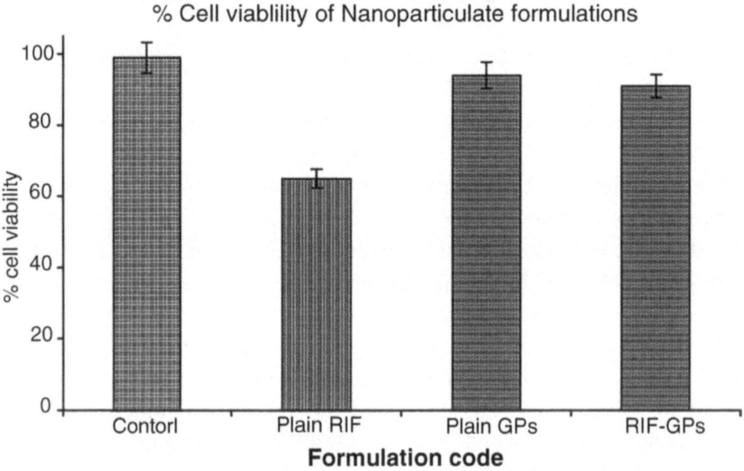

Fig. 2 Cell viability studies on J774 cells after 72 h exposure (values represent mean ± S.D., $n = 3$) [39]. Reused with permission from Elsevier

concentration. A significant evidence ($P \leq 0.05$) of lesser cytotoxicity was detected for GPs and RIF-GPs as compared to free RIF after 72 h of treatment. These results clearly indicated that GPs, even with a varied degree of modification, were biocompatible and nontoxic to normal J774 cells. They found that the gelatin nanoparticles showed improved chemotherapeutic efficacy of the drug as compared to conventional therapy. Therefore, they concluded that the prepared gelatin nanoparticle may be utilized as potential tool for the delivery of bioactives to the lung tissues leading to minimized side effects and improving the therapeutic efficacy of the drug [39].

Saraogi et al. [40] in the year 2011, prepared the mannosylated gelatin nanoparticles (Mn-GNPs) for the selective delivery of an antitubercular drug, INH, to the alveolar macrophages. They have used gelatin type A (Bloom 300). They have prepared the gelatin nanoparticles using a two-step desolvation method and efficiently conjugated with mannose. The size of nanoparticles (both plain and Mn-GNPs) was found to be in the range of 260–380 nm, and the maximum drug payload was found to be 40–55%. The average particle size of Mn-GNPs was more, whereas drug entrapment was lesser compared to plain GNPs. The organ distribution studies proved the efficiency of Mn-GNPs for spatial delivery of INH to alveolar tissues. Intravenous administration of INH-loaded Mn-GNPs (I-Mn-GNPs) resulted in a significant reduction in bacterial counts in the lungs and spleen of tuberculosis-infected (TB-infected) mice and also a reduction in the hepatotoxicity of the drug. They found that the mannose-conjugated GNPs may be explored as a potential carrier for safer and efficient management of TB through targeted delivery of INH when compared to plain GNPs and free drug [40].

Sharmah et al. [41] used gelatin (type B 75 bloom) as a potential drug carrier for controlled delivery applications. They have used cellulose whiskers (CWs) in controlling the release of the drug because CWs have the capacity to form strong hydrogen bonds. These CWs also provide good strength to the drug carrier material. They have prepared CWs from filter paper cellulose by acid hydrolysis. They have attempted to prepare gelatin-CWs nanoparticles by desolvation method using isoniazid as drug and glutaraldehyde as a crosslinking agent. They found the zeta potential values of the cross-linked gelatin nanoparticles in the range of −10.7 to −21.1 mV. The zeta potential values were decreased with the increase of CWs content. The decrease in surface charge might be owing to an increase in electrostatic interaction between the protonated amino groups of gelatin matrix and –OH groups of CWs. Both the swelling degree (%) and cumulative release (%) decrease with the increased content of CWs. Cytotoxicity study revealed that CWs were nontoxic to human lymphocytes and also gelatin nanoparticles containing CWs were less toxic than CWs-free nanoparticles. Their results suggested that gelatin-CWs nanoparticles have the potential uses in controlled drug delivery [41]. Sarfraz et al. [42], in the year 2016, worked on the immune response to antituberculosis drug-loaded gelatin and polyisobutyl-cyanoacrylate nanoparticles in macrophages. They have loaded anti-TB drugs (moxifloxacin and rifampicin) into gelatin (type B, 225 bloom) and polyisobutyl-cyanoacrylate nanoparticles. They further characterized the prepared nanoparticles. They also determined the cellular immune responses and cellular viability. Finally, they have concluded that the NPs together with the chemotherapeutic drugs might be able to trigger an immune response in macrophages. They found that the combined effect might be able to overcome mycobacteria infections [42].

5.2.2 Albumin-Based Polymers

Bovine serum albumin (BSA) is one of the most useful drugs carriers in the NP form. It is a biodegradable, nontoxic carrier that can be metabolized in vivo with the formation of harmless degradation products that are bioavailable, easily purified, and soluble in water, which enables delivery by injection [43]. Nanoparticle parameters (diameter, polydispersity, bioactive substance loading, and the yield of nanoparticle) are very important for drug transport through the bloodstream. Tazhbayev et al. [43] have used INH as the model drug and they have prepared BSA-INH NPs by an ethanol desolvation of an aqueous protein solution in the drug presence. They found that the properties of nanoparticles are significantly affected by the concentration of BSA, urea, L-cysteine, and the drug. The application of the Taguchi method is used for finding the optimal conditions for BSA-INH NPs [43]. Ma et al. [44] in the year 2022 have worked on the treatment of spinal tuberculosis in rabbits using bovine serum albumin nanoparticles loaded with isoniazid and rifampicin. They found that the INH-RFP-BSA-NPs showed the characteristics of sustained release in vivo and target biodistribution in focus vertebral body. They finally concluded on the basis of their findings that the therapeutic effect of the prepared formulations in rabbit spinal tuberculosis is much better than common INH and RFP [44]. Ge et al. [45] have prepared bovine serum albumin nanoparticles loaded with isoniazid and rifampicin (INH-RFP-BSA-NPs) by a modified self-emulsion solvent diffusion method, with albumin and polylactic acid used as carriers and to form the nanoparticles' structure. After that, they studied the drug release characteristics *in vitro*. They found that the drug loading and drug entrapment efficiencies were high, at 19.8% and 87.8% for isoniazid, respectively, and 20.1% and 98.0% for rifampicin, respectively. Drug release from the prepared nanoparticles was slow and sustained with 97.02% INH cumulative release at 6 days, and full release of RFP requiring 5 days [45].

Joshi and Prabhakar [46] have prepared rifampicin-loaded bovine serum albumin nanoparticles (RIF-BSA NPs) by desolvation method using 1-ethyl-3-(3-dimethylaminopropyl)carbodiimide (EDC) as the cross-linking agent. They found that the use of EDC reduces the time for cross-linking and makes the preparation method simple. RIF-BSA NPs confirmed the enhanced *in vitro* therapeutic efficacy assessed by MTb killing assay compared to the free drug suggesting the possibility of dose reduction. Further, they have found that the FITC-labelled RIF BSA NPs could be efficiently taken up by RAW264.7 cells infected with *Mtb* (H37rv), as confirmed using fluorescence microscopy. Finally, they concluded that the use of RIF-BSA NPs DPI formulation can be a promising strategy for the treatment of pulmonary tuberculosis [46].

5.3 Other Synthetic Polymers

Synthetic polymers used in pharmaceuticals such as polyesters (lactones) and acrylates, PLGA has been the most popular type [4]. Poly (lactic-co-glycolic acid) (PLGA) has been used most successfully for the fabrication of nanoparticles containing antitubercular drugs [4, 47].

5.3.1 PLGA Nanoparticles

Poly (lactic-co-glycolic acid) (PLGA) is one of the most successfully developed biodegradable polymers. Danhier et al. [47] reviewed on PLGA-based nanoparticles, and in their review, they present why PLGA has been chosen to design nanoparticles as drug delivery systems [47]. As patient non-compliance has been a major issue in the successful management of TB. Therefore, a lot more research works were going on the PLGA-based nanoparticles [47–49]. Horváti et al., in the year 2015, worked on antimycobacterial activity of peptide conjugate of pyridopyrimidine derivative against *Mycobacterium tuberculosis* in a series of in vitro and in vivo models. They found that PLGA nanoparticle for encapsulation of a pyridopyrimidine derivative exhibited improved antimycobacterial activity and low toxicity compared to the unencapsulated drug in an infected guinea pig model [49]. Sung et al. [50] have prepared PLGA nanoparticles containing rifampicin using a solvent evaporation process, spray-dried into porous nanoparticle-aggregate particle (PNAPs) containing varying amounts of nanoparticles. They further characterized the physical and aerosol properties of the prepared nanoparticles. *In vitro* release study of the prepared nanoparticles showed an initial burst of rifampicin, with the remainder available for release beyond eight hours [50]. Tripathi et al. [51] have prepared PLGA-based rifampicin nanoparticles using single and double evaporation method, solvent diffusion, and ionic interaction method. They further optimized the processing parameters involved in the method (drug/ polymer ratio, concentration of surfactant, phase ratio (organic phase/aqueous phase) and sonication time) to obtain small nanoparticles with maximum drug entrapment. They found that the release behavior of rifampicin showed a biphasic pattern described by an initial burst (11.26% in 1 days) release followed by a slower and continuous release (more than 30 days). They found that the technology would improve patient compliance, the lack of which is the major reason for the development of multidrug-resistant strains of mycobacterium. Further studies, such as confocal microscopy and study of accumulation of drugs in infected macrophages, can be done and are suggested as future scope of their work [51]. Malathi and Balasubramanian [52] worked on the synthesis of biodegradable polymeric nanoparticles and their controlled drug delivery for tuberculosis. They have attempted to develop a synthetic polymeric anti-TB nanodrug delivery system. A series of PLGA polymers with different molar feed ratios, i.e., 90/10, 75/25, 50/50, were synthesized by using direct melt polycondensation method. They prepared rifampicin-loaded PLGA nanoparticles by the double Emulsion-solvent evaporation method using PVA as a stabilizer. The average diameter of the PVA-coated PLGA-RIF nanoparticles is less than 250 nm. The in vitro release profile of the rifampicin-loaded PLGA nanoparticles showed an initial burst followed by sustained release. They found that the nanoparticles were remarkably advantageous in terms of high drug encapsulation efficiency, low polymer consumption, and better-sustained release profile. These systems could be cost-effective, feasible, and save valuable life and resources in the management of tuberculosis [52].

Hakkimane et al. [2] have prepared a nanoformulation of the two most effective first-line drugs, RIF and INH, which are used in both the phases of the 6-month TB therapy. Since INH is small and a highly hydrophilic molecule, it has low cellular penetration and also low drug loading efficiency in nanoformulation using hydrophobic USFDA-approved polymers like PLGA. This has led to a considerable hindrance in effective treatment with INH. To overcome this issue, they have modified INH into INH benz-hydrazone (IH2) by adding a hydrophobic moiety called benzaldehyde, a commonly used food additive, using Schiff base reaction. The newly formed IH2 is encapsulated in PLGA polymer, and its encapsulation in polymer is increased around 15-fold compared to INH encapsulation in PLGA. They found that RIF and IH2 loaded in NPs release in a slow and sustained manner over a period of 1 month, and they are more stable in NPs formulation compared to the free form. Finally, they have concluded that NP formulations will improve the efficacy of drug delivery for TB treatment [2].

Xie et al. [53] have developed PLGA nanoparticles encapsulating a conventional anti-TB drug (levofloxacin) to design more effective strategies against *Mtb*. They have prepared levofloxacin-nanoparticles using a double emulsification method. They further investigated the average diameter, zeta potential, polydispersity index, morphology, and drug release efficiency in vitro of the prepared LEV-NPs. In this study, the bactericidal effect and mechanism of LFLIU combined with levofloxacin-loaded PLGA nanoparticles on *M. smegmatis* in macrophages are investigated. The results support the potential of LFLIU combined with drug-loaded nanoparticles as a new, noninvasive, safe, and effective method for the treatment of TB [53]. In the same year, Liang et al. [48] prepared rifapentine (RPT)-loaded PLGA and PLGA–PEG NPs using premix membrane homogenization combined with solvent evaporation method. The aim of this work was to develop and characterize RPT-loaded PLGA-based nanoparticles for reducing dosing frequency. Their study revealed that in contrast to free drug, RPT-loaded NPs were more effective against *Mtb* in vitro [48].

6 Conclusion

Different polymers are used for the preparation of nanoparticles in tuberculosis treatment. As conventional treatment of tuberculosis is of long duration, so adherence to the treatment is difficult for the patients. Therefore conventional treatment has some limitations. Limitations of the conventional treatment can be overcome by preparing the polymeric nanoparticles. The polymeric nanoparticles possess good biocompatible and biodegradable characteristics. Therefore, nanoparticles effectively target macrophage. Most of the polymeric nanoparticles are prepared with anti-TB drugs with different manufacturing method, and these works are going on a laboratory scale. Still lot more effort and sophisticated techniques are needed for the development of polymeric nanoparticle for mass production from a laboratory scale. New technologies are necessary so that in near future these polymeric nanoparticles will be available in the market.

References

1. Doroudian M, MacLoughlin R, Poynton F, Prina-Mello A, Donnelly SC. Nanotechnology based therapeutics for lung disease. Thorax. 2019;74:965–76. https://doi.org/10.1136/thoraxjnl-2019-213037.
2. Hakkimane SS, Shenoy VP, Gaonkar SL, Bairy I, Guru BR. Antimycobacterial susceptibility evaluation of rifampicin and isoniazid benz-hydrazone in biodegradable polymeric nanoparticles against *Mycobacterium tuberculosis* H37Rv strain. Int J Nanomedicine. 2018:4303–18. https://doi.org/10.2147/IJN.S163925.
3. Limocon JRA, Madalag LMC, Reliquias PJB, Tionko JVS, Fermin JL, Kee SL, Tan MJT, Jonco MJJ, Pomperada MJF. Small but terrible: utilizing chitosan-based nanoparticles as drug carriers to treat tuberculosis in the philippines. Front Pharmacol. 2021;12 https://doi.org/10.3389/fphar.2021.752107.
4. Tafaghodi M, Khademi F, Shiehzadeh F, Firouzi Z. Polymer-based nanoparticles as delivery systems for treatment and vaccination of tuberculosis. In: Nanotechnology based approaches for tuberculosis treatment. Elsevier; 2020. p. 123–42. https://doi.org/10.1016/B978-0-12-819811-7.00008-4.
5. Stewart GR, Robertson BD, Young DB. Tuberculosis: a problem with persistence. Nat Rev Microbiol. 2003;1:97–105. https://doi.org/10.1038/nrmicro749.
6. Bhatt K, Salgame P. Host innate immune response to Mycobacterium tuberculosis. J Clin Immunol. 2007;27:347–62. https://doi.org/10.1007/s10875-007-9084-0.
7. Muller P, Pieters J. Modulation of macrophage antimicrobial mechanisms by pathogenic mycobacteria. Immunobiology. 2006;211:549–56. https://doi.org/10.1016/j.imbio.2006.06.004.
8. Sherman S, Rohwedder J, Ravikrishnan K, Weg J. Tuberculous enteritis and peritonitis. Report of 36 general hospital cases. Arch Intern Med. 1980;140:506–8. https://doi.org/10.1001/archinte.1980.00330160066028.
9. Global Tuberculosis Report 2017 (2017) 1–147. https://www.who.int/tb/publications/global_report/gtbr2017_main_text.pdf
10. Treatment of Tuberculosis, Guidelines for treatment of drug-susceptible tuberculosis and patient care 2017 Update, 2017
11. Grange JM, Zumla A. The global emergency of tuberculosis: what is the cause? J R Soc Promot Heal. 2002;122:78–81. https://doi.org/10.1177/146642400212200206.
12. Bhowmik D, Chiranjib R, Jayakar B, Kumar K. Recent trends of drug used treatment of tuberculosis. J Chem Pharm Res. 2009;1:127–33.
13. Lalloo UG, Ambaram A. New antituberculous drugs in development. Curr HIV/AIDS Rep. 2010;7:143–51. https://doi.org/10.1007/s11904-010-0054-4.
14. Grotz E, Tateosian N, Amiano N, Cagel M, Bernabeu E, Chiappetta DA, Moretton MA. Nanotechnology in tuberculosis: state of the art and the challenges ahead. Pharm Res. 2018;35 https://doi.org/10.1007/s11095-018-2497-z.
15. Banyal S, Malik P, Tuli HS, Mukherjee TK. Advances in nanotechnology for diagnosis and treatment of tuberculosis. Curr Opin Pulm Med. 2013;19:289–97. https://doi.org/10.1097/MCP.0b013e32835eff08.
16. Patel BK, Parikh RH, Aboti PS. Development of oral sustained release rifampicin loaded chitosan nanoparticles by design of experiment. J Drug Deliv. 2013;2013 https://doi.org/10.1155/2013/370938.
17. Plucinski A, Lyu Z, Schmidt BVKJ. Polysaccharide nanoparticles: from fabrications to applications. J Mater Chem B. 2021;9:7030–62.
18. Pourshahab PS, Gilani K, Moazeni E, Eslahi H, Fazeli MR, Jamalifar H. Preparation and characterization of spray dried inhalable powders containing chitosan nanoparticles for pulmonary delivery of isoniazid. J Microencapsul. 2011;28:605–13. https://doi.org/10.3109/02652048.2011.599437.
19. Garg T, Rath G, Goyal AK. Inhalable chitosan nanoparticles as antitubercular drug carriers for an effective treatment of tuberculosis. Artif Cells Nanomed Biotechnol. 2016;44:997–1001. https://doi.org/10.3109/21691401.2015.1008508.

20. Rawal T, Parmar R, Tyagi RK, Butani S. Rifampicin loaded chitosan nanoparticle dry powder presents: an improved therapeutic approach for alveolar tuberculosis. Colloids Surf B: Biointerfaces. 2017;154:321–30. https://doi.org/10.1016/j.colsurfb.2017.03.044.
21. Petkar KC, Chavhan S, Kunda N, Saleem I, Somavarapu S, Taylor KMG, Sawant KK. Development of novel octanoyl chitosan nanoparticles for improved rifampicin pulmonary delivery: optimization by factorial design. AAPS PharmSciTech. 2018; https://doi.org/10.1208/s12249-018-0972-9.
22. Lu E, Franzblau S, Onyuksel H, Popescu C. Preparation of aminoglycoside-loaded chitosan nanoparticles using dextran sulphate as a counterion. J Microencapsul. 2009;26(4):346–54. https://doi.org/10.1080/02652040802365182.
23. Wardani G, Mahmiah M, Sudjarwo SA. In vitro antibacterial activity of chitosan nanoparticles against Mycobacterium tuberculosis. Pharm J. 2018;10(1):162–6. https://doi.org/10.5530/pj.2018.1.27.
24. Gu X, Cheng Q, He P, Zhang Y, Jiang Z, Zeng Y. Dihydroartemisinin-loaded chitosan nanoparticles inhibit the rifampicin-resistant *Mycobacterium tuberculosis* by disrupting the cell wall. Front Microbiol. 2021;22 https://doi.org/10.3389/fmicb.2021.735166.
25. Junise V, Saraswathi R. Development and characterization of inhaled chitosan nanoparticles loaded with isoniazid. J Pharm Technol Res Manag. 2014;2:159–70. https://doi.org/10.15415/jptrm.2014.22011.
26. Dhamane SP, Jagdale SC. Development of rifampicin loaded hyaluronic acid coated chitosan nanoparticles. Eur J Mol Clin Med. 2020;7(1):3447–58.
27. Debnath SK, Saisivam S, Debanth M, Omri A. Development and evaluation of Chitosan nanoparticles based dry powder inhalation formulations of Prothionamide. PLoS One. 2018;13(1):e0190976. https://doi.org/10.1371/journal.pone.0190976.
28. Chogale MM, Gaikwad SS, Kulkarni SP, Savita P, Patravale VB. Quality-by-design enabled chitosan nanoparticles for antitubercular therapy: formulation, statistical optimization, and *in vitro* characterization. Curr Drug Ther. 2021;16(1):64–82. https://doi.org/10.2174/1574885515666200722150305.
29. Zahoor A, Sharma S, Khuller GK. Inhalable alginate nanoparticles as antitubercular drug carriers against experimental tuberculosis. Int J Antimicrob Agents. 2005;26(4):298–303. https://doi.org/10.1016/j.ijantimicag.2005.07.012.
30. Kumar S, Bhatt DC. Sodium Alginate nanoparticles of isoniazid: preparation and evaluation. Int J Pharm Sci Drug Res. 2019;11(6):382–6. https://doi.org/10.25004/IJPSDR.2019.110616.
31. Shaji J, Shaikh M. Formulation, optimization, and characterization of biocompatible inhalable d-cycloserine-loaded alginate-chitosan nanoparticles for pulmonary drug delivery. Asian J Pharm Clin Res. 2016;9:82–95.
32. Ahmad Z, Pandey R, Sharma S, Khuller K. Alginate nanoparticles as antituberculosis drug carriers: formulation development, pharmacokinetics and therapeutic potential. Indian J Chest Dis Allied Sci. 2006;48:171–6.
33. Scolari IR, Páez PL, Sánchez-Borzone ME, Granero GE. Promising chitosan-coated alginate-tween 80 nanoparticles as rifampicin coadministered ascorbic acid delivery carrier against Mycobacterium tuberculosis. AAPS PharmSciTech. 2019;20:67. https://doi.org/10.1208/s12249-018-1278-7.
34. Soumya RS, Ghosh SK, Abraham ET. Preparation and characterization of guar gum nanoparticles. Int J Biol Macromol. 2010;46:267–9. https://doi.org/10.1016/j.ijbiomac.2009.11.003.
35. Kaur M, Malik B, Garg T, Rath G, Goyal AK. Development and characterization of guar gum nanoparticles for oral immunization against tuberculosis. Drug Deliv. 2015;22(3):328–34. https://doi.org/10.3109/10717544.2014.894594.
36. Goyal AK, Garg T, Rath G, Gupta UD, Gupta P. Chemotherapeutic evaluation of guar gum coated chitosan nanoparticle against experimental tuberculosis. J Biomed Nanotechnol. 2016;12(3):450–63. https://doi.org/10.1166/jbn.2016.2180.
37. Elzoghby AO. Gelatin-based nanoparticles as drug and gene delivery systems: reviewing three decades of research. J Control Release. 2013;172:1075–91. https://doi.org/10.1016/j.jconrel.2013.09.019.

38. Yasmin R, Shah M, Khan SA, Ali R. Gelatin nanoparticles: a potential candidate for medical applications. Nanotechnol Rev. 2017;6(2):191–207. https://doi.org/10.1515/ntrev-2016-0009.
39. Saraogi GK, Gupta P, Gupta UD, Jain NK, Agrawal GP. Gelatin nanocarriers as potential vectors for effective management of tuberculosis. Int J Pharm. 2010;385:143–9. https://doi.org/10.1016/j.ijpharm.2009.10.004.
40. Saraogi GK, Sharma B, Joshi B, Gupta P, Gupta UD, Jain NK, Agrawal GP. Mannosylated gelatin nanoparticles bearing isoniazid for effective management of tuberculosis. J Drug Target. 2011;19(3):219–27. https://doi.org/10.3109/1061186X.2010.492522.
41. Sarmah M, Hussain A, Ramteke A, Maji TK. Isoniazid loaded gelatin-cellulose whiskers nanoparticles for controlled drug delivery applications. J Chem Sci. 2016;128(8):1291–301. https://doi.org/10.1007/s12039-016-1129-6.
42. Sarfraz M, Shi W, Gao Y, Clas S, Roa W, Bou-Chacra N, Lobenberg R. Immune response to antituberculosis drug-loaded gelatin and polyisobutyl-cyanoacrylate nanoparticles in macrophages. Ther Deliv. 2016;7(4):213–28.
43. Tazhbayev Y, Galiyeva A, Zhumagaliyeva T, Burkeyev M, Karimova B. Isoniazid – loaded albumin nanoparticles: Taguchi optimization method. Polymers. 2021;13:3808. https://doi.org/10.3390/polym13213808.
44. Ma R, Zhang J, Chen Z, Ma H, Liu X, Liang S, Wu P, Ge Z. Treatment of spinal tuberculosis in rabbits using bovine serum albumin nanoparticles loaded with isoniazid and rifampicin. Neurol Res. 2022;44(3):268–74. https://doi.org/10.1080/01616412.2021.1979749.
45. Ge Z, Ma R, Xu G, Chen Z, Zhang D, Wang Q, Hei L, Ma W. Development and *in vitro* release of isoniazid and rifampicin-loaded bovine serum albumin nanoparticles. Med Sci Monit. 2018;24:473–8. https://doi.org/10.12659/MSM.905581.
46. Joshi M, Prabhakar B. Development of respirable rifampicin loaded bovine serum albumin formulation for the treatment of pulmonary tuberculosis. J Drug Deliv Sci Technol. 2021;61:102197. https://doi.org/10.1016/j.jddst.2020.102197.
47. Danhier F, Ansorena E, Silva JM, Coco R, Le Breton A, Préat V. PLGA-based nanoparticles: an overview of biomedical applications. J Control Release. 2012;161(2):505–22. https://doi.org/10.1016/j.jconrel.2012.01.043.
48. Liang Q, Xiang H, Li X, Luo C, Ma X, Zhao W, Chen J, Tian Z, Li X, Song X. Development of rifapentine-loaded PLGA-based nanoparticles: in vitro characterisation and in vivo study in mice. Int J Nanomedicine. 2020;15:7491–507. https://doi.org/10.2147/IJN.S257758.
49. Horváti K, Bacsa B, Szabó N, Fodor K, Balka G, Rusvai M, Kiss É, Mező G, Grolmusz V, Vértessy B, Hudecz F, Bősze S. Antimycobacterial activity of peptide conjugate of pyridopyrimidine derivative against Mycobacterium tuberculosis in a series of in vitro and in vivo models. Tuberculosis (Edinb). 2015;95(1):S207–11. https://doi.org/10.1016/j.tube.2015.02.026.
50. Sung JC, Padilla DJ, Garcia-Contreras L, Verberkmoes JL, Durbin D, Peloquin CA, Elbert KJ, Hickey AJ, Edwards DA. Formulation and pharmacokinetics of self-assembled rifampicin nanoparticle systems for pulmonary delivery. Pharm Res. 2009;26(8):1847–55. https://doi.org/10.1007/s11095-009-9894-2.
51. Tripathi A, Gupta R, Saraf SA. PLGA nanoparticles of antitubercular drug: drug loading and release studies of a water in-soluble drug. Int J PharmTech Res. 2010;2:2116–23.
52. Malathi S, Balasubramanian S. Synthesis of biodegradable polymeric nanoparticles and their controlled drug delivery for tuberculosis. J Biomed Nanotechnol. 2011;7:150–1. https://doi.org/10.1166/jbn.2011.1244.
53. Xie S, Li G, Hou Y, Yang M, Li F, Li J, Li D, Du Y. A synergistic bactericidal effect of low-frequency and low-intensity ultrasound combined with levofloxacin-loaded PLGA nanoparticles on *M. smegmatis* in macrophages. J Nanobiotechnol. 2020;18 https://doi.org/10.1186/s12951-020-00658-7.

Solid Lipid Nanoparticles in Tuberculosis

Dwani Rana, Sagar Salave, Ritu Patel, Dignesh Khunt, Manju Misra, Bhupendra Prajapati, Geeta Patel, and Jayvadan Patel

Abstract Tuberculosis (TB) is a deadly disease spread by single infection agents. It creates a huge burden on the healthcare system for the development of effective therapeutics therapy. For the effective treatment of TB, the recent focus is on the avoid multidrug resistance and reducing toxicity. In this context, solid lipid nanoparticles (SLNs) are rising as an alternative and effective strategy to treat TB from the scientific literature data. SLNs are formulated using biodegradable and biocompatible solid lipids. Drugs are embedded into solid lipids which provide the sustain and control release of drugs. Surface modification of SLNs is highly useful for the target delivery of drugs to the lungs and also protection against multiple drug resistance. Still, research is going on in the formulation area for the development of effective delivery platform technology that creates continuous interest in its investigation.

Keywords Tuberculosis · Solid lipid nanoparticles · Pharmacokinetics · Antitubercular drugs · *Mycobacterium tuberculosis*

Abbreviations

M. tuberculosis	*Mycobacterium tuberculosis*
TB	Tuberculosis
SLNs	Solid lipid nanoparticles

D. Rana · S. Salave
NIPER-Ahmedabad, Gandhinagar, Gujarat, India

R. Patel · D. Khunt · M. Misra
Graduate School of Pharmacy, Gujarat Technological University, Gandhinagar, Gujarat, India

B. Prajapati (✉) · G. Patel
Shree S. K. Patel College of Pharmaceutical Education and Research, Ganpat University, Mahesana, Gujarat, India
e-mail: bhupen27@gmail.com

J. Patel
Nootan Pharmacy College, Sankalchand Patel University, Visnagar, Gujarat, India

R. Shegokar, Y. Pathak (eds.), *Tubercular Drug Delivery Systems*,
https://doi.org/10.1007/978-3-031-14100-3_6

NPs	Nanoparticles
AIDS	Acquired immunodeficiency syndrome
INH	Isoniazid
RIF	Rifampicin
PYZ	Pyrazinamide
ETB/EMB	Ethambutol
MDR	Multi-drug resistance
SDR	Single-drug resistance
XDR	Extensively drug resistance
PIT	Phase inversion temperature
SCF	Supercritical fluid
HPH	High-pressure homogenization
%EE	Percentage entrapment efficiency
%DR	Percentage drug loading
PDI	Polydispersity index
SEM	Scanning electron microscopy
TEM	Transmission electron microscopy
DSC	Differential scanning calorimetry
GI	Gastrointestinal
BBD	Box–Behnken design
STRS	Streptomycin sulfate
PBS	Phosphate buffer saline
SGF	Simulated gastric fluid
SIF	Simulated intestinal fluid
RFB	Rifabutin
GMS	Glyceryl monostearate
MN	Mannose
DPI	Dry powder inhaler

1 Introduction

Mycobacterium tuberculosis (M. tuberculosis) is the responsible bacteria for tuberculosis (TB) which attacks the lungs and also other organs such as the brain, kidney, and spine. TB is known to spread through the air and has a higher chance of contagion in AIDS (acquired immunodeficiency syndrome) [1, 2]. In the world, TB comes in the top 10 deadliest diseases from a single infectious agent [3]. TB is usually categorized as latent and active. Bacteria are present in the patient's body with no symptoms and without infectiousness known as latent TB. Active TB is communicable and has dynamic bacteria, having symptoms such as more than 3 days of coughing, chest pain, night sweats, blood in sputum, mass loss, fatigue, chills, and fever [1]. Conventional therapy includes various antituberculosis drugs such as isoniazid (INH), rifampicin (RIF), pyrazinamide (PYZ), and ethambutol (ETB). Apart from these, thioacetazone, kanamycin, and aminosalicylic acid are also used as treatment modalities [4]. However, conventional therapy possesses various drawbacks such as systemic side effects over the period, poor patient compliance owing to the long duration of therapy [3]. The bacteria have also acquired resistances *viz.*, multi-drug resistance (MDR), single-drug resistance (SDR), and extensive drug resistance (XDR), worsening management of the disease [5].

Various studies are carried out in this regard and the rising trends in delivery systems viz., microparticles, nanoparticles, microspheres liposomes, solid lipid nanoparticles, novel implants, carrier-based drug delivery systems, etc. These advancements may provide targeted drug delivery and sustained and controlled release, reducing the dosing frequency intending to improve patient compliance [2].

Solid lipid nanoparticles (SLNs) are extensively studied owing to their ability to improve bioavailability, provide a sustained release or controlled release, site-specific delivery, no carrier toxicity, hydrophilic and lipophilic drugs absorption enhancement, improvement of drug stability, and difficulty for scale-up [2]. SLNs are composed of solid lipid at room temperature which is stabilized by surfactant and the core is loaded with drug [6].

The system may consist of a solid core matrix of monoglycerides like glycerol monostearate, diglycerides like glycerol behenate, triglycerides like tristearin, steroids like cholesterol, fatty acids like stearic acid, and waxes like cetyl palmitate, surfactants for stabilizing the lipid dispersion, and emulsifiers for avoiding particle agglomeration [2].

2 Nanotechnology in TB

Nanotechnology has been proved for its valuable and unique role in the diagnosis of several diseases and effective drug delivery. Nanoscale dosage form enhances the stability and carrier capacity with the considerable enhancement of drug bioavailability. Other merits of nanoscale dosage forms are the sustained and controlled release of anti-TB drug. Ultimately, it reduce the dosage of anti-TB drug which is the major benefit for the chronic treatment of TB [7].

In the majority of cases, diagnosis of TB has required the higher-end analytical laboratoriesl which require intensive skilled labor. Despite this, currently, Mycobacterium DNA detection in clinical samples using nanoparticles has been established. Recently, Silva et al. has fabricated moveable and cheap optoelectronic platform amalgamate double color tuned light-emitting diode with amorphous/nanocrystalline silicon photodetector, flat spectral response, and integrating electronic for signal acquisition. With suitable software, the fabricated device could lessen the detection time and human error [8].

Some ultrasensitive techniques which utilize methodologies for detection such as Raman scattering, calorimetric, fluorometric, and electrochemical detection have been employed to detect the gold nanoparticle in a sputum sample. Thiol-modified oligonucleotides have been developed to detect the causative agent simply and efficiently. These nanoparticles help *M. tuberculosis* to differentiate from other members of *Mycobacterium* species [9].

3 Solid Lipid Nanoparticle

In the 1990s, SLNs were introduced as a substitute for nanoparticles, liposomes, and emulsion systems [9]. SLNs are nanoparticles that are composed of lipids and fall under the novel category of drug delivery systems. They also have a potential role in controlled and site-specific drug delivery and can incorporate various drugs by acting as nanocarriers. SLNS are spherical colloidal particles, composed of a lipidic core, surfactant, and water [3]. A combination of emulsifiers is used to stabilize the lipidic core and prevent its agglomeration [10].

3.1 Composition of SLN

SLNs are colloidal carrier-based matrix systems that are composed of solid lipid, aqueous, and surfactants or emulsifiers. Lipids constitute fatty acids, steroids, triglycerides, and waxes. Commonly, sodium cholate and phosphatidylcholine are used as a stabilizer in the formulation. SLNs comprising solid lipid and emulsifiers are generally safe concerning toxicity and biocompatibility. The solid lipid is made of either natural or chemical which is biodegradable [11].

3.2 Method of Preparation of SLN

SLNs are either prepared via a dispersion system that acts via a template or precursor or use of particular instrumentation [12]. Various techniques are: (A) HOT homogenization, (B) phase inversion temperature (PIT) method, (C) microemulsion dilution or cooling, (D) coacervation, (E) solvent injection, (F) solvent evaporation, (G) solvent diffusion, (H) SCF extraction of emulsions, (I) membrane contactor techniques, and (J) electrospray (Fig. 1).

Emulsions are a major category of precursors used for the preparation of SLNs [13]. The category of the emulsion can be direct, reversed, or can be multiple. The basic rationale behind using emulsions is that emulsions are solid at room temperature; when heated at a certain temperature about its melting point, they liquefy, which when emulsified with water at the same temperature and further cooling results in solidification of droplets to form SLNs [14].

3.2.1 High-Pressure Homogenization (HPH)

HPH is a well-known technique for the formulation of lipidic nanoparticles. This method was first applied by Muller and Lucks. This process engages dissolving or dispersing the drug in the liquefied lipid by maintaining the temperature greater

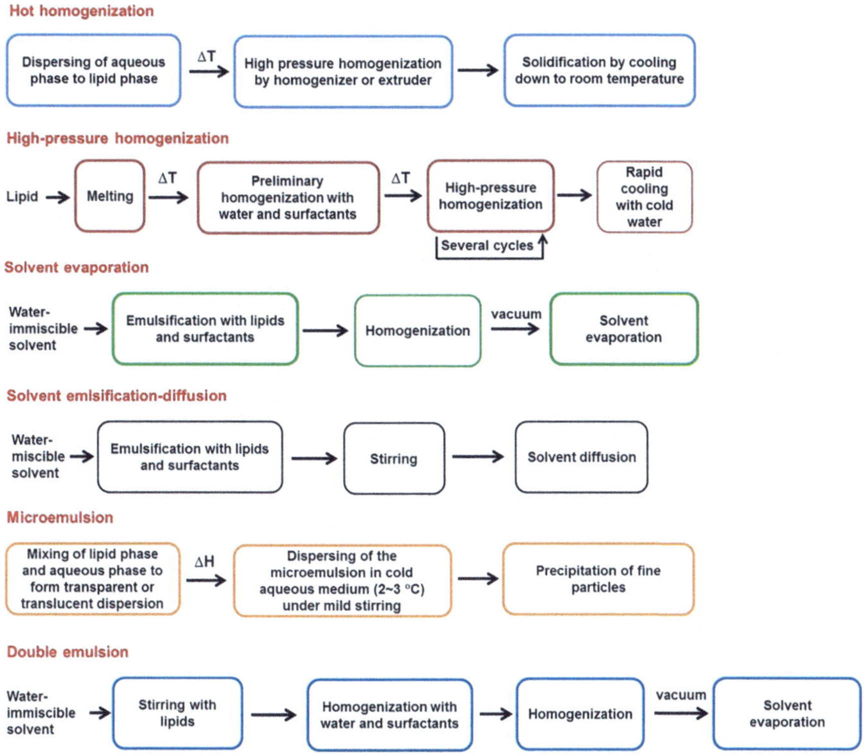

Fig. 1 Different types of methods for the formulation of solid lipid nanoparticles (Reproduced from Chih-Hung et al. (2017) [11] (2014), an open-access article distributed under the Creative Commons Attribution License that permits unrestricted use, distribution, and reproduction in any medium)

than the 5–10 °C of their melting point. Melted lipid is added to the HOT aqueous surfactant solution with continuous stirring. This dispersion of drug melt with an aqueous surfactant is called a pre-emulsion. Piston gap homogenizer is used for homogenizing the pre-emulsion. High pressure of up to 100 to 2000 bar helps to move the fluid in the narrow gap. When cooled, the lipid tends to recrystallize and form SLN [15]. Particle size is reduced due to a reduction in viscosity at higher temperatures [16].

High shear and ultrasound homogenization are some of the older techniques that were used for the production of SLNs. Compared to the homogenization technique, this method is easier to perform, easier to handle, however can compromise on the quality of SLNs that are formed. One of the key limitations is the broader particle size distribution which may lead to stability issues. If ultrasounds are used, the risk of metal contamination is added.

3.2.2 Phase Inversion Temperature (PIT) Method

A temperature at which a surfactant has equal affinity for both aqueous and lipid phases which called as PIT [17]. This method is based on temperature-provoked inversion of water in oil to oil to water emulsion and conversely. PIT is commonly used for the formulation of nanoemulsions. The rationale behind inversion is that on heating the ethoxy groups get dehydrated and further the lipophilicity of the surfactant increases and there is a decrease in HLB values. The use of non-ionic polyoxyethylated surfactants is a prerequisite of this method and must have temperature-dependent properties. This technique is based on the affinity of surfactants for oil and water-based on temperature [13]. Modifications of temperatures above or below the PIT can favor the inversion of emulsion. Oil, water, and surfactants are heated at a temperature that is greater than PIT under constant stirring which forms without emulsions. Both the phases need to be heated above the PIT and the aqueous phase has to be added to the oil phase under constant agitation. [18] These formed emulsions are rapidly cooled under constant stirring that precipitates the lipid. This method offers nanoparticles with narrow particle size distribution and size range.

3.2.3 Microemulsion Dilution

The first investigator to use this microemulsion template was Gasco and the method was developed in the early 1990s [19]. In this technique, a nanoemulsion is formed by diluting a microemulsion. The process involves the addition of drugs into the melted lipidic phase by exposing the lipids to a temperature above their melting point; an aqueous phase containing surfactant and water that is under isothermal conditions is added under constant agitation to the first phase, forming a thermodynamically stable microemulsion.

Further, this microemulsion is added in the cold aqueous phase (2–10 °C) with slow mixing which has 25–50 times more volume than that of HOT emulsion [20]. A nanoemulsion is formed on dilution as lipids tend to crystallize to form SLNs.

3.2.4 Coacervation Method

Battaglia et al. used the coacervation for preparation of SLNs [21]. A homogenous nanoparticle suspension is formed by coupling or interaction of a micellar soap (fatty acid alkaline salt) with an acidic solution which is a coacervating solution based on proton exchange. This interaction is carried out using various types of amphiphilic polymer which act as stabilizing agents [22]. A micellar soap solution is formed at a temperature above its Krafft point. Micellization is enhanced by the pre-dissolving drug in ethanol and then is dissolved in lipids. Coacervating solution is added dropwise to the solutions causing the lipids to precipitate. Acidification takes place at a temperature of 40–50 °C above the Krafft point of fatty acid sodium. The suspension is further cooled to a temperature of 15 °C to complete the precipitation of SLNs. The advantage of this method is that a homogenous stable nanosuspension can be formed by adjusting the concentration of lipid or by modifying the concentration of the micellar solution.

3.2.5 Supercritical Fluid-Based Method

A newer technique for the production of SLNs has the advantages of being solvent-free and is a method of interest for the past few years. This method offers particles as dry powder instead of suspensions. SCF is above the critical temperature and pressure of the fluid and can be modulated by changing pressure. CO_2 is a widely used SCF as it has a lower critical point of 31 °C and a critical pressure of 74 bar. Four main processes under SCF technology are: (1) rapid expansion of supercritical solutions, (2) GAS (gas anti-solvent) process, (3) particles from gas-saturated solutions/suspensions, and (4) supercritical fluid extraction of emulsions [13].

3.2.6 Membrane Contactor Technique

This technique produces SLNs with help of a membrane contactor. The module consists of two phases: lipid phase and the water phase, which are separated by a porous Kerasep ceramic membrane (0.1–0.45 μm pore size). The membrane has an active ZrO_2 layer on an ALO_2- TiO_2. The membrane allows water to circulate tangentially inside the module and takes away or detaches the particles formed at the outlet of the pore. Initially, the lipid phase is heated above its melting point and is pressed through the pores of the membrane which forms small droplets. The water dispersion is finally cooled to form SLNs. Pressure is created by nitrogen for the liquid phase to move [23]. The formation of SLN depends on the concentration of the lipid phase and the cross velocity of the aqueous phase. The surfactants added to the formulation will also affect the particle size of the nanoparticles [24].

3.2.7 Spray Drying Method

This method is a substitute for lyophilization and is comparatively a cheaper method and causes the transformation of an aqueous dispersion into a drug product. The particles are aggregated due to higher temperatures and by shear forces. Lipids with a melting point greater than 70 °C were recommended by Mullera and Freitas [14].

3.2.8 Drying of SLNs

In comparison to liquid forms, solid forms are more stable and preferable such as dry powder inhalers. Again dry powder is dispersed into the aqueous media when to use products. Generally dying can be achieved using spray drying, lyophilization, and electrospinning. Drying is ideal for the solid oral and topical SLNs formulation but for the pulmonary and parenteral route, SLNs must be sterilized. Heating-based sterilization is impacted on the release or leaching of the drug so not suitable. For sterilization of SLNs, liquid solutions filter out using membrane filter and it is a good alternative option [25].

3.3 Characterization of SLN

Physicochemical characterizations are requisite after the formulation of SLNs for quality control (Tables 1 and 2). Because formulation complexity, particle size, and dynamic nature of the delivery system are the prime concern for the characterization of SLNs. The majority of tools only measure the dilute sample, so the probable effect of sample preparation needs to be taken into consideration. SLNs are evaluated for the degree of crystallinity, particle size, zeta potential entrapment efficiency (% EE), and surface morphology [26]. Table 2 summarizes the various methods and characterization of developed SLNs of different drugs used in the treatment of TB. Similarly, in Table 3 various pharmacokinetic parameters are tabulated.

Table 1 Different characterization parameters of SLNs

Sr. no	Parameter	Commonly used techniques
1.1.1.	Size	Photon correlation spectroscopy, Dynamic light scattering, Quasi-elastic light scattering, Laser diffraction, Coulter counter
	PDI	
	Zeta potential	
2.	Surface morphology	Scanning electron microscopy (SEM), Transmission electron microscopy (TEM), Cryo-electron microscopy, Atomic force microscopy
3.	Degree of crystallinity	Differential scanning calorimetry, Powder X-ray diffractometry
4.	Encapsulation efficacy	Centrifugation, Micro-centrifugation techniques
5.	Interaction or excipients to drug compatibility studies	Differential scanning calorimetry
6.	Drug release studies	In-vitro release of drug

Table 2 Tabular representation of method and characterization of developed SLNs, NLCs, lipid capsules of different TB drugs

Anti-TB drugs	Method of preparation	Route of drug delivery	Characterization parameters	Refs.
Rifampicin/Isoniazid	High shear homogenization/ultrasonication	Oral	PS: 150 nm	[27]
			ZP: −26 mV	
			%EE:	
			RIF: 84%	
			INH: 74%	
Isoniazid	Microemulsion/precipitation	Oral	PS: 132 nm	[28]
			PDI: 0.3	
			%EE: 70%	
			%DL: 95%	
Rifampicin	Microemulsion/high-pressure homogenization	Oral	PS: 460 nm	[29]
			%EE: 85%	
			%DL: 16%	
Rifampicin	Modified lipid film hydration	Inhalatory	PS: 830 nm	[30]
	Lyophilization		PDI: 0.2	
			%EE: >85%	
Rifabutin	High shear homogenization/Lyophilization	Inhalatory	PS: 100–300 nm	[31]
			PDI: 0.2	
			%ZP: −18 to −28 mV	
			%EE: 58–70%	
			%DL: 4.2–6.8%	
Isoniazid	High shear homogenization/Lyophilization	Oral and Intravenous	PS: 167 nm	[30]
			PDI: 0.25	
			%EE: 98%	

(continued)

Table 2 (continued)

Anti-TB drugs	Method of preparation	Route of drug delivery	Characterization parameters	Refs.
Streptomycin sulfate	Microemulsification-dilution	Intranasal	PS: 140 nm	[32]
			PDI: 0.3	
			%EE: 55%	
Isoniazid	High-pressure homogenization/ultrasonication/ Lyophilization	Not reported	PS: 120–185 nm	[30]
			PDI: 0.2–0.3	
			ZP: −30 to −35 mV	
			%EE: 62–73%	
Ethambutol hydrochloride	HOT homogenization and ultrasonication, Spray drying	Inhalatory	PS: <100 nm	[33]
			PDI: 0.2–0.5	
			%EE: >98%	
			%DL: 10–30%	
Rifabutin	Solvent injection method	Inhalatory	PS: 250 nm	[30]
			PDI: 0.4	
			ZP: 3.4 mV	
			EE: 88%	
Rifabutin	Solvent diffusion evaporation method	Oral	PS: 350 nm	[34]
			PDI: 0.3	
			%EE: 60%	
Rifampicin, isoniazid, pyrazinamide	Emulsion solvent diffusion/centrifugation/vacuum dried	Oral/ Nebulization	%EE:	[35]
			Rifampicin ~50%, isoniazid 40%, pyrazinamide 40%	
Isoniazid	Microemulsification	Ocular	PS: 150 nm	[36]
			PDI: 0.15	
			ZP: −0.35 mV	
			%EE: 65%	

Anti-TB drugs	Method of preparation	Route of drug delivery	Characterization parameters	Refs.
Rifampicin	Emulsion solvent diffusion Centrifugation – Lyophilization	Oral	PS: 450 nm	[37]
			PDI: 0.4	
			%EE: 30–79%	
Rifampicin	HOT ultrasonication	Inhalatory	PS: 160 nm	[38]
	Lyophilization		PDI: <0.2	
			ZP: – 25 mV	
			%EE: 75%	
			%DL: 2.7%	
Isoniazid	Double emulsion technique	Inhalatory	PS: 500 nm	[38]
			PDI: 0.4	
			ZP: +27 to +39 mV	
			%EE: 35–50%	
			%DL: 1.5–2.3%	
Rifabutin	HOT high shear homogenization/Spray drying	Inhalatory	PS: 4.3–5.2 mm	[31]
			EE: 80–90%	
			DL: 6–9%	
Rifampicin	Melt emulsifying technique by sonication Lyophilization	Inhalatory	PS: 0.47–1.72 mm	[39]
			PDI: 0.3–0.9	
			ZP: −40 mV to −50 mV	
			%EE: 43–61%	
			%DL: 11–16%	

(continued)

Table 2 (continued)

Anti-TB drugs	Method of preparation	Route of drug delivery	Characterization parameters	Refs.
Isoniazid	HOT ultrasonication	Inhalatory	PS: 236 ± 9 nm	[40]
	Lyophilization		PDI: 0.240 ± 0.012	
			ZP: −19 ± 2 mV	
			%EE: 75.13 ± 0.97%	
			%DL: 10.46 ± 1.24%	

Reproduced from Gordillo-Galeano A, Ospina-Giraldo LF, Mora-Huertas CE. Lipid nanoparticles with improved biopharmaceutical attributes for tuberculosis treatment. International Journal of Pharmaceutics. 2021 Mar 1;596:120321. with permission of the copyright holder, Elsevier, Amsterdam) [3]
**PS* particle size, *PDI* polydispersity index, *ZP* zeta potential, *EE* entrapment efficiency, *DL* drug loading

Table 3 Different Pharmacokinetics parameter of developed anti-TB SLNs from reported literature works

Anti-TB drug carrier	Route of drug delivery	T_{max} (h)	C_{max} (mg/mL)	AUC_{0-t} (mg h/mL)	Refs.
RIF	Oral	3.33 ± 0.58	8.34 ± 0.84	32.51 ± 1.45	[27]
RIF SLNS		4.66 ± 1.15	8.43 ± 0.19	294.44 ± 29.95	
INH	Oral	2.34 ± 0.58	10.92 ± 1.62	34.32 ± 2.67	[28]
INH-SLNs		3.33 ± 1.15	11.25 ± 2.58	170.52 ± 16.01	
STRS	Intranasal	4.00 ± 0.03	0.02 ± 0.01*	0.49 ± 0.01**	[32]
STRS-SLNs		2.00 ± 0.03	0.22 ± 0.02*	3.78 ± 0.60***	
RFD	Oral	2	3.9	21.37	[34]
RFD_SLNs		4	3.44	73.57	
INH	Ocular	0.25	15.09 ± 3.10	44.00 ± 4.62	[36]
INH-SLNs		0.25	23.31 ± 3.10	188.13 ± 37.10	

Reproduced from Gordillo-Galeano A, Ospina-Giraldo LF, Mora-Huertas CE. Lipid nanoparticles with improved biopharmaceutical attributes for tuberculosis treatment. International Journal of Pharmaceutics. 2021 Mar 1;596:120321. with permission of the copyright holder, Elsevier, Amsterdam [3]

*Cmax expressed as % radioactivity/g

**AUC expressed as % radioactivity·h/g

***Pharmacokinetic parameters measured in ocular aqueous humor

T_{max} time taken to reach peak plasma concentration, C_{max} maximum concentration, AUC_{0-t} area under the curve of a plasma (or ocular aqueous humor) concentration versus time profile

4 Anti-TB Therapy of SLN

4.1 Oral Delivery

SLN protects the drug against chemical and enzymatic attacks in the gastrointestinal (GI) tract. It increases the residence time and permeation of the drug across the GI epithelium. Figure 2 shows the different mechanism by which SLN cross the GI epithelium.

Obinu et al. formulated SLNs as a strategy for increasing oral permeability of a new compound-SS13, having potential against multidrug-resistant tuberculosis. For the formulation of SLN, witepsol and/or gelucire were used for improving the solubility of SS13. SLNs were formulated using the solvent evaporation method having a size range of 200–450 nm. The percentage drug loading of 86.84 ± 2.49% and 100.00 ± 3.11% was obtained with witepsol and gelucire, respectively. The fabricated system exhibited desired intestinal absorption and improved permeation across intestinal mucosa comparison to the free drug [41].

Pandey et al. developed oral SLN using the emulsion solvent diffusion technique and antitubercular chemotherapy incorporating isoniazid (INZ), rifampicin (RIF), and pyrazinamide (PYZ). Drug encapsulation efficiency was found to be 51 ± 5%, 45 ± 4%, and 41 ± 4% for rifampicin SLN, isoniazid SLN, and pyrazinamide SLN,

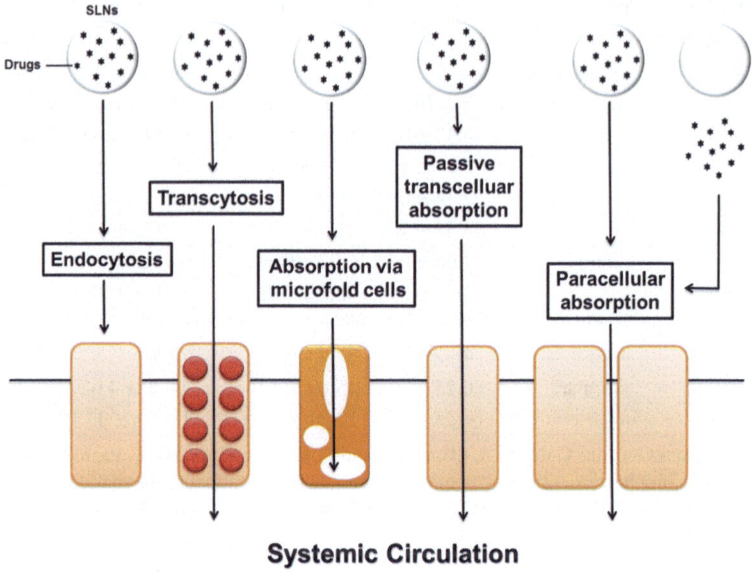

Fig. 2 Different mechanism by which SLN cross the GI epithelium (Reproduced from Chih-Hung et al. (2017) [11] (2014), an open access article distributed under the Creative Commons Attribution License that permits unrestricted use, distribution, and reproduction in any medium)

respectively. Incorporation of PVA served as a barrier to diffusional release and hence sustained release was observed following single oral delivery [35].

Similarly, Khatak et al. formulated SLNs of antituberculosis drugs RIF, INZ, and PYZ for oral administration to treat *M. marinum* infection. SLNs were formulated using the microemulsion technique and optimization was carried out using central composite design (CCD). The %EE was found to be 86.40 ± 0.274, 83.84 ± 0.269, and 81.43 ± 0.576 of optimized IF, INH, and PYZ loaded SLN, respectively. The *in vitro* drug release study confirmed the maximum release of 85.21%, 92.47%, and 88.30% for RIF, INH, and PYZ, respectively, in PBS (pH 6.8) after 24 h. Two times growth inhibition of *M. marinum* was observed by SLNs compared to standard anti-tubercular drugs in modified MMT assay on murine macrophage cell line (RAW 264.7) [42].

Rifampicin possesses certain limitations such as poor bioavailability, hepatic first-pass metabolism, and GI instability when orally administered with isoniazid. To overcome these challenges, Chokshi et al. developed rifampicin-embedded SLNs using the design of experiment for the treatment of tuberculosis. In the study, SLNs were formulated using the HPH method. Effect of Independent factor *viz.*, concentration of drug, concentration of emulsifiers, and homogenization pressure, on dependent factor *viz.*, particle size, % loading of a drug (%DL), and %EE, was studied using 3^3 Box–Behnken design (BBD). The developed RIF-SLNs displayed 456 ± 11 nm, 84.12 ± 2.78%, and 15.68 ± 1.52% of mean diameter, %EE, and %DL, respectively. SLNs prepared using poloxamer 188 revealed an anti-lipolytic effect.

The result of *in vitro* dissolution at different pH (1.2, 4.5, 6.8, and 7.4) was suggested that biphasic drug release profile for drug-loaded SLNs with the stability of developed formulation throughout the GIT. Fabricated SLNs were stable following accelerated stability studies and the system could be employed for efficient management of tuberculosis [29].

Banerjee et al. developed SLNs loaded with rifampicin (RIF) and isoniazid (INZ) for the management of tuberculosis. The human macrophage cell line (THP-1) was used for the *in vitro* internalization evaluation. The authors fulfilled that formulated systems could be a potential tactic for delivery of drug with enhanced permeation through intestinal, lymphatic, and cellular uptake. As a result of this, oral bioavailability is improving an ultimately better antitubercular effect [27].

Singh et al. developed SLNs of hydrophilic streptomycin sulfate (STRS) showing improved oral bioavailability and enhanced macrophage uptake. The SLNs were prepared to employ the cold HPH technique giving drug loading of 30% and entrapment efficiency of 51.17 ± 0.95%. Incorporation of polyethylene glycol 600 imparted mucus-penetrating property and attributed to the small particle size of nanoparticles (218.1 ± 15.46 nm). The developed system was found to be stable in SGF/SIF and displayed zero-order drug release. Intracellular uptake was 60 times enhancement experiential in THP-1 and Pgp expressing LoVo and DLD-1 cell lines with the bits of help of fluorescein-SLNs. In comparison to free drugs, STRS-SLNs demonstrated a three-fold reduction in MIC against *M. tuberculosis H37RV (256182)*. The developed system also depicted enhanced oral absorption and bioavailability validating its potential in the treatment of TB [43].

Nirbhavane et al. developed a sustained release oral rifabutin (RFB). SLN was formulated using solvent diffusion evaporation method and glyceryl monostearate (GMS) as solid lipid. The developed RFB-SLNs were showing 345 ± 17.96 nm and 0.321 ± 0.09 particle size and PDI, respectively. Stability studies of RFB-SLNs in different GI fluids suggested that developed SLNs were well stable. Drug release of SLN-loaded formulation in SIF and PBS was showing up to 48 h and 7 days sustain release, respectively. RFB-SLN was showing a five-time increment of oral bioavailability in comparison to the free drug [34].

Chokshi et al. investigated the development of surface engineered rifampicin-loaded lipid nanoparticulate system for the management of pulmonary tuberculosis. They fabricated mannose-appended rifampicin containing SLN (Mn-RIF-SLNs). The developed Mn-RIF-SLNs showed particle size of 479 ± 13 nm and percentage entrapment efficiency of 79.41 ± 2.42%. The developed nanoparticles were found to be non-toxic and safe compared to the free drug in the cytotoxicity performed on the J774A.1 cell line. The intracellular uptake was also significantly enhanced by 1.79-folds and bioavailability by around 17-folds via oral routes. The biodistribution studies signified the 1.8 times higher lung accumulation of Mn-RIF-SLNs in comparison to RIF SLN, which shows the effectiveness of Mn-RIF-SLNs for the treatment of TB [44] shown in Fig. 3.

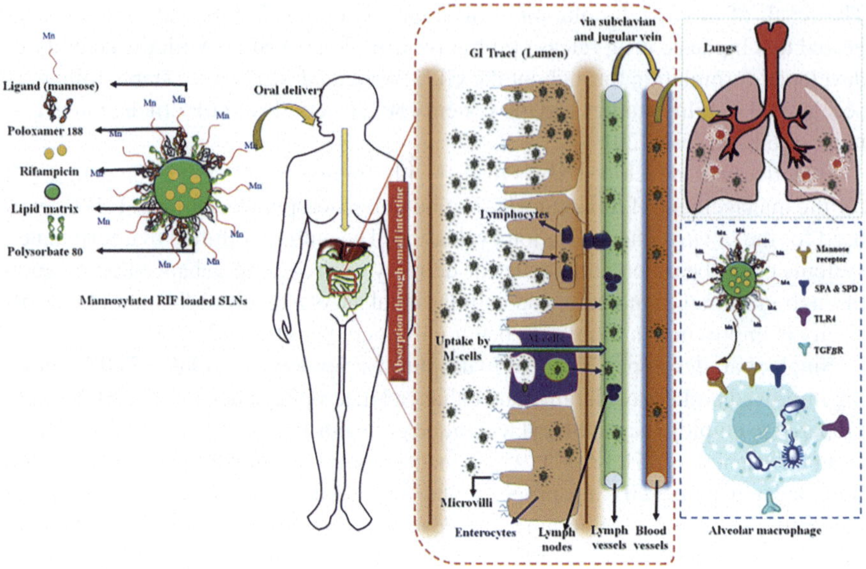

Fig. 3 Diagrammatic representation of targeted mannose rifampicin SLN for the treatment of TB via oral routes (Reproduced from Chokshi, Nimitt V., et al. "Fabrication and Characterization of Surface Engineered Rifampicin Loaded Lipid Nanoparticulate Systems for the Potential Treatment of Tuberculosis: An In Vitro and In Vivo Evaluation." Journal of Pharmaceutical Sciences 110.5 (2021): 2221–2232. [44] with permission of the copyright holder, Elsevier, Amsterdam)

4.2 Mucoadhesive Drug Delivery

Vieira et al. developed mucoadhesive chitosan-coated SLNs (C-SLNs) loaded with RIF for pulmonary delivery across alveolar epithelial cells. The developed SLNs had a smooth spherical shape with a size range of 245–344 nm. Zeta potential was found to be −30 mV for SLNs and +40 mV for C-SLNs. Chitosan-SLNs depicted elevated *in vitro* mucoadhesive properties and higher permeability in alveolar epithelial cells A549 comparison to uncoated SLNs, indicating that the developed chitosan-SLNs can be used as a promising carrier for safer and efficient management of TB [38].

4.3 Ocular Delivery

Singh et al. developed SLN of isoniazid for ocular delivery of highly water-soluble isoniazid (INH) by micro emulsification. The developed SLNs system was showing the particle size of 149.2 ± 4.9 nm, the zeta potential of −0.35 ± 0.28 mV, PDI of 0.15 ± 0.02, entrapment efficiency of 65.2 ± 2.2%, and extended-release of 48 h. Further, fluorescein-labeled SLNs displayed considerable *in vitro* and *in vivo* uptake, enhanced corneal permeability (1.6 times), and five-times lower MIC. Further, the system exhibited efficient targetability and improved localized concentration of INH in aqueous humor displaying an effective ocular delivery system [36].

4.4 Inhalation Drug Delivery

Nemati et al. developed ethambutol-loaded SLNs as dry powder inhaler (DPI) formulation for tuberculosis therapy. Given orally, ethambutol leads to cellular toxicity whereas direct administration into lungs (i.e., pulmonary administration) might overcome this side effect. EMB-loaded SLNs were formulated by HOT homogenization and ultrasonication. DPI formulations of EMB-loaded SLNs with and without mannitol were formulated using a spray dryer. The developed system had a particle size of sub-100 nm and encapsulation efficiency higher than 98%. Biocompatibility and non-toxicity of EMB-loaded SLNs were confirmed via MTT assay. The above-mentioned study evidenced that EMB-loaded SLNs as DPI preparation could be used in direct drug delivery to the lung [33].

Gaspar et al. developed a hybrid platform for pulmonary delivery of antibiotics using RFB-loaded SLN (glyceryl dibehenate and glyceryl tristearate lipids) embedded in mannitol and trehalose. Encapsulation was carried out spray-drying method, which resulted in dry powders with desired characteristics for pulmonary administration. The liposomes developed using lipid-glyceryl dibehenate had particle size of 108 ± 5 nm, % EE of 89.9 ± 5.1%, and DL of 9.0 ± 0.5%, whereas glyceryl tristearate containing SLNs depicted particle size of 191 ± 7 nm, % EE of 81.0 ± 9.6%, and DL of 6.0 ± 0.7%. Both these formulations used tween 80 as the surfactant component displayed suitable polydispersity index values (<0.2). Results showed that both the formulations had the desired biodistribution and reached deep into the lungs. This RFB-SLN microencapsulation approach could become a potential strategy for treating TB [31].

Intramacrophagic delivery of antitubercular drugs can be achieved by DPI via the pulmonary route. Similar type studies were carried out by Maretti et al. wherein they developed rifampicin-loaded SLNs (SLNs) by the melt emulsifying technique and go after freeze-drying. Design of experiments (DoE) was applied for evaluating the effect of pre-freezing conditions and a >50% SLNs-respirable fraction was found with adequate yields [39].

Active drug targeting is known to overcome the challenges of conventional therapy. Maretti et al. also attempted the formulation of inhalable lipid NPs for the delivery of drugs to the infected tissue. They developed RIF-loaded SLNs surface modification with novel mannose derivatives for TB targeted delivery via inhaled route by DPI. Mannose receptors are overexpressed on the membranes of infected alveolar macrophages (AM), a preferred site of *M. tuberculosis*, and in this regard, mannose was considered a relevant ligand for achieving active drug targeting. The fabricated SLNs demonstrated a suitable drug payload, *in vitro* release, and more efficient ability to enter macrophages (≅80%) compared to bare RIF (≅20%) and non-functionalized SLNas (≅40%). This technology was regarded as a green process, suitable for scale-up development [45].

Cheng Ma et al. developed mannose SLN (MAN-IC-SLN) encapsulated the pH-sensitive prodrug of INH for the treatment of latent tuberculosis infection. 6-Octadecylimino-hexane-1,2,3,4,5-pentanol (MAN-SA) decorated of SLNs was demonstrated a 97.2% cell uptake in macrophages comparison to unmodified SLNs (42.4%). The result of *in vivo* antibiotic efficacy test is showing 83% decrease in the colony-forming unit with SLN formulation in comparison to the free INH group (60% decreases). The study exhibited that macrophage targeting and pH-sensitive SLNs could be used as a potential platform for latent TB infection [40].

Mannosylated SLNs for the selective delivery of rifampicin to macrophages was also developed by Vieira et al. A sustained release profile was achieved in simulated pulmonary fluids, with more than 80% of RIF remaining entrapped inside the SLNs lipid matrix after 8 h. The developed formulation could increase the drug's bioavailability by effective target delivery and a high degree of internalization in human macrophages [31]. Vieira et al. also attempted to target rifampicin to alveolar macrophages by developing chitosan-coated SLNs, promoting mucoadhesion and permeation. The developed formulation had a monodisperse population having a diameter of around 250–500 nm adequate for lung deposition, zeta potential between −31 mV for uncoated SLNs and +33 mV for coated SLNs, the % EE of approximately 90%, and loading capacity (LC) 4.5% [46].

Kumar et al. developed streptomycin sulfate SLNs (STRS-SLNs) for non-invasive intranasal (IN) delivery. STRS-SLNs were developed using the nano colloidal aqueous dispersion technique, achieving particle size of 140.1 ± 7.0 nm and entrapment efficiency of 54.83 ± 2.1%. *In vivo* studies following intranasal delivery of STRS-SLNs showed significant biodistribution in the brain and blood and lower levels of the drug in the liver and spleen of mice compared to free STRS. The developed system owing to its smaller size, lipophilic enclosure, and presence of tween 80 imparted inhibition of Pgp-efflux and better bioavailability [32].

4.5 Advantages of SLNs for Sustaining the Delivery of TB Drugs and Its Industrial Feasibility

SLNs are versatile carriers, and the incorporation of drugs into SLNs improves their bioavailability and stability [35]. SLNs are known to demonstrate a longer retention time in the lungs owing to their hydrophobicity. SLNs exhibit industrial feasibility and capability of large-scale production due to the highly developed techniques, abundant material supply, and systematic product development. Moreover, precise particle size control and possibilities of surface modification of SLNs have opened great possibilities for achieving the target delivery of anti-TB agents [40].

Possibility of Surface Modification

Several studies have been carried out for developing surface-modified SLNs. Costa et al. developed INH-loaded SLN, reinforced with stearyl amine and surface-functionalized with mannose. The fabricated functionalized SLNs were devoid of toxicity when tested in a human lung epithelial cell line (NCI-H441) and reduced intrinsic cytotoxicity of INH upon incorporation into SLN. These functionalized nanocarriers may signify a beneficial platform for targeting alveolar macrophages to deliver anti-infective drugs [47]. Vieira et al. developed SLNs using mannose as a lectin receptor-ligand conjugate for macrophage targeting and to increase the therapeutic index of RIF. The developed SLNs yielded particle size in the range of 160–250 nm, and drug encapsulation efficiency was above 75%. Results suggested that mannosylation improved internalization in macrophages, and the surface-modified formulation could be a promising tool for targeted tuberculosis therapy [48]. Similarly, Truzzi et al. developed respirable SLN assemblies (SLNas) loaded with RIF for an alveolar macrophage passive targeting. In this study, mannosylated SLNas were exploited for alveolar macrophage active targeting to mannose receptors, located on the macrophage membrane, and overexpressed in the case of *Mycobacterium tuberculosis* infection. For achieving this, SLNas were surface-decorated with a newly synthesized mannosylated derivative acting as both surfactant, required for the producing nanoparticle, and functionalized agent for the alveolar macrophage active targeting. Results suggested the ability of mannosylated SLNas to interact with mannose receptors on J774 and MH-S macrophage cell lines improving cell internalization ability in comparison with non-functionalized SLNas and bare RIF [49].

NLCs/Lipid Capsule Research for TB and Close TB Drugs

Nanostructured lipid carriers (NLCs) are second-generation SLNs developed using a mixture of liquid and solid lipids at room temperature that form an imperfect lipid matrix. Moreover, NLCs can incorporate a wider concentration of surfactants, co-surfactants, and lipophilic drugs compared to SLNs. NLCs exhibits improved physical and chemical stability. They favor controlled/sustained release of drugs and expulsion during storage is also minimized. Moreover, toxicity and adverse effects are also reduced. Owing to such features, research has been carried out of NLCs for delivering anti-TB drugs. Sato et al. developed copper (II) complex-loaded NLCs for the treatment of *M. tuberculosis*. Evaluation of the antimicrobial activity depicted the activity and selectivity of the copper (II) complexes against *Staphylococcus aureus*, *Escherichia coli*, and *M. tuberculosis*. Moreover, increase in solubility of the active substance was achieved and a controlled-release system, maintaining the drug plasma concentration within the therapeutic range for several hours or days, was created, decreasing the toxicity and increasing patient compliance [50]. Nemati et al. formulated Ethambutol hydrochloride-loaded NLCs using HOT homogenization followed by ultrasonication technique. The particle size and encapsulation efficiency EMB-loaded NLCs were below 100 nm and >98%, respectively. The developed NLCs exhibited low toxicity and desirable biocompatibility [51]. Pinheiro et al. designed, developed, and characterized NLCs for the selective delivery of RFB to alveolar macrophages. The NLCs were synthetized using high-shear homogenization and ultrasonication techniques. The developed NLCs exhibited both passive and active targeting strategies to be efficiently internalized by the alveolar macrophages, traffic to the acidified phagosomes and phagolysosomes, and release bactericidal concentrations of the antituberculosis drug intracellularly. The system was further functionalized with mannose and characterized. Functionalized NLCs depicted pH-sensitive drug release with a faster drug release at acidic pH than at neutral pH. The nanocarrier was found to be stable and effective for managing tuberculosis [52].

5 Limitations to Consider on SLN

Due to compose of the solid lipid, SLNs are more prone to the stability and degradation problem. Different issue related to these are the high pressure-prone drug degradation, crystallization of lipid, and presence of other colloidal carriers.

High pressure-prone drug degradation is the main issue related to higher molecular weight and long-chain atoms drug such as protein and DNA in comparison to low molecular weight drugs. [53].

Crystallization of lipid is a crucial problem and study form the many years. Now day determination of lipid modifications methods is well developed such as Differential Scanning Calorimetry and X-ray diffractometry. The focus was given on the drug entrapped in the SLNs. Surfactants are present on the surface and also inside the core of lipids. The additionally available surfactant may form the micelles and embed the drugs inside. It is one of the important factors for drug release and also stability. The solution to these types of problems is increasing the viscosity of the lipid matrix, so redistribution of drugs can be inhibited [54].

6 Conclusion

SLNs are made of the solid lipid matrix using scalable methods. SLNs are of great benefit for the targeting of TB due to high and frequent dosing, low solubility, and targeting properties. One of the bigger benefits of SLNs is it is highly useful against multiple resistance TB in comparison to conventional therapy. One of the key areas to modify the surface of SLN for pulmonary targeting leads to the development of novel anti-TB treatments. For completion of these, need to the establishment of extensive preclinical pharmacokinetics with pharmacodynamic studies for supportive documents. Results of preclinical studies of animals are extrapolated into human pharmacokinetic and pharmacodynamic studies. Proving the safety with the efficacy of SLNs leads to successful market introduction of SLN for the treatment of TB.

References

1. Ahmed El-Harati A, Charcosset C, Fessi H. Influence of the formulation for solid lipid nanoparticles prepared with a membrane contactor. Pharm Dev Technol. 2006;11(2):153–7.

2. Anon. Formulation and characterization of ethambutol loaded nanostructured lipid carrier
3. Banerjee S, Roy S, Bhaumik KN, Pillai J. Mechanisms of the effectiveness of lipid nanoparticle formulations loaded with anti-tubercular drugs combinations toward overcoming drug bioavailability in tuberculosis. J Drug Target. 2020;28(1):55–69.
4. Banyal S, Malik P, Tuli HS, Mukherjee TK. Advances in nanotechnology for diagnosis and treatment of tuberculosis. Curr Opin Pulm Med. 2013;19(3):289–97.
5. Battaglia L, Gallarate M, Cavalli R, Trotta M. Solid lipid nanoparticles produced through a coacervation method. J Microencapsul. 2010;27(1):78–85.
6. Battaglia L, Gallarate M, Panciani PP, et al. Techniques for the preparation of solid lipid nano and microparticles. In: Application of nanotechnology in drug delivery; 2014.
7. Bhandari R, Kaur IP. Pharmacokinetics, tissue distribution and relative bioavailability of isoniazid-solid lipid nanoparticles. Int J Pharm. 2013;441(1–2):202–12.
8. Bhattacharya S, Prajapati BG. Recent excavations on cationic solid lipid nano particles. J Chem Pharm Res. 2015;7(10):559–63.
9. CDC, Centers for Disease Control and Prevention. Basic TB facts | TB | CDC. Centers Dis Control Prev. 2016:1.
10. Charcosset C, El-Harati A, Fessi H. Preparation of solid lipid nanoparticles using a membrane contactor. J Control Release. 2005;108(1):112–20.
11. Chaturvedi SP, Kumar V. Research journal of pharmaceutical, biological and chemical sciences production techniques of lipid nanoparticles: a review. Science. 2012;3(3):525–41.
12. Chokshi NV, Khatri HN, Patel MM. Formulation, optimization, and characterization of rifampicin-loaded solid lipid nanoparticles for the treatment of tuberculosis. Drug Dev Ind Pharm. 2018;44(12):1975–89.
13. Chokshi NV, Rawal S, Solanki D, et al. Fabrication and characterization of surface engineered rifampicin loaded lipid nanoparticulate systems for the potential treatment of tuberculosis: an in vitro and in vivo evaluation. J Pharm Sci. 2021;110(5):2221–32.
14. Chuan J, Li Y, Yang L, et al. Enhanced rifampicin delivery to alveolar macrophages by solid lipid nanoparticles. J Nanopart Res. 2013;15(5):1–9.
15. Costa A, Sarmento B, Seabra V. Mannose-functionalized solid lipid nanoparticles are effective in targeting alveolar macrophages. Eur J Pharm Sci. 2018;114:103–13.
16. Du Toit LC, Pillay V, Danckwerts MP. Tuberculosis chemotherapy: current drug delivery approaches. Respir Res. 2006;7(1):1–18.
17. Duan Y, Dhar A, Patel C, et al. A brief review on solid lipid nanoparticles: part and parcel of contemporary drug delivery systems. RSC Adv. 2020;10(45):26777–91.
18. Duong VA, Nguyen TTL, Maeng HJ. Preparation of solid lipid nanoparticles and nanostructured lipid carriers for drug delivery and the effects of preparation parameters of solvent injection method. Molecules. 2020;25(20):1–36.
19. Gasco MR. Method for producing solid lipid microspheres having a narrow size distribution, US Patent 5,250,236. 1993
20. Gaspar DP, Gaspar MM, Eleutério CV, et al. Microencapsulated solid lipid nanoparticles as a hybrid platform for pulmonary antibiotic delivery. Mol Pharm. 2017;14(9):2977–90.
21. Gordillo-Galeano A, Ospina-Giraldo LF, Mora-Huertas CE. Lipid nanoparticles with improved biopharmaceutical attributes for tuberculosis treatment. Int J Pharm. 2021;596:120321.
22. Kathe N, Henriksen B, Chauhan H. Physicochemical characterization techniques for solid lipid nanoparticles: principles and limitations. Drug Dev Ind Pharm. 2014;40(12):1565–75.
23. Kaur M, Garg T, Narang RK. A review of emerging trends in the treatment of tuberculosis. Artif Cells Nanomed Biotechnol. 2016;44(2):478–84.
24. Khatak S, Mehta M, Awasthi R, et al. Solid lipid nanoparticles containing anti-tubercular drugs attenuate the Mycobacterium marinum infection. Tuberculosis. 2020;125
25. Kumar M, Kakkar V, Mishra AK, Chuttani K, Kaur IP. Intranasal delivery of streptomycin sulfate (STRS) loaded solid lipid nanoparticles to brain and blood. Int J Pharm. 2014;461(1–2):223–33.

26. Lin C-H, Chen C-H, Lin Z-C, Fang J-Y. Recent advances in oral delivery of drugs and bioactive natural products using solid lipid nanoparticles as the carriers. J Food Drug Anal. 2017;25(2):219–34.
27. Ma C, Wu M, Ye W, et al. Inhalable solid lipid nanoparticles for intracellular tuberculosis infection therapy: macrophage-targeting and pH-sensitive properties. Drug Deliv Transl Res. 2021;11(3):1218–35.
28. Maretti E, Rustichelli C, Romagnoli M, et al. Solid Lipid Nanoparticle assemblies (SLNas) for an anti-TB inhalation treatment – a design of experiments approach to investigate the influence of pre-freezing conditions on the powder respirability. Int J Pharm. 2016;511(1)
29. Maretti E, Costantino L, Buttini F, et al. Newly synthesized surfactants for surface mannosylation of respirable SLN assemblies to target macrophages in tuberculosis therapy. Drug Deliv Transl Res. 2019;9(1)
30. Mehnert W, Mäder K. Solid lipid nanoparticles: production, characterization and applications. Adv Drug Deliv Rev. 2012;64(Suppl):83–101.
31. Montenegro L, Sarpietro MG, Ottimo S, Puglisi G, Castelli F. Differential scanning calorimetry studies on sunscreen loaded solid lipid nanoparticles prepared by the phase inversion temperature method. Int J Pharm. 2011;415(1–2):301–6.
32. Montenegro L, Campisi A, Sarpietro MG, et al. In vitro evaluation of idebenone-loaded solid lipid nanoparticles for drug delivery to the brain. Drug Dev Ind Pharm. 2011;37(6):737–46.
33. Müller RH, Mäder K, Gohla S. Solid lipid nanoparticles (SLN) for controlled drug delivery – a review of the state of the art. Eur J Pharm Biopharm. 2000;50(1):161–77.
34. Mutyam Pallerla S, Prabhakar B. A review on solid lipid nanoparticles. Int J Pharm Sci Rev Res. 2013;20(2):196–206.
35. Nemati E, Mokhtarzadeh A, Panahi-Azar V, et al. Ethambutol-loaded solid lipid nanoparticles as dry powder inhalable formulation for tuberculosis therapy. AAPS PharmSciTech. 2019;20(3)
36. Nirbhavane P, Vemuri N, Kumar N, Khuller GK. Lipid nanocarrier-mediated drug delivery system to enhance the oral bioavailability of rifabutin. AAPS PharmSciTech. 2017;18(3):829–37.
37. Obinu A, Porcu EP, Piras S, et al. Solid lipid nanoparticles as formulative strategy to increase oral permeation of a molecule active in multidrug-resistant tuberculosis management. Pharmaceutics. 2020;12(12):1–24.
38. Pandey R, Sharma S, Khuller GK. Oral solid lipid nanoparticle-based antitubercular chemotherapy. Tuberculosis. 2005;85(5–6):415–20.
39. Patel AN, Prajapati BG. Design and development of solid lipid nanoparticles (SLNs) of zolmitriptan for the treatment of migraine. Eur J Biomed Pharm Sci. 2016;3:253–9.
40. Pinheiro M, Ribeiro R, Vieira A, Andrade F, Reis S. Design of a nanostructured lipid carrier intended to improve the treatment of tuberculosis. Drug Des Devel Ther. 2016;10:2467.
41. Sato MR, Oshiro Junior JA, Machado RTA, et al. Nanostructured lipid carriers for incorporation of copper(II) complexes to be used against Mycobacterium tuberculosis. Drug Des Devel Ther. 2017;11:909.
42. Seung KJ, Keshavjee S, Rich ML. Multidrug-resistant tuberculosis and extensively drug-resistant tuberculosis. Cold Spring Harb Perspect Med. 2015;5(9):a017863.
43. Shah B, Khunt D, Bhatt H, Misra M, Padh H. Application of quality by design approach for intranasal delivery of rivastigmine loaded solid lipid nanoparticles: effect on formulation and characterization parameters. Eur J Pharm Sci. 2015;78:54–66.
44. Silva LB, Veigas B, Doria G, et al. Portable optoelectronic biosensing platform for identification of mycobacteria from the Mycobacterium tuberculosis complex. Biosens Bioelectron. 2011;26(5):2012–7.
45. Singh M, Guzman-Aranguez A, Hussain A, Srinivas CS, Kaur IP. Solid lipid nanoparticles for ocular delivery of isoniazid: evaluation, proof of concept and in vivo safety & kinetics. Nanomedicine. 2019;14(4):465–91.

46. Singh M, Schiavone N, Papucci L, et al. Streptomycin sulphate loaded solid lipid nanoparticles show enhanced uptake in macrophage, lower MIC in Mycobacterium and improved oral bioavailability. Eur J Pharm Biopharm. 2021;160:100–24.
47. Sivadasan D, Madavan B, Penmatsa SD, Bathini ST. Formulation and characterization of solid lipid nanoparticles of rifampicin. Correspondance. 2012;
48. Soo PC, Horng YT, Chang KC, et al. A simple gold nanoparticle probes assay for identification of Mycobacterium tuberculosis and Mycobacterium tuberculosis complex from clinical specimens. Mol Cell Probes. 2009;23(5):240–6.
49. Tekade RK, Maheshwari R, Tekade M, Chougule MB. Solid lipid nanoparticles for targeting and delivery of drugs and genes. Elsevier; 2017.
50. Tekade RK, Maheshwari R, Tekade M, Chougule MB. Solid lipid nanoparticles for targeting and delivery of drugs and genes. In: Nanotechnology-based approaches for targeting and delivery of drugs and genes. Elsevier; 2017. p. 256–86.
51. Truzzi E, Nascimento TL, Iannuccelli V, et al. In vivo biodistribution of respirable solid lipid nanoparticles surface-decorated with a mannose-based surfactant: a promising tool for pulmonary tuberculosis treatment? Nanomaterial. 2020;10(3):568.
52. Vieira ACC, Chaves LL, Pinheiro S, et al. Mucoadhesive chitosan-coated solid lipid nanoparticles for better management of tuberculosis. Int J Pharm. 2018;536(1):478–85.
53. Vieira ACC, Chaves LL, Pinheiro M, et al. Mannosylated solid lipid nanoparticles for the selective delivery of rifampicin to macrophages. Artif Cells Nanomed Biotechnol. 2018;46(1):653–63.
54. Vieira ACC, Chaves LL, Pinheiro M, et al. Lipid nanoparticles coated with chitosan using a one-step association method to target rifampicin to alveolar macrophages. Carbohydr Polym. 2021;252

Dendrimers-Based Drug Delivery in Tuberculosis

Dhaval J. Patel, Sunita Chaudhary, Ankit B. Chaudhary, and Jayvadan K. Patel

Abstract Tuberculosis (TB) is a disease caused by a microorganism (*Mycobacterium tuberculosis*) which invades the lungs and initiates the disease. Currently, treatment with anti-tubercular (anti-TB) drugs becomes the only option available because Bacillus Calmette-Guérin (BCG) vaccination and chemoprophylaxis appear to be unsatisfactory. Currently available treatment is oral medications which are toxic to the patients and especially to the HIV patients leads to death. It was observed that resistant strains, such as multi-drug resistant (MDR-TB) and extensively drug resistant (XDR-TB), to all major anti-TB drugs have the average treatment periods of 6–32 months against TB, MDR-TB, and XDR-TB. Carrier or delivery systems such as liposomes, nanoparticles, microspheres, solid lipid nanoparticles, and dendrimers have been developed for the sustained delivery of anti-TB drugs and have demonstrated better chemotherapeutic efficacy when investigated in animal models. Dendrimer is a drug attachment platform that can bind and release pharmaceuticals via a variety of methods. Due to the tight and globular structure of dendrimers as well as the number of surface functional groups, drug molecules can be encapsulated both inside and linked to the surface groups. Dendrimers have been used to successfully administer drugs directly into target sites, mimic a sustained release formulation, and maintain high efficacy for anti-TB drugs. Most of the current anti-TB drugs are associated with low solubility, limited permeability, and inadequate bio-distribution-associated side effects. Many scientists have suggested the use of dendrimers for the current anti-TB drugs to enhance their efficacy and overcome some of the drawbacks. Various analytical procedures for analyzing the physico-chemical characteristics of dendrimers have been mentioned. It includes spectro-

D. J. Patel (✉) · A. B. Chaudhary
Saraswati Institute of Pharmaceutical Sciences, Gandhinagar, Gujarat, India
e-mail: dhaval6668@gmail.com

S. Chaudhary
Arihant School and Bio Research Institute, Gandhinagar, Gujarat, India

J. K. Patel
Nootan Pharmacy College, Sankalchand Patel University, Visnagar, Gujarat, India

R. Shegokar, Y. Pathak (eds.), *Tubercular Drug Delivery Systems*,
https://doi.org/10.1007/978-3-031-14100-3_7

scopic, rheological, microscopic, chromatographic, calorimetric, and electrophoretic characterization. Hopefully, new drug delivery strategies will improve the future and, in turn, lessen the negative effects associated with the use of dendrimers in therapy of tuberculosis.

Keywords Tuberculosis · Novel drug delivery · Dendrimers

1 Introduction

Tuberculosis (TB) is a widespread disease and presents difficulties that transcend the conventional medical approach. Tuberculosis is a very old scourge. Its mortality and morbidity keeps increasing because it is a global health problem. It is the most ordinary opportunistic infection in acquired immunodeficiency syndrome. TB is a disease caused by a microorganism (*Mycobacterium tuberculosis* (*M. tuberculosis or Mtb*)) which invades the lungs and initiates the disease [1].

As per the latest global report, more than 75% of the world population is infected by Mtb and around 2 million deaths were caused due to TB in 2017. Currently available treatment is orally given medications which are toxic to the patients, especially to the HIV patient's leads to death. It was observed that resistant strains, such as multi-drug resistant (MDR-TB) and extensively drug resistant (XDR-TB), to all major anti-TB drugs have the average treatment periods of 6–32 months against TB, MDR-TB, and XDR-TB [2].

1.1 Current Anti-tuberculosis Drug Therapy

Currently, treatment with anti-tubercular (anti-TB) drugs becomes the only option available because Bacillus Calmette-Guérin (BCG) vaccination and chemoprophylaxis appear to be unsatisfactory. The goals of treatment are to ensure cure without relapse, to prevent death, to impede transmission, and to prevent the emergence of drug resistance. Long-term treatment with a combination of drugs is required. Treatment of active TB with a single drug should never be attempted, and a single drug should never be added to a failing regimen, the result being the development of MDR TB [3].

As suggested by WHO (define), treatment of TB and drug-resistant cases requires multi-drug therapy, comprising the following [1]:

I. An initial severe phase of Rifampicin (RIF), Isoniazid (INH), Pyrazinamide (PYZ), and Ethambutol (ETB) administered daily for 2 months.

Table 1 Classification of antitubercular drugs

Group	First-line oral anti-TB agents	Injectable anti-TB agents	Newer second-line drugs (Fluoroquinolones)	Oral second-line anti-TB Drugs	Agents with an unclear role in the treatment of drug-resistant TB
Drugs	Isoniazid	Streptomycin	Levofloxacin	Thiacetazone	Clofazimine
	Pyrazinamide	Kanamycin	Ciprofloxacin	P-amino salicylic acid	Linezolid
	Ethambutol	Amikacin	Oflaxacin	Ethionamide	Thioacetazone
	Rifampicin	Capreomycin	Moxifloxacin	Cycloserine	Clarithromycin
	Streptomycin	Vincomycin	Gatifloxacin		

II. A continuation phase of RIF and INH for an extra 4 months, daily or 3 (three) times per week, to be administered. INH eliminates most of the rapidly replicating bacilli in the first 2 weeks of treatment, together with streptomycin and ETB. Table 1 represents the various examples of anti-TB drugs.

1.2 Novel Drug Delivery Systems for the Treatment of TB

The critical problem with the current TB chemotherapy is that when the drug is taken intravenously or administered orally, it is distributed throughout the body via systemic blood circulation, and a majority of molecules do not reach their targets and, consequently, stay in the body causing adverse side effects. Drugs have a short plasma half-life and rapid clearance, which limit their effectiveness [4]. A number of novel implants like microparticulate and a variety of other carrier-based drug delivery systems incorporating the principal anti-tuberculosis agents target the site of tuberculosis infection or reduce the dosing frequency with the aim of improving patient outcomes (Fig. 1). Recent trends in controlled drug delivery have seen microencapsulation of pharmaceutical substances in biodegradable polymers as an emerging technology. Carrier or delivery systems such as liposomes, nanoparticles, microspheres, solid lipid nanoparticles, and dendrimers have been developed for the sustained delivery of anti-TB drugs and have demonstrated better chemotherapeutic efficacy when investigated in animal models (e.g., mice) [1, 3]. Dendrimers are responsible nanocarriers that can retain anti-TB medicine in the blood at pH 7.4 as well as drug release effectively in an acidic environment. Dendrimers can be an effective and attractive candidate for encapsulation of anti-TB drug due to their unique structure and diverse route of administration. Mainly anti-TB drugs like RIF, INH, PYZ, streptomycin, gentamycin, sparfloxacin, amikacin, clofazimine, rifabutin and capreomycin formulated in liposomes and RIF is formulated in niosomes. Drugs like INH, RIF, PYZ, ETB, streptomycin, moxifloxacin, and econazole are formulated in nanoparticles and microparticles.

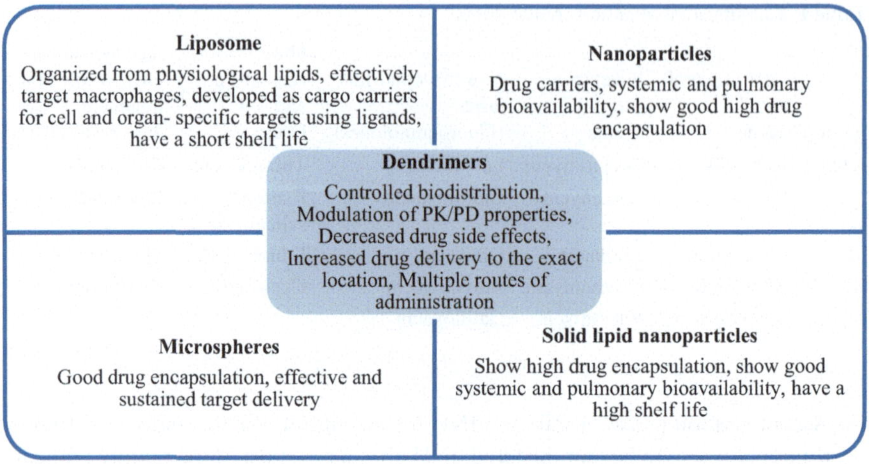

Fig. 1 Prominent features of drug carriers that have been used for the treatment of tuberculosis [1]

2 Dendrimer: Definition, Origin, and Properties

The word dendrimer comes from the word dendron, meaning "branch," and meros, meaning "part" [2]. Dendrimers exhibit a highly branched structure which consists of some specific components like central core, interior layer which is formed by repetitive units of polymer joined by core, and finally peripheral area to provide surface functionality. The typical structure of dendrimers is shown in Fig. 2, which covers all the parts: core, branches, and peripherals [5, 6]. Dendrimers are synthesized by repetition of chemical reaction and grown out of focal point, i.e., core. Layer is formed around the core which is known as "generation" number of dendrimers. Thus, a dendrimer having five concentric layers when going from the center to the periphery is denoted as the 5.0 G generation dendrimer [7]. Dendrimers possess asymmetric shape in first, second, and third generations as compared with higher generations. During the synthesis of the dendrimer, when it reaches its fourth generation or higher generations, it attains highly branched and globular structures [7]. Flexibility is higher with low-generation dendrimers while denser and highly rigid compound is observed in higher generation compounds [2].

Dendrimers' 3D structure provides specific and unique properties such as surface functionality that can be "tailored" to attach a variety of drugs, genes, and targeting agents, and the inner core that can act as a "host" for "guests" (that is, drug molecules). Many types of dendrimers are developed which exhibit key potential, like PAMAM dendrimers (Starbust), poly-etherhydroxyl-amine (PEHAM) dendrimers (Priostar), PPI dendrimers (Astramol) carbosilane dendrimers, and phosphorus-based dendrimers developed by Majoral JP and Caminade AM. [2] From 1980 to 1990, many research groups have developed various designs of

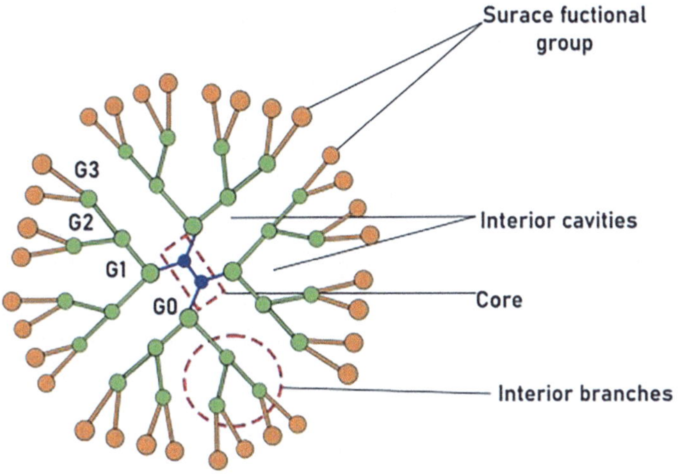

Fig. 2 Representation of a typical dendrimer

Table 2 Dendrimer types and associated research groups

Research groups	Type of dendrimers
Tomalia and co-workers	Polyamidoamine (PAMAM)
Denkewalter and co-workers	Poly(L-lysine) (PLL)
Newkome and co-workers	Polyamide
Grinstaff and co-workers	Polyester (PGLSA-OH)
Vogtle and co-workers	Polypropylenimine (PPI)
Hult and co-workers	Poly (2,2-bis (hydroxy methyl) propionic acid (bis-MPA)
Frechet and Meijer	Polyether

different dendrimers, and currently many new designs of dendrimers are in development [5]. From the time then, more than 100 diverse dendrimer structures have been recognized. Few known dendrimers are listed in Table 2 [8].

Most of the current anti-TB drugs are associated with low solubility, limited permeability, and inadequate bio-distribution-associated side effects. Many scientists have suggested the use of dendrimers for the current anti-TB drugs to enhance their efficacy and overcome some of the drawbacks. It becomes the need of researchers to understand the drug dendrimers interaction to study the solubility enhancement property of dendrimers. Many different types of drug dendrimers interaction have been developed to date, and they are further classified as follows:

1. Drug encapsulation in dendritic structure with formation of hydrogen bonds, hydrophobic interactions, and electrostatic interactions
2. Peripheral interaction of drug and dendrimers by covalent bond formation

Dendrimers have been used as nanocarriers treating a variety of disorders, including Alzheimer's disease, inflammation, HIV, herpes simplex virus, cancer and employing scaffolds of prodrugs using different routes of administration [9]. This chapter mainly focuses on dendrimers as a drug delivery system in the treatment of tuberculosis.

3 Physicochemical Characterization of Dendrimers

Characterization of dendrimers is critical for predicting and understanding their characteristics, morphology, and interactions. Because the characteristics of dendrimers cannot be defined using a single technique, it is necessary to approach them from a multifocus perspective. Dendrimer characterization procedures must be rigorous in order to create regulatory strategies for assuring nanomaterial safety. Various analytical procedures for analyzing the physicochemical characteristics of dendrimers have been documented in the literature [10, 11]. It includes spectroscopic, rheological, microscopic, chromatographic, calorimetric, and electrophoretic characterization (Table 3).

Table 3 Various techniques employed for the physicochemical characterization of dendrimers

Sr. no.	Characterization methods	Purpose of the methods to be used
1.	*Spectroscopy and spectrometry*	
	Nuclear magnetic resonance (1D, 2D, and 3D NMR studies)	Determining the structure and the dynamics of molecules suspended in dendrimer
	Mass spectroscopy technique	Determining molecular mass and its generations
	Matrix-assisted laser desorption time-of-flight (MALDI-TOF)	
	Electrospray ionization (ESI)	
	Fourier-transform infrared spectroscopy and RAMAM spectroscopy	For determination of synthesis progress, surface modification and identification of interactions of dendrimers–dendrimers, dendrimers–drugs, or dendrimers conjugated with other molecules
	UV–visible spectroscopy	Provides a proof of synthesis and surface modification of dendrimers
	Fluorescence spectroscopy	Provides valuable information regarding size, surface, interaction between the drug and dendrimers, and shape of dendrimers

(continued)

Table 3 (continued)

Sr. no.	Characterization methods	Purpose of the methods to be used
2.	*Rheology and Physical Properties*	
	Rheology	Identification of macroscopic properties of dendrimer
	Differential scanning calorimetric (DSC)	For determination of solid-state characterization (e.g., glass transition temperature)
	Thermal analysis	Determination of dendrimer's thermal stability before and after surface modification
	Microscopy	SEM and TEM confirm the molecular size and morphology of dendrimers AFM studying the structure and interaction of dendrimers with lipid bilayer
	Scanning electron microscopy (SEM), transmission electron microscopy (TEM), atomic force microscopy (AFM)	
	Scattering techniques	Determination of internal structure of dendrimers
	Small-angle X-ray scattering (SAXS) analysis	
	Dynamic light scattering (DLS)	
3.	*Chromatography*	
	High-performance liquid chromatography (HPLC), ultra-high-performance liquid chromatography (UPLC), gel permeation chromatography (GPC), size exclusion chromatography (SEC)	For purification of dendrimer and molecular weight distribution of dendrimers (using SEC)
4.	*Electrophoresis*	
	Gel electrophoresis	Used for analysis of the purity of dendrimers' synthesis
5.	*X-ray diffraction*	
	Single-crystal X-ray diffraction (XRD)	Structural information (crystalline or amorphous) at the atomic level

4 Pharmacokinetic and Drug Release Behavior from Dendrimers

Controlling the distribution of drugs in multiple organs after systemic administration is a major difficulty in drug delivery. The lack of adequate drug carriers, which can influence not only its pharmacokinetic profile but also its biodistribution, causes inefficient drug transport to the target tissue and possible side effects after administration. Dendrimers as drug delivery systems have several advantages, including extended drug circulation duration, drug protection from the environment, increased drug stability and possibly its effectiveness, and the capacity to target diseased tissue [12].

In case of dendrimers, parenteral administration has been utilized mostly suggesting the potential for dendrimer absorption across various epithelial barriers, including the intestine and the skin [13]. The volume of distribution (Vd), plasma elimination half-life ($t_{1/2}$), the area under the curve (AUC), and finally the maximum concentration of the drug (C_{max}) (peak concentration) and the time to peak (T_{max}) are all affected by the incorporation, encapsulation, or even conjugation of a bioactive molecule with dendrimers. It is important to notice that the physicochemical characteristics of dendrimers such as their size, their shape, their active surface groups, as well as the kind of linkers between dendrimers and drugs influence the pharmacodynamic and pharmacokinetic behavior of drugs. Surface-decorated dendrimers are reported to exhibit improved pharmacokinetic profiles than plain dendrimers [12, 13]. Polycationic PAMAM dendrimers exhibit fast clearance from the bloodstream upon intravenous or intraperitoneal administration and accumulate either in the liver, kidney, spleen, or pancreas. The pharmacokinetic profiles of dendrimers are for the most part, determined by surface charge and dendrimer molecular weight. The liver absorption is reduced when the PAMAM surface is modified with hydrophilic polyethylene oxide chains or by acetylation, likely due to steric stabilization of the dendrimer surface and/or lowering of the positive charge. PAMAM dendrimers with a negatively charged periphery have significantly longer blood circulation durations, while liver accumulation still occurs to some extent [14]. Several drugs that are complexed with or conjugated to dendrimeric structures have been reported in the literature, and their pharmacodynamic and pharmacokinetic behavior has been examined using parameters such as Tmax, Cmax, and AUC.

PAMAM dendrimers have been considered a promising drug delivery system that accommodates drug molecules either within the shell voids or at the outer surface. Two main approaches can attain drug delivery by PAMAM dendrimers:

(a) The first approach takes place due to changes in physical circumstances such as pH and temperature. These variations could trigger a conformational change of PAMAM dendrimers, which will facilitate the release of conjugated drug molecules outside the polymer.
(b) The second approach happens for the covalently linked molecules; the release can only occur through appropriate in vivo enzymatic cleavage of the covalent bonds.

The fifth-generation PEGylated PAMAM dendrimer was studied for its pharmacokinetic characteristics in the Wistar albino rats, and the study compared the C_{max} (μg/mL), T_{max} (h), AUC(μg/L*h), $t_{1/2}$(h), and Mean Residence Time (MRT) (h) between RIF and RIF-loaded PEGylated PAMAM dendrimer. The pharmacokinetic profile of 5G EDA-PAMAM dendrimers loaded with RIF and RIF (0.5–120 h) is shown in Table 4. RIF was no longer present in plasma after 6 h; however, it was still present in 5G EDA-PAMAM dendrimers after 120 h. For PEGylated 5G EDA-PAMAM dendrimers-loaded RIF, lower and longer release was observed, resulting in greater AUC, $t_{1/2}$ and MRT values compared to free RIF [15, 16].

Different ways for loading drugs into dendrimers allow for drug release from the dendrimers based on therapeutic needs. In addition, the type of linking employed to conjugate the drug to the surface of dendrimers can have a significant impact on the regulated distribution of bioactives from the dendrimeric scaffold.

Table 4 Comparison of pharmacokinetic characteristics of RIF and RIF-loaded PEGylated PAMAM dendrimer [2, 15]

	Pharmacokinetics parameters				
	C_{max} (μg/mL)	T_{max} (h)	AUC (μg/L*h)	$t_{1/2}$ (h)	MRT (h)
RIF alone	19.81	02	1154	2.14	4.01
RIF encapsulated in the -fifth generation EDA-PAMAM dendrimer	47.85	48	71,451	66.3	90.18

5 Dendrimer-Based Cell Line Studies

Dendrimers have recently been gaining the interest of biomedical researchers due to their medicinal potential. The development of a dendrimeric nano-architecture with well-defined size, shape, and regulated external activity has promise in biological and pharmaceutical applications such as drug administration, solubilization, DNA transfection, and diagnostics. Dendrimers have been tested for the transport of various bioactives inside cells. Cell cytotoxicity assays, cell uptake investigations using fluorescence microscopy, cell line studies, flow cytometry, gamma scintigraphy, and confocal microscopy are among the techniques that have been used to investigate dendrimer safety and efficacy for over a decade [17].

Ex vivo characterization of dendrimers is commonly done using cell line experiments. A number of cell lines such as HeLa cells, KB cells, Caco-2 cells, Molt-3 leukemia cells, PAM 212 cells, HEK-293 cells, A549 cells, LM3 cells, and many more are currently employed in characterization of bioactives and provide a basis for correlation with in vivo studies. The different assays performed on cell lines are Cell Cytotoxicity Assay, XTT [2,3-bis (2-methoxy-4-nitro-5-sulfophenyl)-5-[(phenylamino)carbonyl]-2H-tetrazolium hydroxide], MTT [3-(4, 5-di-methylthiazol-2yl)-2, 5-diphenyltetrazolium bromide], Sulforhodamine B Assay, Luciferase Assay, Flow Cytometric Analysis, Cell Uptake Study, Single Cell Gel Electrophoresis (COMET) assay, Dihydrofolatereductase inhibition assay, and Receptor Blockade Cytotoxicity Assay [23].

In many countries, TB is still a severe problem. Noncompliance with prescribed regimens is a serious issue, owing to the fact that TB therapy necessitates regular, ongoing multiple-drug dosage. RIF is the most important drug in TB treatment. The mannosylated dendrimers of RIF showed negligible cytotoxicity in Vero cells (ATCC-CCL-81e, *Cercopithecus aethiops* Kidney) possibly due to shielding of the internal cationic charges by the surface hydroxyl group. A significant difference was observed ($p = 0.0112$) with cytotoxicity in case of plain mixtures as compared to drug-loaded mixtures [18, 19].

6 Dendrimers as Drug Carriers

Dendrimers are hydrophobic from internal structure because of hydrophobic interaction and hydrogen bonding. Dendrimers are a preferable carrier for encapsulating hydrophobic drugs/bio-actives [36]. Dendrimers can act as an excellent carrier for drug solubilization due to their various properties like compatible host and guest chemistry, 3D-branched geometry, higher water solubility, and encapsulation efficiency. It was observed that when dendrimer is used for drug solubilization, then the changes in the properties and performance of dendrimers were found based on generation change. Encapsulation efficiency for hydrophobic drug is increased in higher generation dendrimers like G4 and above. Therefore, drug entrapment in dendrimer carried out by non-covalent bonding is a preferred technique for drug solubilization [6]. Poly(amidoamine) (PAMAM) and poly(propyleneimine) (PPI) are the most explored dendrimers in the field of drug delivery. In addition to their ability to carry a considerable load of small molecules, they have been shown to incorporate into biomembranes and cells and are thus capable of delivering small active compounds in the intracellular region. In the following section, we will discuss the encapsulation/conjugation of anti-TB drugs with dendrimers which could be improved by their therapeutic effectiveness against tuberculosis.

6.1 Rifampicin Encapsulation in PAMAM Dendrimers

Rifampicin (RIF) is an antibiotic class of drug having bactericidal activity and belongs to rifamycin family. Rifampicin is one of the most essential therapeutic drugs in the first-line TB regimen. Rifampicin has good clinical outcomes when used at recommended concentrations. Rifampicin shows limited clinical effect due to its poor water solubility.

Bellini et al. have carried out the interaction of Rifampicin with PAMAM dendrimer of G4 generation using molecular dynamics simulation. Molecular dynamics simulation is a powerful tool to study various types of association system in dendrimers. It was observed in one study that one molecule of G4-PAMAM dendrimer can load maximum 20 RIF molecules per G4-PAMAM dendrimer. Generally, at neutral and low pH, molecular dynamic simulations were generated depending upon docking of RIF molecules within the G4 PAMAM cavities. It was shown that the stability of the complex is measurable at neutral pH, and at low pH, RIF molecules were rapidly and simultaneously expelled to the solvent bulk. These outcomes of experiments describe the role of PAMAM dendrimers as nanocarriers to deliver drugs in acidic cellular structures such as alveolar macrophages.

Researchers reported that cationic PAMAM and PPI dendrimers are not suitable candidates as drug carriers because of their hemolytic and cytotoxic properties at even relatively low concentrations as well as having short half-life and are cleared quickly from the body. Upon reduction of positive charge on PAMAM and PPI, cationic dendrimer either by the addition of lipid or by PEGylation has shown to decrease the toxicities [38].

Another attempt was to prepare and explore the efficiency of a surface-modified 4.0 G PAMAM using dendrimer as a novel delivery system for RIF. The 4.0 G PAMAM dendrimer having various concentrations of polyethylene glycol (PEG 2000) was synthesized using 4-nitrophenyl chloroformate as an activator. Fourier-transform infrared and proton nuclear magnetic resonance analyses were used to confirm the PEGylation of 4.0 G PAMAM dendrimer. Simple dissolution solvent evaporation method was used for loading RIF into dendrimers. The polymer encapsulation efficiency (EE%) was determined using a validated HPLC method. In vitro drug release was studied at pH 7.4 using dialysis bag method. The MTT technique was used to assess the cytotoxicity of the dendrimer formulations against raw 264.7 cell lines. RIF molecules were encapsulated into dendrimers in varying proportions and enhanced drug loading was noticed after dendrimer PEGylation. MTT results confirmed that the toxicity of the native G4 dendrimer was concentration-dependent and a significant decrease in toxicity was noticed after dendrimer PEGylation. The complex between drug and PEGylated-PAMAM dendrimers upon evaluation showed promising biocompatible delivery system for RIF with high loading capability and prolonged release behavior [39].

Furthermore, when RIF was loaded with PAMAM dendrimers G5, whose PEGylation was done using polyethylene glycol 2000 and epichlorohydrin was used to link both the polymers, led to decrease in toxicity. Also, an increase in the loading capacity of the drug was noted with the above-mentioned structural design and hence proved suitable for sustained delivery of the drug rifampicin [40].

Dineshkumar et al. recently developed RIF-loaded fifth-generation PEGylated EDA-PAMAM dendrimers. The PEGylation of 5G PAMAM dendrimers was confirmed by Fourier-transform infrared spectrophotometry and NMR spectra. Around 98% RIF entrapment was found in PEGylated EDA-PAMAM dendrimer. The prolonged release behavior of RIF was observed for PEGylated (81% in 120 h) while comparing with non-PEGylated (98% in 72 h) PAMAM dendrimers. The hemolytic studies showed PEGylated PAMAM dendrimer showed lower toxicity effects (less than 2.5%) than non-PEGylated PAMAM dendrimer (11.6–25.3%) [15].

Rajabnezhad et al. formulated co-spray-dried microspheres using three different generations of PAMAM dendrimers (G1, G2, and G3) and RIF as a model anti-TB drug for pulmonary drug delivery. The PAMAM G3-rifampicin microspheres displayed the smallest particle size (6.21 μm) and bulk density (0.024 g/ml) compared to other two generations. The release rate of drug from PAMAM G1-G3 microspheres was studied. PAMAM G3-rifampicin microspheres exhibited slow and sustained drug release up to 72 h at pH 7.4 compared to other two generations of microspheres at the same dose (5 mg). PAMAM G3-rifampicin microspheres showed 4- and 30-fold improvement in plasma drug concentration and mean residence time compared to intravenous administration of drug alone. In summary, PAMAM G3 dendritic microspheres were identified as suitable drug carriers for the pulmonary delivery of rifampicin into lung tissues. [41]

6.2 *Rifampicin Encapsulation in PPI Dendrimers*

As PAMAM dendrimers, PPI dendrimers are also utilized in a wide range of applications in drug delivery. Kumar et al. developed mannosylated fifth-generation (5G) PPI dendrimeric nanoparticles for delivery of RIF to macrophages. Drug encapsulation is primarily based on hydrophobic interactions and hydrogen bonding, both of which contribute to the drug's physical binding to the core. Amine-terminated PPI dendrimers show a high level of cytotoxicity and hemolytic effect. Hemolytic toxicity was significantly reduced with mannose molecules (30D-mannose molecules) from 15.6% to 2.8%. Also, RIF-containing dendrimers reduced hemolytic effect of free RIF from 9.8% to 6.5%. Moreover, RIF dendrimers improved the survival in epithelial cell line of kidney compared to free RIF (from 50% to 85%). In alveolar macrophages taken from rat lungs, phagocytic absorption of RIF and RIF-loaded dendrimers resulted in a noticeable rise in RIF intracellular content [42, 43].

Carrying the same approach further, the outcomes of fourth- and fifth-generation PEGylated PPI dendrimers were examined. They exhibited effects similar to the sustained release form of RIF. The PEGylation improved both the entrapment and the release of RIF from the PPI fourth- and fifth-generation dendrimer. The hemolytic activity of PEG-grafted dendrimers (1–3%) was lower than that of NH_2-terminated dendrimers (14–20%) [42].

6.3 *Isoniazide Encapsulation in PAMAM Dendrimers*

Isoniazid, another significant anti-TB medicine, has also been laced with dendrimers. Singh et al. used the dialysis approach to load Isoniazid into 1.5-generation PAMAM dendrimers. UV spectroscopy and FTIR techniques confirmed the loading of IND. This formulation demonstrated that 93.25% of IND was delivered continuously for up to 24 h following zero-order kinetics [44].

6.4 *Ethambutol-Jeffamine Conjugate*

Sardari S and coworker proposed the combination of ethambutol with third generation of poly(propyleneoxide) amines (Jeffamine) dendrimer which also possesses antimycobacterial effect along with self-target delivery property. This dendrimer-ethambutol conjugate targeted tolectin receptors on macrophages using mannose molecules as ligands on the outer surface of dendrimers. The authors believe that putting this technique into practice could make tuberculosis treatment easier and reduce multiple drug resistance [38].

7 Dendrimers and Toxicity

Before a new medicine may be utilized in human treatments and diagnostics, its properties must be thoroughly documented. Dendrimers are biocompatible nanoparticle macromolecules that improve the activity and efficiency of active pharmaceutical compounds while minimizing their toxicity [45]. Injected polymers are not eliminated as easily, especially if they are not readily degraded into smaller units or are too large to be filtered via the kidneys. Long-term accumulation of low-molecular-weight compounds is not often a problem because they are excreted in the urine or in the feces after metabolism. Thus, for dendrimers, which can be classified as low molecular weight or polymeric depending on their generation, acceptable biocompatibility must be accompanied by a reasonably fast renal elimination rate or biodegradation rate [14, 46].

It has been shown that the cytotoxicity of the dendrimer depends on the generation to which it belongs and also on the nature of its surface, given by terminal functional groups. Cytotoxicity was highlighted in cationic and amine dendrimers. Cationic dendrimers can interact with negatively charged cell membranes, causing them to break down [47, 48]. Cationic dendrimers with terminal primary amino groups, such as PAMAM and polypropyleneimine (PPI) dendrimers, have been shown to have concentration-dependent toxicity and hemolysis, whereas dendrimers containing only neutral or anionic components have been shown to be much less toxic and hemolytic. Thus, reducing cytotoxicity is closely linked to surface modification of cationic dendrimers in order to neutralize or completely modify them to anion [49, 50]. The study of various structural modulations in the peripheral area of dendrimers was used to investigate the occurrence and modulation of dendrimer cytotoxicity. Attachment of carbohydrates, acetyl, and polyethylene glycol (PEG) derivatives can significantly affect cell viability while retaining other beneficial properties [45]. The use of mannosylated dendritic architecture for the selective delivery of RIF, an anti-tuberculosis drug, to alveolar macrophages was developed and investigated by Jain and coworkers. They found that mannosylation of the dendrimers reduced RBC hemolysis by inhibiting RBC interaction with the charged quaternary ammonium ions present in the amine-terminated dendrimers [19].

In vitro and in vivo toxicity are quite closely correlated. Mice tolerate low dosages of positively charged PAMAM dendrimers administered intraperitoneally [51]. Generally, 5.0G PPI and native PPI dendrimers get accumulated in the liver and cause liver problems, but on the other hand PEGylated dendrimers do not harm the liver [52]. After injecting doses greater than 1 g/kg intraperitoneally or intravenously, no acute or subchronic toxicity was seen when 50% of the cationic groups of a structurally identical dendrimer were substituted with neutral polyethylene oxide chains [14, 53]

Glycodendrimers are a newer form of dendrimer, and these modulations result in a considerable reduction in cytotoxicity [54]. The interaction of liposomes and human serum albumin (HSA) with glucose-modified carbosilane dendrimers was studied from the first to the third generation. Interestingly, in contrast to other dendritic type polymers, the extent of the dendrimers interaction with biological models was not related to dendrimers generation. Because of the strong interactions with liposomes and the weak interactions with HAS, cancer cells can be targeted by overexpression of glucose transporters demonstrating that glucose-modified carbosilane dendrimers can be used as drug delivery carriers in cancer therapy [55].

8 Clinical Outlook

Compared to other nanoparticles such as liposomes, few dendrimers are undertaking clinical trials, despite the fact that the number of publications on dendrimer research has steadily increased over the last two decades [56, 57]. Dendrimer-based nanomedical products were first introduced in the late 1990s. Cardiac diagnostics (Stratus) and DNA gene vectors (Superfect, Qiagen, Hilden, Germany) are commercial products before the year 2000 focused on simple dendrimer compositions; applications relied on their structure-controlled nanoscale sizes and surface chemistry. Several dendrimer-based products emerging from the period 2000–2010 included (i) organic light-emitting diodes (Cambridge Display/Sumitomo, Tokyo, Japan); (ii) the antiviral topical nanopharmaceutical SPL7013 (Starpharma), which is presently in Phases II–III clinical trials as a microbicide for treatment of bacterial vaginosis; (iii) MRI agents (Gadomer-17, Bayer Schering Pharma AG); (iv) siRNA delivery vector (Priofect, Merck KGaA, Darmstadt, Germany); (iv) cardiac diagnostics (Stratus); (v) protein detection amplifiers (UltraAmp, Genisphere, Inc.); and (vi) ocular/surgical adhesives (OcuSeal/Adherus, HyperBranch Medical Technology, Inc., Durham, NC, USA) [58].

Various clinical trials related to "dendrimers" are registered and reachable on the Europe portal (EU Clinical Trials Register [https://www.clinicaltrialsregister.eu/]) and on US FDA portal (ClinicalTrials.gov [https://clinicaltrials.gov/]). The pharmaceutical application of dendrimers has recently undergone clinical translation since the first dendrimer called Vivagel® has been translated from research studies to clinical trials as a new FDA drug application. Vivagel®, developed by Starpharma (Melbourne, Australia), is a formulation of polyanioniclysine G4 dendrimers with an anionic surface of naphthalene-disulfonate (SPL7013) in a Carbopol® gel. It exhibits antiviral activity against HIV and HSV for the treatment of sexually transmitted infections and is given by intravaginal administration. Complete preclinical and phase I clinical studies have shown that 0.5–3% SPL7013 formulations are safe and well tolerated after seven-day vaginal application without systemic absorption. A phase II clinical trial of Vivagel® is ongoing with fast-track status [59–61]. Currently DEP-docetaxel is under active development by Starpharma and AstraZeneca as a tumor-targeting nanotherapy which has exhibited enhanced effectiveness against breast, prostate, lung, and ovarian cancer compared to docetaxel alone. Another dendrimer undergoing preclinical study is the multi-antigenic peptide PHSCN-lysine dendrimer for inhibition of invasion and growth of breast cancer cells via $\alpha 5\beta_1$ integrin-selective recognition in a metastatic murine cancer model [58].

Extensive application of dendrimers as nanocarriers has been described essentially in the oncology domain. In the dendrimer domain, a few articles have highlighted the encapsulation of anti-TB drugs such as RIF (Table 5). Currently there is no marketed pharmaceutical product where dendrimer has been used for tuberculosis; however, several other dendrimer-based products are commercially available for various applications (Table 6). Despite the promise of dendrimers for the solubilization/encapsulation of hydrophobic medications, clinical and commercial uses of this approach in the field of drug development and delivery have yet to be established [6, 7, 11].

Table 5 Cell line studies of various dendrimers and therapeutic moieties

Drug	Dendrimer	Cell line used	Refs.
Methotrexate	PAMAM	KB	[20]
		CCRF-CEM cell line	[21]
		Human umbilical vein endothelial cells (HUVEC)	[22]
		EGFR-expressing rat glioma cell line F98(EGFR)	[23]
Camptothecin	Poly (glycerol succinic acid)	Human breast adenocarcinoma (MCF-7), colorectal adenocarcinoma (HT-29), non–small cell lung carcinoma (NCI-H460), and glioblastoma (SF-268)]	[24]
Paclitaxel	PAMAM	Human ovarian carcinoma A2780 cell line	[25]
		Prostate cancer cell line (PC-3M)	[26]
Doxorubicin	PAMAM	DMS114 (lung cancer)	[27]
	Polyester dendrimer	Murine B16F10 melanoma cells and the human breast cancer cell lines MDA-MD-231 and MDA-MD-435	[28]
Methylprednisolone	PAMAM	A549 human lung epithelial carcinoma cells	[29]
Estrogen	PAMAM	MCF-7 breast cancer cells	[30]
Dexamethasone	PAMAM	Human embryonic kidney 293 cells and mouse neuroblastoma Neuro2A cells	[31]
Efavirenz	PPI	Hepatoma (Hep G2) cell lines	[32]
Lamivudine	Mannosylated PPI	MT2 cell lines (human cord leukocyte cell line)	[33]
Rifampicin	Mannosylated PPI	*Vero* cells (ATCC-CCL-81e, *Cercopithecus aethiops* Kidney)	[19]
Ibuprofen	PAMAM	Human lung epithelial carcinoma (A549 cells)	[34]
Naproxen	PAMAM	Human intestinal adenocarcinoma cells (Caco-2)	[35]

Table 6 Various drugs studied using dendrimer platform

Sr no	Drug used	Dendrimer used	Application	Refs.
1.	Naproxen	PAMAM	High permeability across Caco-2 cell	[13]
2.	Rifampicin	PAMAM	Improved solubility and tissue uptake	[19, 37]
		Mannosylated PPI	Sustained drug release and targeted delivery	
3.	Methotrexate	PAMAM	Targeted drug delivery and increment in cytotoxicity	[21]
4.	Doxorubicin	PAMAM Polylysine	Improved cytotoxicity	[22]
			Prolonged plasma exposure and diminished drug toxicity	
5.	Lamivudine	Mannosylated PPI	Prolonged drug release	[33]
6.	Amphotericin	PAMAM	Improved solubility	[62]
7.	Imatinib Mesylate	PPI	Enhanced solubility	[63]
8.	Haloperidol	PAMAM	Improved solubility	[64]
9.	Docetaxel and Paclitaxel	Dendrimer–TPGS mixed micelles	Enhanced solubility and cytotoxicity	[65]
10.	Beclomethasone dipropionate	PAMAM	Improved solubility	[66]
11.	Paclitaxel	Poly(butylene oxide)-poly(ethylene oxide) block copolymer PAMAM dendrimer	Enhanced solubility and cytotoxicity	[67]
12.	Ibuprofen, Ketoprofen, and Diflunisal	Polypropylene oxide cored PAMAM	Improved drug permeation through the skin	[68]
13.	Candesartan cilexetil, Silybin, Risperidone	PAMAM	Enhanced solubility	[69–71]
14.	Famotidine	PPI	Improved solubility	[72]
15.	Efavirenz and Zidovudine	PPI	Targeted drug delivery	[73, 74]
16.	Etoposide	PAMAM	High loading capacity	[75]
	Erythromycin	PAMAM	Sustained release and improve activity	[76]
17.	Pilocarpine	PAMAM	Prolonged corneal residence time	[77]

9 Conclusion

Tuberculosis remains a leading cause of mortality worldwide even in the twenty-first century. Multiple antibiotics, such as isoniazid, rifampicin, pyrazinamide, and ethambutol, must be given for a long time to kill bacteria in the treatment of

Table 7 Dendrimer-based marketed formulations

Marketed product	Type of dendrimer	Name of company	Application	Current status
VivaGel	Poly-L-Lysine	Starpharma	Mucoadhesive gel for bacterial vaginosis	Phase-III trial
SuperFect	PAMAM	Qiagen	Cell transfection and gene delivery	Phase-IV trial
PrioFect	PAMAM	Starpharma	Cell transfection and gene delivery	Phase-III trial
Stratus CS	PAMAM	Siemens Healthcare diagnostics	Measurement of cardiac biomarkers	Marketed
DEP docetaxel	ND	Starpharma	Anti-cancer agent	Phase-II trial
Starburst	PAMAM	Starpharma	Commercially available PAMAM dendrimers	Marketed
Priostar	PEHAM/PEA	Starpharma	Commercially available poly-lysine-based dendrimers for crop protection	Marketed
Alert ticket	PAMAM	US Army Lab	Anthrax-detecting agent	Marketed

ND not defined

tuberculosis. Many antimicrobial drugs are complicated to administer because of the limited water solubility, rapid breakdown, significant cytotoxicity to tissues, and clearance in the blood circulation. In recent years, the encapsulation of antimicrobial drugs in all carrier systems (nanoparticle, liposomes, dendrimers, solid lipid nanoparticles, and microspheres) has emerged as an innovative and promising change that increases therapeutic efficiency and reduces undesirable side effects of the drugs (Table 7).

Dendrimer is a drug attachment platform that can bind and release pharmaceuticals via a variety of methods. Due to the tight and globular structure of dendrimers as well as the number of surface functional groups, drug molecules can be encapsulated both inside and linked to the surface groups. Dendrimers have been used to successfully administer drugs directly into target sites, mimic a sustained release formulation, and maintain high efficacy for anti-TB drugs. Dendrimers could be used to provide drugs via the oral, parenteral, intraocular, and nasal routes. A variety of dendrimers, including PPI and PAMAM dendrimers, are introduced to compare their uses in drug administration. Dendrimer characterization procedures must be rigorous in order to create regulatory strategies for assuring nanomaterial safety. The combination of MALDI-TOF-MS, NMR, SEC, and electron microscopy techniques, on the other hand, can provide comprehensive information on dendrimer characterization and derivatives. The high cost of procurement and maintenance of these machines is a significant disadvantage.

In vitro cell line investigations are important not only for characterization of bioactives, but also as a foundation for in vivo studies. Dendrimers may have applications in disease targeting and diagnosis, based on the examples shown in this article. According to the literature, PAMAM dendrimer has been used most frequently in cell line investigations. Other dendrimer types under investigation have yet to be thoroughly investigated for their potential. In summary, dendrimer has limitations too, e.g., it could be toxic beyond certain concentration levels. Hopefully, new drug delivery strategies will improve the future and, in turn, lessen the negative effects associated with the use of dendrimers in the therapy of tuberculosis.

References

1. Kaur M, Garg T, Narang RK. A review of emerging trends in the treatment of tuberculosis. Artif Cells Nanomed Biotechnol. 2016;44(2):478–84.
2. Mignani S, Tripathi RP, Chen L, Caminade AM, Shi X, Majoral JP. New ways to treat tuberculosis using dendrimers as nanocarriers. Pharmaceutics. 2018;10(3):105.
3. Du Toit LC, Pillay V, Danckwerts MP. Tuberculosis chemotherapy: current drug delivery approaches. Respir Res. 2006;7(1):1–8.
4. Nasiruddin M, Neyaz M, Das S. Nanotechnology-based approach in tuberculosis treatment. Tuberculosis Res Treat. 2017;22:2017.
5. Tomalia DA, Baker H, Dewald J, Hall M, Kallos G, Martin S, Roeck J, Ryder J, Smith P. A new class of polymers: starburst-dendritic macromolecules. Polym J. 2002;34(5):132–47.
6. Choudhary S, Gupta L, Rani S, Dave K, Gupta U. Impact of dendrimers on solubility of hydrophobic drug molecules. Front Pharmacol. 2017;8:261.
7. Singh J, Jain K, Mehra NK, Jain NK. Dendrimers in anticancer drug delivery: mechanism of interaction of drug and dendrimers. Artif Cells Nanomed Biotechnol. 2016;44(7):1626–34.
8. Mehta P, Kadam S, Pawar A, Bothiraja C. Dendrimers for pulmonary delivery: current perspectives and future challenges. New J Chem. 2019;43(22):8396–409.
9. Noriega-Luna B, Godínez LA, Rodríguez FJ, Rodríguez A, Zaldívar-Lelo de Larrea G, Sosa-Ferreyra CF, Mercado-Curiel RF, Manríquez J, Bustos E. Applications of dendrimers in drug delivery agents, diagnosis, therapy, and detection. J Nanomater. 2014;2014
10. Valdes Lizama O, Vilos C, Durán-Lara E. Techniques of structural characterization of dendrimers. Curr Org Chem. 2016;20(24):2591–605.
11. Madaan K, Kumar S, Poonia N, Lather V, Pandita D. Dendrimers in drug delivery and targeting: drug-dendrimer interactions and toxicity issues. J Pharm Bioallied Sci. 2014;6(3):139.
12. Villalonga-Barber C, Micha-Screttas M, Steele BR, Georgopoulos A, Demetzos C. Dendrimers as biopharmaceuticals: synthesis and properties. Curr Top Med Chem. 2008;8(14):1294–309.
13. Kesharwani P, Jain K, Jain NK. Dendrimer as nanocarrier for drug delivery. Prog Polym Sci. 2014;39(2):268–307.
14. Lee CC, MacKay JA, Fréchet JM, Szoka FC. Designing dendrimers for biological applications. Nat Biotechnol. 2005;23(12):1517–26.
15. Dineshkumar P, Panneerselvam T, DeeptiBrundavani K, Selvaraj K, Vijayaraj KP. Formulation of rifampicin loaded PEGylated 5.0 G EDA-PAMAM dendrimers as effective long-duration release drug carriers. Curr Drug Ther. 2017;12(2):115–26.
16. Ahmed RM. Development of rifampicin loaded in surface-modified 4.0 G PAMAM dendrimer as a novel antituberculosis pulmonary drug delivery system. University of Western Cape; 2020. http://etd.uwc.ac.za/xmlui/handle/11394/8073
17. Kesharwani P, Gajbhiye V, Tekade K, Jain N. Evaluation of dendrimer safety and efficacy through cell line studies. Curr Drug Targets. 2011;12(10):1478–97.

18. Telenti A, Imboden P, Marchesi F, Matter L, Schopfer K, Bodmer T, Lowrie D, Colston MJ, Cole S. Detection of rifampicin-resistance mutations in Mycobacterium tuberculosis. Lancet. 1993;341(8846):647–51.
19. Kumar PV, Asthana A, Dutta T, Jain NK. Intracellular macrophage uptake of rifampicin loaded mannosylated dendrimers. J Drug Target. 2006;14(8):546–56.
20. Patri AK, Kukowska-Latallo JF, Baker JR. Targeted drug delivery with dendrimers: comparison of the release kinetics of covalently conjugated drug and non-covalent drug inclusion complex. Adv Drug Deliv Rev. 2005;57(15):2203–14.
21. Gurdag S, Khandare J, Stapels S, Matherly LH, Kannan RM. Activity of dendrimer– methotrexate conjugates on methotrexate-sensitive and-resistant cell lines. Bioconjug Chem. 2006;17(2):275–83.
22. Shukla R, Thomas TP, Desai AM, Kotlyar A, Park SJ, Baker JR. HER2 specific delivery of methotrexate by dendrimer conjugated anti-HER2 mAb. Nanotechnology. 2008;19(29):295102.
23. Wu G, Barth RF, Yang W, Lee RJ, Tjarks W, Backer MV, Backer JM. Boron containing macromolecules and nanovehicles as delivery agents for neutron capture therapy. Anti Cancer Agents Med Chem. 2006;6(2):167–84.
24. Morgan MT, Nakanishi Y, Kroll DJ, Griset AP, Carnahan MA, Wathier M, Oberlies NH, Manikumar G, Wani MC, Grinstaff MW. Dendrimer-encapsulated camptothecins: increased solubility, cellular uptake, and cellular retention affords enhanced anticancer activity in vitro. Cancer Res. 2006;66(24):11913–21.
25. Khandare JJ, Jayant S, Singh A, Chandna P, Wang Y, Vorsa N, Minko T. Dendrimer versus linear conjugate: influence of polymeric architecture on the delivery and anticancer effect of paclitaxel. Bioconjug Chem. 2006;17(6):1464–72.
26. Devarakonda B, Judefeind A, Chigurupati S, Thomas S, Shah GV, Otto DP, de Villiers MM. The effect of polyamidoamine dendrimers on the in vitro cytotoxicity of paclitaxel in cultured prostate cancer (PC-3M) cells. J Biomed Nanotechnol. 2007;3(4):384–93.
27. Papagiannaros A, Dimas K, Papaioannou GT, Demetzos C. Doxorubicin–PAMAM dendrimer complex attached to liposomes: cytotoxic studies against human cancer cell lines. Int J Pharm. 2005;302(1-2):29–38.
28. Padilla De Jesús OL, Ihre HR, Gagne L, Fréchet JM, Szoka FC. Polyester dendritic systems for drug delivery applications: in vitro and in vivo evaluation. Bioconjug Chem. 2002;13(3):453–61.
29. Khandare J, Kolhe P, Pillai O, Kannan S, Lieh-Lai M, Kannan RM. Synthesis, cellular transport, and activity of polyamidoamine dendrimer– methylprednisolone conjugates. Bioconjug Chem. 2005;16(2):330–7.
30. Kim SH, Funk CC, Erdogan ZM, Katzenellenbogen JA. Estrogen dendrimer conjugates that preferentially activate extranuclear, nongenomic versus genomic. Mol Endocrinol. 2006;20(3):491–502.
31. Choi JS, Ko KS, Park JS, Kim YH, Kim SW, Lee M. Dexamethasone conjugated poly (amidoamine) dendrimer as a gene carrier for efficient nuclear translocation. Int J Pharm. 2006;320(1-2):171–8.
32. Dutta T, Agashe HB, Garg M, Balasubramanium P, Kabra M, Jain NK. Poly (propyleneimine) dendrimer based nanocontainers for targeting of efavirenz to human monocytes/macrophages in vitro. J Drug Target. 2007;15(1):89–98.
33. Dutta T, Jain NK. Targeting potential and anti-HIV activity of lamivudine loaded mannosylated poly (propyleneimine) dendrimer. BiochimicaetBiophysicaActa (BBA)-General Subjects. 2007;1770(4):681–6.
34. Kolhe P, Khandare J, Pillai O, Kannan S, Lieh-Lai M, Kannan RM. Preparation, cellular transport, and activity of polyamidoamine-based dendritic nanodevices with a high drug payload. Biomaterials. 2006;27(4):660–9.
35. Najlah M, Freeman S, Attwood D, D'Emanuele A. In vitro evaluation of dendrimer prodrugs for oral drug delivery. Int J Pharm. 2007;336(1):183–90.

36. Medina SH, El-Sayed ME. Dendrimers as carriers for delivery of chemotherapeutic agents. Chem Rev. 2009;109(7):3141–57.
37. Bellini RG, Guimarães AP, Pacheco MA, Dias DM, Furtado VR, de Alencastro RB, Horta BA. Association of the anti-tuberculosis drug rifampicin with a PAMAM dendrimer. J Mol Graph Model. 2015;60:34–42.
38. Ghavami G, Sardari S. New chimeric anti-tubercular dendrimers with self-delivering property. Afr J Microbiol Res. 2011;5(18):2575–82.
39. Ahmed R, Aucamp M, Ebrahim N, Samsodien H. Supramolecular assembly of rifampicin and PEGylated PAMAM dendrimer as a novel conjugate for tuberculosis. J Drug Deliv Sci Technol. 2021;66:102773.
40. Shukla R, Sethi A, Handa M, Mohan M, Tripathi PK, Kesharwani P. Dendrimer-based drug delivery systems for tuberculosis treatment. In: Nanotechnology based approaches for tuberculosis treatment. Academic; 2020. p. 163–74.
41. Rajabnezhad S, Casettari L, Lam JK, Nomani A, Torkamani MR, Palmieri GF, Rajabnejad MR, Darbandi MA. Pulmonary delivery of rifampicin microspheres using lower generation polyamidoamine dendrimers as a carrier. Powder Technol. 2016;291:366–74.
42. Vijayaraj Kumar P, Agashe H, Dutta T, Jain NK. PEGylated dendritic architecture for development of a prolonged drug delivery system for an antitubercular drug. Curr Drug Deliv. 2007;4(1):11–9.
43. Shegokar R, Al Shaal L, Mitri K. Present status of nanoparticle research for treatment of tuberculosis. J Pharm Pharm Sci. 2011;14(1):100–16.
44. Singh N, Gautam SP, Singh HL, Dhiman A, Siddiqui G, Verma A. Isonizidloded dendrimer based nano carriers for the delivery of anti-tuberculosis. Indian Res J Pharm Sci. 2016;3:519–29.
45. Chis AA, Dobrea C, Morgovan C, Arseniu AM, Rus LL, Butuca A, Juncan AM, Totan M, Vonica-Tincu AL, Cormos G, Muntean AC. Applications and limitations of dendrimers in biomedicine. Molecules. 2020;25(17):3982.
46. Drobník J, Rypáček F. Soluble synthetic polymers in biological systems. Polym Med. 1984:1–50.
47. Abd-El-Aziz AS, Agatemor C. Emerging opportunities in the biomedical applications of dendrimers. J Inorg Organomet Polym Mater. 2018;28(2):369–82.
48. Janaszewska A, Lazniewska J, Trzepiński P, Marcinkowska M, Klajnert-Maculewicz B. Cytotoxicity of dendrimers. Biomolecules. 2019;9(8):330.
49. Malik N, Wiwattanapatapee R, Klopsch R, Lorenz K, Frey H, Weener JW, Meijer EW, Paulus W, Duncan R. Dendrimers:: relationship between structure and biocompatibility in vitro, and preliminary studies on the biodistribution of 125I-labelled polyamidoamine dendrimers in vivo. J Control Release. 2000;65(1-2):133–48.
50. Jevprasesphant R, Penny J, Jalal R, Attwood D, McKeown NB, D'emanuele A. The influence of surface modification on the cytotoxicity of PAMAM dendrimers. Int J Pharm. 2003;252(1-2):263–6.
51. Roberts JC, Bhalgat MK, Zera RT. Preliminary biological evaluation of polyamidoamine (PAMAM) StarburstTM dendrimers. J Biomed Mater Res. 1996;30(1):53–65.
52. Kaur D, Jain K, Mehra NK, Kesharwani P, Jain NK. A review on comparative study of PPI and PAMAM dendrimers. J Nanopart Res. 2016;18(6):1–4.
53. Chen HT, Neerman MF, Parrish AR, Simanek EE. Cytotoxicity, hemolysis, and acute in vivo toxicity of dendrimers based on melamine, candidate vehicles for drug delivery. J Am Chem Soc. 2004;126(32):10044–8.
54. Augustus EN, Allen ET, Nimibofa A, Donbebe W. A review of synthesis, characterization and applications of functionalized dendrimers. Am J Polym Sci. 2017;7(1):8–14.
55. Wrobel D, Müllerová M, Strašák T, Růžička K, Fulem M, Kubíková R, Bryszewska M, Klajnert-Maculewicz B, Malý J. Glucose-modified carbosilane dendrimers: Interaction with model membranes and human serum albumin. Int J Pharm. 2020;579:119138.
56. Wijagkanalan W, Kawakami S, Hashida M. Designing dendrimers for drug delivery and imaging: pharmacokinetic considerations. Pharm Res. 2011;28(7):1500–1.

57. Zolnik BS, Sadrieh N. Regulatory perspective on the importance of ADME assessment of nanoscale material containing drugs. Adv Drug Deliv Rev. 2009;61(6):422–7.
58. Kannan RM, Nance E, Kannan S, Tomalia DA. Emerging concepts in dendrimer-based nanomedicine: from design principles to clinical applications. J Intern Med. 2014;276(6):579–617.
59. McCarthy TD, Karellas P, Henderson SA, Giannis M, O'Keefe DF, Heery G, Paull JR, Matthews BR, Holan G. Dendrimers as drugs: discovery and preclinical and clinical development of dendrimer-based microbicides for HIV and STI prevention. Mol Pharm. 2005;2(4):312–8.
60. Patton DL, Cosgrove Sweeney YT, McCarthy TD, Hillier SL. Preclinical safety and efficacy assessments of dendrimer-based (SPL7013) microbicide gel formulations in a nonhuman primate model. Antimicrob Agents Chemother. 2006;50(5):1696–700.
61. Rupp R, Rosenthal SL, Stanberry LR. VivaGel™(SPL7013 Gel): a candidate dendrimer–microbicide for the prevention of HIV and HSV infection. Int J Nanomedicine. 2007;2(4):561.
62. Jose J, Charyulu N. Solubility enhancement of an antifungal agent by association with dendrimers. Indian J Res Pharm Biotechnol. 2015;3:2320–3471.
63. Karthikeyan R, Vijayarajkumar P. PEGylated nanoarchitechture mediated solubility enhancement of tyrosine-kinase inhibitor. Inventi Rapid: Novel Excipients. 2015;2:1–4.
64. Katare YK, Daya RP, Sookram Gray C, Luckham RE, Bhandari J, Chauhan AS, Mishra RK. Brain targeting of a water insoluble antipsychotic drug haloperidol via the intranasal route using PAMAM dendrimer. Mol Pharm. 2015;12(9):3380–8.
65. Pooja D, Kulhari H, Singh MK, Mukherjee S, Rachamalla SS, Sistla R. Dendrimer–TPGS mixed micelles for enhanced solubility and cellular toxicity of taxanes. Colloids Surf B: Biointerfaces. 2014;121:461–8.
66. Nasr M, Najlah M, D'Emanuele A, Elhissi A. PAMAM dendrimers as aerosol drug nanocarriers for pulmonary delivery via nebulization. Int J Pharm. 2014;461(1-2):242–50.
67. Zhou Z, D'Emanuele A, Attwood D. Solubility enhancement of paclitaxel using a linear-dendritic block copolymer. Int J Pharm. 2013;452(1-2):173–9.
68. Koç FE, Şenel M. Solubility enhancement of Non-Steroidal Anti-Inflammatory Drugs (NSAIDs) using polypolypropylene oxide core PAMAM dendrimers. Int J Pharm. 2013;451(1-2):18–22.
69. Gautam SP, Verma A. PAMAM dendrimers: novel polymeric nanoarchitectures for solubility enhancement of candesartan cilexetil. Pharm Sci. 2012;1:1–4.
70. Huang X, Wu Z, Gao W, Chen Q, Yu B. Polyamidoamine dendrimers as potential drug carriers for enhanced aqueous solubility and oral bioavailability of silybin. Drug Dev Ind Pharm. 2011;37(4):419–27.
71. Prieto MJ, Temprana CF, del Río Zabala NE, Marotta CH, del Valle AS. Optimization and in vitro toxicity evaluation of G4 PAMAM dendrimer–risperidone complexes. Eur J Med Chem. 2011;46(3):845–50.
72. Gupta U, Agashe HB, Jain NK. Polypropylene imine dendrimer mediated solubility enhancement: effect of pH and functional groups of hydrophobes. J Pharm Pharm Sci. 2007;10(3):358–67.
73. Dutta T, Garg M, Jain NK. Targeting of efavirenz loaded tuftsin conjugated poly (propyleneimine) dendrimers to HIV infected macrophages in vitro. Eur J Pharm Sci. 2008;34(2-3):181–9.
74. Gajbhiye V, Ganesh N, Barve J, Jain NK. Synthesis, characterization and targeting potential of zidovudine loaded sialic acid conjugated-mannosylated poly (propyleneimine) dendrimers. Eur J Pharm Sci. 2013;48(4-5):668–79.
75. Wang F, Bronich TK, Kabanov AV, Rauh RD, Roovers J. Synthesis and evaluation of a star amphiphilic block copolymer from poly (ε-caprolactone) and poly (ethylene glycol) as a potential drug delivery carrier. Bioconjug Chem. 2005;16(2):397–405.
76. Bosnjakovic A, Mishra MK, Ren W, Kurtoglu YE, Shi T, Fan D, Kannan RM. Poly (amidoamine) dendrimer-erythromycin conjugates for drug delivery to macrophages involved in periprosthetic inflammation. Nanomedicine. 2011;7(3):284–94.
77. Vandamme TF, Brobeck L. Poly (amidoamine) dendrimers as ophthalmic vehicles for ocular delivery of pilocarpine nitrate and tropicamide. J Control Release. 2005;102(1):23–38.

Liposomes for Delivery of Antitubercular Drugs

Priya Shrivastava, Laxmikant Gautam, Sonal Vyas, and Suresh P. Vyas

Abstract TB is a severe infectious disease that is yet to be eradicated and controlled. The ongoing therapy of TB (DOTS strategy) is associated with certain limitations including a complicated therapeutic regimen, prolonged therapy with a multidrug combination, patient noncompliance, off-target distribution of the bioactive(s), subtherapeutic levels at the target site, appearance of several adverse effects, and development of multidrug resistance (MDR). The liposomes-based nanomedicines have been developed to offer distinctive benefits. They can potentially address the challenges associated with conventional therapeutic options. These nanoconstructs allow an increase in the bioavailability of the bioactive(s). Consequently, the frequency of administration of drugs and duration of therapy are reduced. The results obtained with such nanoplatforms, however preliminary, are promising. They may be extremely useful alternatives to current therapy as they improve the targeting ability of the payloads at the site of infection and cell-specific delivery. Furthermore, integrating the liposomal nanoconstructs with the inhalational or pulmonary route of administration provides the most promising approach in anti-TB drug therapy. This chapter outlines the potential role of liposomes in addressing the challenges associated with TB chemotherapy.

Keywords Nanoconstructs · Liposomes · Macrophage targeting · Tuberculosis · Passive drug targeting · Active drug targeting · Antitubercular therapy

---

P. Shrivastava · L. Gautam · S. P. Vyas (✉)
Drug Delivery Research Laboratory, Department of Pharmaceutical Sciences,
Dr. Harisingh Gour Vishwavidyalaya, Sagar, Madhya Pradesh, India
e-mail: vyas_sp@rediffmail.com

S. Vyas
Director Sampoorna Path Care Labs, Sagar, Madhya Pradesh, India

R. Shegokar, Y. Pathak (eds.), *Tubercular Drug Delivery Systems*,
https://doi.org/10.1007/978-3-031-14100-3_8

1 Introduction

Tuberculosis (TB) continues to be a leading infectious disease globally [1]. It is the second most infectious disease after HIV (human immunodeficiency virus) infection. Pulmonary TB is one of the most prevalent manifestations of TB. Furthermore, the co-infection between Mtb (*Mycobacterium tuberculosis*) infection and HIV has dramatically accelerated deaths and morbidity in developing and developed nations. HIV infection remains the most significant risk factor; so far, the activation to the active disease state from the latent stage is concerned [2]. Nevertheless, co-infected (TB + HIV) patients are at a higher risk of developing active TB. Conclusively, TB contributes to the primary cause of mortality in co-infected patients. That is why TB is gaining intense attention from research scientists globally in recent years. Several bioactive(s) have demonstrated enormous potential in the therapy of TB in the past; however, most of the therapeutic moieties are associated with certain limitations such as poor metabolic stability, poor solubility, low permeability, and hence low bioavailability [3, 4]. Other limitations include a complicated therapeutic regimen and prolonged therapy that leads to patient noncompliance, off-target distribution of bioactive(s), subtherapeutic levels at the target site, development of MDR, etc. These challenges result in high dose requirements that further lead to drug-associated toxicities. Due to these aforementioned reasons, conventional or standard treatment regimens are ineffective in the complete eradication of infection.

In addition to these factors, TB bacilli develop reservoirs in the alveolar macrophages, specifically by inhibiting phagosomes and lysosomes fusion. Consequently, the pathogen/antigen processing and presentation, T helper cell involvement, and as a result immunological consequences including bactericidal activity are inhibited. Furthermore, the infected macrophages when exposed to a subtherapeutic concentration of bioactive(s) develop MDR by activation and remodeling of responsible bio-constituents mainly associated with the membrane or cytosolic machinery. In addition, under the modified set of modified bio-condominium, the infectious agent, i.e., Mtb, also tends to mutate into a strain that averts pharmacodynamic activity of the bioactive(s) [5]. Furthermore, Mtb strains develop resistance to multiple antitubercular bioactive(s) during the therapy resulting in therapeutic failure. These reservoirs of tubercle bacilli together with drug-resistant Mtb strains (MDR and XDR strains) are challenging in effective treatment [6]. However, if bacilli are not treated, they can spread to other areas of the body and cause secondary TB infection, also referred to as post-primary infection [7]. Other factors include enzymatic degradation and metabolism of orally administered medicament before reaching the site of action or systemic circulation that renders the drug with compromised bioactivity and poor bioavailability [8]. These drawbacks of conventional antitubercular therapy entail for a quest for newer improved drug delivery approaches for the delivery of currently approved novel drugs for safe and effective treatment [9, 10].

To circumvent these challenges, liposomal-based therapeutics are being developed with antitubercular bioactive(s), which might be extremely useful alternatives to the current therapy. Liposomes are nanobiotechnological modules that possess the potential to be employed in the delivery of therapeutic moieties for TB chemotherapy [11, 12]. They are biocompatible, safe, and may enhance the therapeutic index of antitubercular drugs. Furthermore, they can be loaded with either type of drug(s), i.e., hydrophilic and lipophilic. The characteristics of these delivery modules, including surface charge, composition, size, and presence of ligands, can significantly extend the site specificity to such carriers [13]. The development of targeted and controlled release formulations of antitubercular bioactive(s) that are either intravenously delivered or inhaled/aerosolized directly to the lungs appears to be the putative and efficacious therapeutic alternatives. The intravenous route is of particular interest in extrapulmonary TB, whereas the inhalational route is advantageous for delivering bioactive(s) directly to the lungs (primary organ of pulmonary TB) [14]. The chapter focuses on the role of liposomes as bioactive(s) delivery modules for antitubercular bioactive(s) for targeting the alveolar macrophages.

2 Role of Nanomedicines in TB Chemotherapy

Nanomedicine-based therapeutics have gained a focal attention over the past years as superior alternatives to the conventional therapy regimens in TB chemotherapy [15, 16]. This is due to their potential to deliver multiple bioactive simultaneously, improved selectivity, high efficacy, ability to address MDR, etc. Moreover, they decrease drug-associated side effects by delivering the bioactive(s) to the site of action and avoiding its nontarget accumulation. They increase the stability of the bioactive(s) by shielding them from external environmental conditions. They extend the residence time of the bioactive(s) by blocking its efflux resulting in greater therapeutic benefits.

In addition to this, pulmonary nanomedicine possesses specific therapeutic benefits over systematically or orally administered nanotherapeutics owing to the localized higher concentration of the antitubercular bioactive(s)-loaded delivery modules to lung tissues where Mtb infection occurs and develops [17]. The pulmonary administration prevents first-pass metabolism of anti-TB drugs and facilitates rapid absorption owing to the broad alveolar surface area. They avoid nontargeted distributions of the bioactive(s) that lead to the reduction in the liver and renal toxicity-related clinical concerns and improve the therapeutic efficacy with maximum patient compliance. They can directly penetrate alveolar macrophages and deep lung tissues resulting in greater deposition and drug accumulation, and also consistency in localized drug delivery at the target site. Additionally, aerosolized pulmonary nanotherapeutics possess superior penetration into the alveolar macrophages. Consequently, drug absorption and retention in the lungs are improved. The intracellular delivery of nanotherapeutics by localized delivery and targeting of drug release to the infected cells holds the potential to circumvent the problems associated with conventional antitubercular therapy. Surface-functionalized nanotherapeutics with stimuli-responsive release of the bioactive(s) or macrophage targeting could be a promising alternative to the existing therapy. These approaches may improve the therapeutic efficacy of the anti-TB bioactive(s) against resistant Mtb strains [18]. Some potential advantages of nanomedicines in TB chemotherapy are discussed below.

2.1 Role in Improving Drug Permeability Through the Thick Mycolic Acid Cell Wall

The major problem associated with conventional TB chemotherapy is the poor penetration of antitubercular bioactive(s) in the Mtb-infected alveolar macrophages (target site). This is due to the presence of a thick hydrophobic outer envelope that inadvertently affects the absorption and hence the desired drugs concentration at the target site [19]. The thick outer coat of Mtb is lipidic. It is composed of lipophilic lipoarabinomannan (LAM) and mycolic acid (a form of hydroxyl fatty acid) covalently linked to arabinogalactan. This covalent linkage bestows significant hydrophobic characteristics to the outer cell coat of Mtb. This hydrophobic character renders the cell wall of Mtb impervious to the hydrophilic anti-TB bioactive(s). This results in the development of resistance to several chemotherapeutics and antibiotics [20]. Poor permeability and limited solubility are some other drawbacks associated with the antitubercular bioactive(s). Consequently, bioactive(s) delivery modules that facilitate selective drug internalization vis-à-vis improved penetration as well as antimicrobial activity, as a result, may significantly increase therapeutic effectiveness [21]. Previous works have reported that the carriers of the nanometric size range could potentially improve the permeability of a bioactive across the thick hydrophobic bacterial envelope, consequently reducing bacterial burden, thereby improving therapeutic outcomes [22, 23].

2.2 Role in Addressing Drug Resistance Due to Bacterial Microenvironment

The efflux pumps (systems) operate in both the cells, i.e., the host cell as well as in bacterial cells. Therefore, the required therapeutic concentration fails to reach its optimum level. The host cells, particularly T cells and macrophages which are the common cellular tropics of infection, are equipped with P-glycoprotein (P-gp) pumps [24, 25]. 170 kd plasma membrane P-glycoprotein (P-gp) functions as a metabolically active drug efflux pump in human T cells and macrophages. It accounts for the majority of drug-resistant infections. These systems pump solute/drugs out of the cell (Fig. 1). Thus, they allow the Mtb to regulate its internal environment by removing toxic substances such as antibacterial agents, metabolites, etc. The Mtb genome encodes for a majority of different efflux transporters. Out of these efflux transporters, primarily ABC and MFS superfamily proteins are involved in

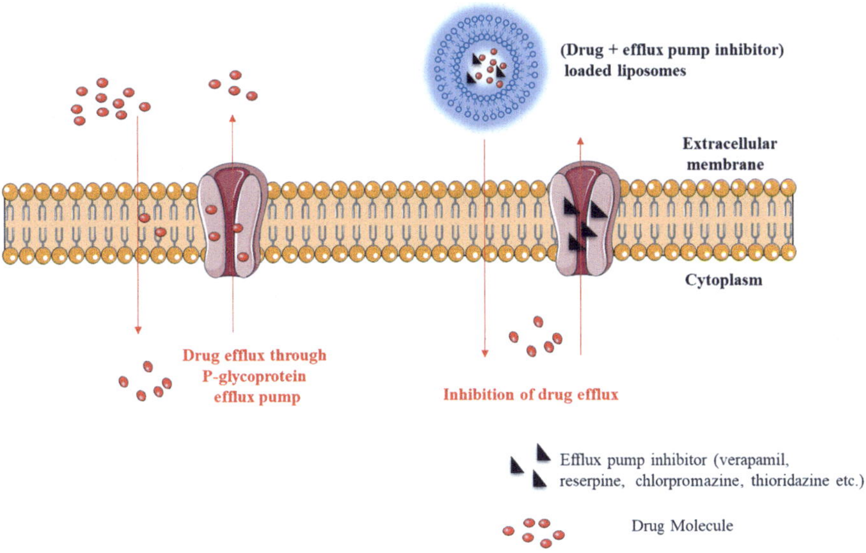

Fig. 1 Schematic representation of role of nanocarrier in addressing drug resistance due to bacterial microenvironment (self-drawn)

bioactive(s) resistance. The use of efflux pump inhibitors could be the potential approach to combat drug efflux. These nonantibiotic chemicals are small molecules that potentiate a co-administered antimicrobial agent by blocking bacterial efflux systems [26]. A variety of efflux pump inhibitors such as calcium channel blocker Verapamil, the alkaloid reserpine, and the phenothiazine derivatives chlorpromazine and thioridazine are well established for use in Mtb infection. Moreover, a nanomedicine-based approach may be able to effectively circumvent the challenges relating to drug efflux particularly operating in cellular tropics of infection. Demitto et al. [27] worked on Verapamil and Rifampicin combination for the treatment of MDR-TB. This combination was used against $MtbH_{37}R_V$ strain MDR Mtb clinical isolates. The Verapamil and Rifampicin combination demonstrated synergism in MDR Mtb clinical isolates. The co-adjunct therapy with Verapamil inhibited Mtb efflux pumps. This renders the Mtb more susceptible to Rifampicin [27].

2.3 Role in Controlled Drug Release

The delivery of anti-TB drugs using nanocarrier-based systems offers distinct advantages over free anti-TB drugs. These include prolonged or extended circulation, improved access of the therapeutic payload to the Mtb-infected macrophages, and hence enhancing the efficacy of the therapy [28]. Apart from this, the prolonged release of the bioactive(s) from nanocarriers builds a consistent therapeutic concentration of the bioactive(s) over a prolonged period [29]. This improves the pharmacokinetic profile of the therapeutic moieties and ensures a lower dose requirement and less frequent dosage regimens. Rajan et al. demonstrated that Rifampicin (RIF)-loaded chitosan nanoparticles could increase the delivery of the bioactive(s) by improving the solubility of RIF. Furthermore, the release of the RIF was also controlled [30]. The pegylation of nanoparticles could prolong the systemic circulation of carriers and control the release of the bioactive(s) for a better antitubercular effect.

2.4 Role in Reducing the Dosage Size and Dosing Frequency

Nanotherapeutics-based formulations could minimize the dose, frequency of dosing along with toxic effects associated with the bioactive(s). It is due to site-specific accumulation of the bioactive(s) that enhances the therapeutic efficacy [31]. This would improve patient compliance. Zaru et al. reported the efficacy of RIF-loaded liposomes for alveolar macrophage targeting. The author worked on a passive drug-targeting strategy. Freeze-dried, redispersible RIF-loaded liposomes were prepared by using the thin film hydration technique and extensively characterized. The formulations were administered in rats following aerosol inhalation. The developed nanosystems (liposomes) were stable in the presence of mucus. They were avidly internalized and taken up by the alveolar macrophages and also deposited in the lower airways. The growth of *Mycobacterium avium* was inhibited even at low doses of RIF [32]. Rawal et al. worked on RIF-loaded freeze-dried chitosan nanoparticles, developed for the treatment of TB. The formulation was prepared by using the ionic gelation method and extensively characterized. The mass median aerodynamic diameter (MMAD) was determined by using an Anderson Cascade Impactor (ACI) and was found to be 3.3 μm. *In vitro* drug deposition study results demonstrated that the fine particle fraction (FPF) of the nanoparticles contain 33% of the loaded drug dose. The large FPF value indicated that the deposition of the bioactive(s) in the lungs was satisfactory. The formulations demonstrated sustained and localized delivery of the bioactive(s) together with the targeting of the nanoparticles to the alveolar macrophages. The accumulation of the bioactive(s) in the infected macrophages was improved. The dose and dosing frequency and drug-related toxicity were reduced [33].

2.5 Role in Macrophage Targeted Chemotherapy

Targeted nano-delivery systems offer multiple advantages in the treatment of chronic human infectious diseases. They possess highly functional corona that could be modified by using specific targeting moieties to enable target-oriented delivery of drug(s) [34]. They have great potential to effectively deliver the loaded bioactive(s) to the infected macrophages and other host cells of Mtb. They minimally release the bioactive(s) in the blood circulation; thus, prevent off-target distribution and accumulation; while sustain and maintain the therapeutic concentration of the bioactive(s) at the infected site. This could improve the potency of the bioactive(s) against tubercle bacilli in infected host cells (macrophages) [35]. Consequently, the dose requirement would be minimized. Apart from this, they offer protection to the loaded bioactive(s) from renal clearance, hepatic degradation, and first-pass metabolism in order to improve the pharmacokinetic characteristics of the loaded drug.

The *tubercle bacilli* replicate within the phagosomes of the host macrophages. The poor penetration of the bioactive(s) into the host cell (macrophages) is one of the major challenges in effective TB chemotherapy. This leads to poor therapeutic outcomes. The design and development of nanotherapeutics for the delivery and accumulation of bioactive(s) selectively in the infected macrophages may result in an optimum/higher therapeutic concentration. This may lead to improvement in the antitubercular action of the loaded bioactive(s). Nanoconstructs for targeting macrophages take advantage of targeting moieties in their assembly that identify/recognize the receptors highly expressed on Mtb-harboring macrophages. Basha and coworkers reported nanoparticles-mediated dual delivery of anti-TB drugs to infected macrophages. RIF and Levofloxacin combination was used in the study. The drugs were complexed with cyclodextrin. The developed complexes were then crosslinked onto curdlan nanoparticles by using epichlorohydrin. Raw 264.7 and L929 cells were used for the cell line studies. The developed formulation was found to be nontoxic to these cell lines. The curdlan nanoparticles were selectively recognized by a dectin-1 receptor expressed on the surface of the macrophages. The investigation revealed that curdlan nanoparticles demonstrated 1.8 times higher uptake by macrophages and killed more than 95% of Mtb in 4 h [36].

Nanoconstructs could be surface functionalized to attain targeted delivery of the bioactive(s) through an active drug targeting approach. The actively targeted nanoplatforms-based formulations take advantage of targeting moieties that possess an affinity for a specific receptor for precise delivery of the bioactive(s) to the infected macrophages or desired site. Chono et al. developed mannose-conjugated liposomes to exploit the mannose receptors that exist on the surface of macrophages and facilitate the delivery to the target site upon pulmonary administration. The particle size of liposomes was between 100 and 2000 nm. The developed formulations were administered in rats through the pulmonary route. The cellular uptake (both *in vitro* and *in vivo*) by rat alveolar macrophages was significantly higher in the case of mannose-conjugated liposomes compared to plain liposomes [37]. The developed mannosylated liposomal formulation appears to be the potential carrier for macrophage-targeted TB chemotherapy.

3 Liposomal Nanoconstructs

Liposomes are self-spherical, artificial vesicles composed of bilayers of phospholipids that enclose an aqueous phase in their core [38–40]. These are like cells; however, the sizes are much smaller. These vesicular constructs are produced as a result of self-assemblies of the amphiphiles in an aqueous compartment forming unilamellar or multilamellar concentric bilayers. Hydrophobic hydrocarbon tails or fatty acids tails form the lipophilic/hydrophobic core of the bilayers that are shielded from the aqueous compartment whereas the polar head groups are exposed to the aqueous medium and form the outer hydrophilic surface [41]. They are spontaneously formed on the hydration of a dry phospholipid film above the glass transition temperature of phospholipids. They are of varying sizes ranging from 30 nm to several micrometers [42]. The thickness of the phospholipid bilayer membrane ranges from 4 to 7 nm. They are classified primarily in the following categories on the basis of their size and the number of bilayers forming the vesicles (lamellarity) – small unilamellar vesicles (SUVs) (30–100 nm); large unilamellar vesicles (LUVs) (>100 nm); giant unilamellar vesicles (>1000 nm); multilamellar vesicles (MLVs) (>500 nm); and multivesicular vesicles (1000 nm to several micrometers). The liposomes produced by using the lipid thin film hydration technique are multilamellar vesicles (MLVs) [43]. These MLVs are composed of several concentric lipid bilayers arranged in an onion-like structure (hydrated multilayers) with diameters ranging from 500 nm to several micrometers. The MLVs can be further processed by sonication or extrusion techniques, through a filter to generate unilamellar vesicles (ULVs) [44] (Fig. 2). These ULVs are liposomes with single lamellae (membrane bilayer). Based on their size, these ULVs can be further classified as small

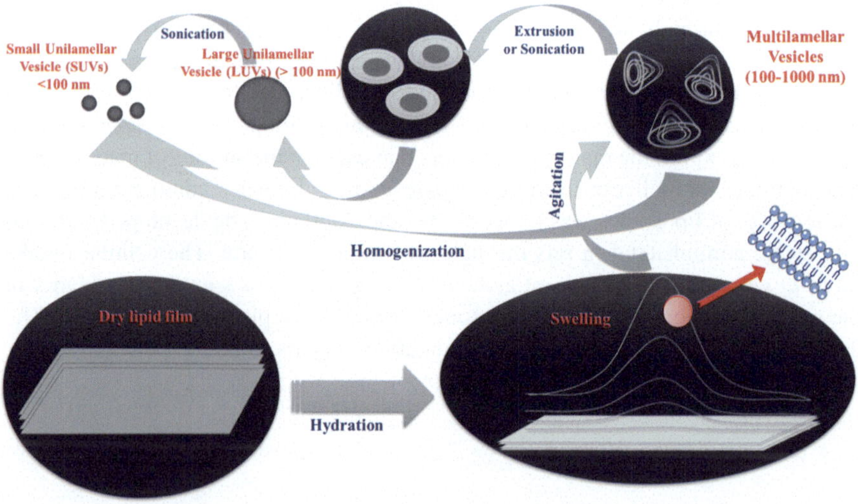

Fig. 2 Schematic representation of methods of preparation of liposomes (self-drawn)

unilamellar vesicles (SUVs) and large unilamellar vesicles (LUVs). The difference in the release rates of the bioactive(s) strongly depends on the phospholipids that the bioactive has to cross during the release process [45].

On the basis of their composition and constitution, they may also be classified as conventional, anionic, cationic, long-circulating, stimuli-responsive, and immunoliposomes. Moreover, there are various other vesicles-based systems considered as modified liposome(s) vesicles. The examples include ethosomes, sphyngosomes, virosomes, transferosomes, emulsomes, pharmacosomes, and enzymosomes. These are analogous to lipid-based liposomes. Besides this, niosomes, aquasomes, and bilosomes are nonlipid-based liposome analogs [46].

4 Benefits and Limitations of Liposomes as Drug Delivery Modules

Liposomes have sparked considerable interest since their discovery by Bengham and colleagues in 1965 due to their flexible organization and versatility in biophysical and physicochemical characteristics [47]. The realization of the remarkable potential of the phospholipids to self-assemble substantiated previous findings that all intracellular vesicles and plasma membranes are comprised of bilayers of phospholipids. It encouraged the extensive use of liposomes as the primary model for studying the biophysical and physicochemical features of the cell/biological membranes [48, 49]. One of the distinct merits of liposomes is their chemical biphasic nature as they contain both hydrophilic and hydrophobic regions. Therefore, their applications have enormously increased from merely membrane/biomimetic modules to attractive vehicles for the delivery of both hydrophobic and hydrophilic bioactive(s). Consequently, in 1970, just after 5 years of their development, liposomes emerged as well-acceptable drug delivery modules. Furthermore, liposomes were among the first generations of nanocarriers approved by the USFDA in 1995 for study in clinical trials. Initially, they were used for the delivery of doxorubicin for the treatment of ovarian cancer and AIDS-related Kaposi's sarcoma [50].

Today, liposomes are the most extensively employed carrier for bioactive(s) delivery. They are also widely exploited as delivery modules for antimicrobials. They offer numerous advantages including safety and biocompatibility; nonimmunogenicity; the potential to carry large drugs payloads; the ability for self-assembly; and a vast array of biophysical and physicochemical characteristics that could be altered to engineer their carrier characteristics [51]. Encapsulation within liposomes protects the therapeutic moieties from early inactivation, degradation, and dilution in the circulation. Despite their merits, they still suffer some limitations such as instability in plasma. When drugs-loaded liposomes enter the bloodstream, during their circulation, they interact with the plasma proteins such as high-density lipoproteins (HDLs), low-density lipoproteins (LDLs), and opsonins (immunoglobulin, fibronectin, lipoproteins) [52]. The opsonins bind to the liposomes followed by adsorption of opsonins onto the phospholipid membrane and signaling for their presence. This signal is detected by the mononuclear phagocyte system (MPS)/ reticuloendothelial system (RES) cells which engulf liposomes and eliminate them from the blood circulation through the opsonization process. This leads to the elimination of the bioactive(s) at the hepatic level following its metabolism by the Kupffer cells. Another process of elimination involves the metabolism of liposomes by the MPS. The elimination or clearance process by the MPS cells via plasma proteins opsonization is influenced by a number of variables including size, surface charge, and colloidal stability. In general, negatively charged large liposomes are eliminated more rapidly from the blood circulation than small, positively charged, or neutral liposomes [53]. Thus, negatively charged liposomes have a shorter half-life in blood circulation. However, this probable drawback of liposomes has been used for the effective therapeutic delivery of the antibacterial/antimicrobial bioactive(s) to treat opportunistic intracellular infections thriving within the MPS such as TB.

Nevertheless, when the target site is other than the MPS, the use of liposomes that are able to escape passive uptake by MPS is justified in order to allow for longer circulation. The search for long-circulating liposomes thus began with the development of certain lipids, i.e., pegylated lipids; the PEG-coated stealth liposomes that are sterically stabilized. The protection of liposomes from scavenging by the MPS is offered by the pegylation that constitutes a steric barrier [54]. The liposome surfaces could be further altered/decorated with targeting moieties such as peptides, antibodies, carbohydrates, etc. It improves the specificity of these nanoconstructs to recognize and bind to target cells. Consequently, the improvement in the specificity results in a favored accumulation of these nanocarriers at the infected or target sites and relatively lower concentrations in nontarget cells or healthy tissues saving them from eventual toxic effects of the drug(s) [55, 56]. Conclusively, there are promising and potential merits of liposomes to be a carrier of choice for drugs (Fig. 3).

5 Routes of Administration of Liposomes in the Treatment of TB

Liposomes have previously been evaluated as an anti-TB drug delivery module for the treatment of both pulmonary and extrapulmonary TB [57]. The success and efficacy of the therapy, however, are strongly dependent on the routes of administration of these nanoconstructs. It is well-established that liposomes are susceptible to intestinal lipases; therefore, their oral application seems limited. The permeability and stability of vesicular systems could be increased for oral drug delivery by modulating the compositions of the phospholipid bilayers, by adding targeting moieties (ligands) or modifications using polymers. Routes other than the oral route have also been explored for their administration [58]. The intravenous (IV) route of administration has also been studied for the delivery of antitubercular drug(s). Several factors including leakage of the encapsulated content from liposomes in the

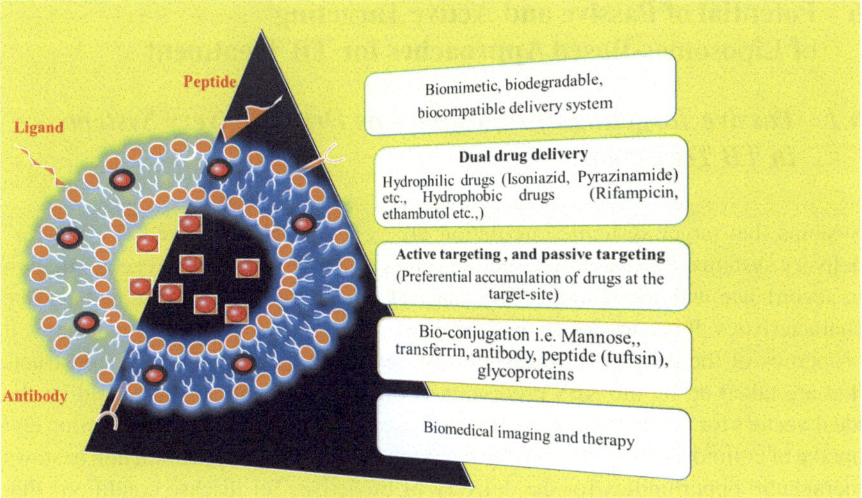

Fig. 3 Potential advantages of liposomes

plasma compartment before reaching the infected or target cells/tissues, rapid clearance from the blood circulation, and uptake of the liposomes by the liver and spleen macrophages have been the concern of research and studies.

Gaspar and colleagues reported that preferential accumulation of antitubercular drugs-loaded liposomes occurs in the spleen and liver after IV administration, rather than their accumulation in the lungs. This may be a potential approach for the treatment of extrapulmonary TB [59]. In contrast to this, if the surface of the liposomes is functionalized by using polymers or targeting moieties, the lung specificity could be achieved [60]. In addition to the IV route, the noninvasive routes such as nasopulmonary route and inhalatory route have equally been explored. It would provide convenience to the patient and the possibility of effective localized delivery of antimicrobials. This may be a very interesting and attractive alternative strategy, particularly for the therapy of pulmonary TB [61]. Thus, the administration of liposomal nanoconstructs loaded with anti-TB drugs through inhalatory or nasopulmonary route may also be further explored. However, the size of the liposomes must be tuned because the area in the respiratory tract where these nanoconstructs are deposited principally depends on their size. Deposition of 1 nm particles occurs largely in the upper airways (nose, pharynx, and larynx). The particles of approx. 5 nm are considered appropriate for deposition into the trachea and bronchi regions while particles of 20 nm sizes are optimally deposited into the deeper alveolar regions of the lungs. Similarly, in the case of particles greater than 15 μm in size, they are reportedly retained in the throat and get swallowed in the gut [62].

6 Potential of Passive and Active Targeting of Liposomes-Based Approaches for TB Treatment

6.1 Passive Targeting of Liposomes as Drug Delivery Systems in TB Treatment

Systems that target systemic circulation are generally characterized as "passive" delivery systems. Passive drug targeting refers to the distribution and accumulation in accordance with the nature of the material of construct and its size and surface characteristics. It occurs due to the body's natural response to the physicochemical properties of the drug or drug-carrier systems [63]. The colloidal nanoconstructs that are taken up by the RES predominant organs (liver and spleen) could be the ideal vectors for passive drug targeting to these compartments. The intervention and uptake of colloidal nanoconstructs by macrophages by passive phenomenon bestows therapeutic opportunities for the delivery of bioactives for disease conditions that involve macrophage cells of the RES, e.g., leishmaniasis. Targetable modules in this class include drug-loaded bilayer vesicular modules as well as cellular nanoconstructs of micron or submicron size range.

6.1.1 Nanocarriers Characteristics Affect Passive Drug Targeting Strategy

The physicochemical characteristics of the developed nanoconstruct and the pathophysiological characteristics of the tumor microenvironment influence the biodistribution, pharmacokinetics, and toxicity profiles of the loaded bioactive(s). Previous studies have reported that the size and surface charge of the nanosystems affect the pathway of their cellular uptake and consequently their therapeutic effectiveness. These parameters influence the adhesion of the particles and their interaction with cells [64].

One of the major applications of nanomedicines is in the selective and effective delivery of several antitumor bioactives based on extravasation owing to nanosize of the modules and their subsequent retention due to nonavailability of the lymphatic drainage system. The microvasculature of healthy tissues varies by tissue type, but in most tissues including the heart, brain, and lungs, there are intercellular junctions with less than 10 nm intercellular spaces; however, tumor microvasculature contains pores, fenestrae ranging from 100 to 1000 nm in diameter. Therefore, a tumor within these tissues could be targeted selectively by designing and developing bioactive(s) delivery modules that are larger than the intercellular gap of the healthy tissue but smaller than the pores found in tumor vasculature. Therefore, the nanocarriers of size-ranging from 10 to 1000 nm are considered to be appropriate for the targeted delivery of various antitumor bioactives. The size of the nanoconstructs is a pivotal parameter for their permeation and retention in the tumor and thus is due to the fenestrations in tumor vessels. The accumulation of liposomes or large molecules in cancerous tissues/tumors is the result of a "leaky" microvasculature and impaired lymphatics, supporting the tumor area (also known as enhanced permeability and retention effect) [65, 66]. Previously it was reported that the enhanced permeability and retention phenomenon may also operate in the infected and inflamed tissues. Thus, as a general rule, this phenomenon could be advantageous for improvising nanosystems accumulation at the site of infection or infected area, particularly in the granuloma in the case of Mtb infection.

EPR phenomenon is found associated with etiology where cancer and inflammation develop due to hypervenulated (fenestrated) blood vessels. This leads to extravasation and subsequent retention. This has been reported to be of benefit in the case of carcinoma and also other inflammatory diseases such as myocardial infarction and arthritis inflammation due to infection.

The extravasation through the discontinuous endothelium of the tumor microvasculature is a principal pathway for the transport/movement of liposomes into the tumor interstitium. Even the size of the liposomes in the passive targeting approach determines the degree or extent of extravasation from the normal vasculature. Once in the tumors, nontargeted liposomes are localized in the interstitium surrounding the tumor cells; however, a limited accumulation and distribution of liposomes within the tumor interstitium results due to typically high interstitial pressure (which though helps in trapping the liposomes within the tumor area but prevents the access of bioactive into the necrotic zone) and a large interstitial space compared with normal body tissues [67, 68].

The surface chemistry and charge of the nanosystems also play a crucial role in the circulation time. Lipophilic/hydrophobic or charged modules are opsonized rapidly by the cells of the MPS. Therefore, the nanosystems are prepared with hydrophilic, slightly anionic charged, and sterically stabilized surfaces. Furthermore, surface stabilization is affected by using polyethylene glycol (PEG), a hydrophilic polymer that is adsorbed or grafted on the surface of nanocarriers [52, 68]. PEG coatings on nanosystems shield the surface from aggregation. Additionally, pegylation also prevents nonspecific interactions by changing the surface charge and hydration of the nanosystems. It has been reported that 5–9 mol% of DSPE-PEG2000 facilitates the optimal loading of PEG-modified lipids in the liposomes. Consequently, each polymer chain forms a mushroom-like globular structure with a slight overlay between distinct polymers at this lipid concentration. Furthermore, it also ascertains complete "stealth" coverage of the nanosystems surface [69].

It has been reported in a study that the first injection of pegylated liposomes was demonstrated to produce PEGs-specific IgM resulting in rapid clearance and increased hepatic uptake of the second dose of pegylated liposomes. This is referred to as an accelerated blood clearance phenomenon. It is a major impediment to the pharmacokinetics and pharmacodynamics of pegylated liposomes and other PEG-coated nanosystems. Furthermore, PEG corona may act as a steric barrier averting the nanosystems from being efficiently internalized into the cancerous tissue/cells. This problem is referred to as the "PEG dilemma." Consequently, while designing a drug delivery module, an acceptable balance/adjustment between prolonged circulation time and effective cellular uptake should be sought. The potential solutions could be the use of shorter PEG chains (i.e., Mw < 1000). The PEG attachment by enzyme-cleavable bond or grafting tumor-specific targeting moieties on nanosystems surface could be advantageous in circumventing the challenges [65]. However, pegylation is a clinically accepted method to ameliorate the surface characteristics of nanosystems and develop "stealth" bioactive(s) delivery modules. Furthermore, it offers an opportunity to chemically attach/link/conjugate a targeting moiety on nanosystems surface, which improves their intracellular uptake. This so called "active drug targeting" approach will be discussed in Sect. 6.2.

In regard to TB, another approach that appears advantageous is the avid uptake of the drug delivery modules by the mononuclear phagocytic system (macrophages, monocytes, and dendritic cells) via phagocytosis or endocytosis. The approach might be successful when infectious agents dwell within macrophagic tropics; which not only harbor the infection but also cluster together to constitute granulomas. The enhanced uptake could be achieved by tuning or altering their properties such as size, shape, porosity, elasticity, charge, lipophilicity, rigidness, etc. Passive targeting in TB particularly in pulmonary TB is of enormous interest when it is attempted through the inhalatory route [70]. Localized administration of nanotherapeutics through the inhalatory route prevents the gastrointestinal degradation of the bioactive(s) and first-pass metabolism in the liver; enhances accumulation and retention of bioactive(s) in the alveoli; prevents penetration of bioactive(s) into the blood circulation. This results in a higher accumulation of drug(s) in the lungs, and a high therapeutic effect simultaneously minimizes the adverse effects. This type of targeting solely depends on the natural propensity of the phagocytic cells to engulf exogenous particles [71]. This approach possesses the ability to deliver the bioactive(s) or nanotherapeutics to the desired site of action. Vyas et al. developed liposomal aerosols for improved delivery of Rifampicin to infected alveolar macrophages. *In vitro* penetration efficiency of the developed liposomal formulation was recorded to be 1.5–1.8 times higher in comparison to free drug solution-based aerosol. A higher concentration of the bioactive(s) in the lungs was observed in the case of ligand-anchored liposomes-based aerosols in comparison to free drug and plain liposomes-based aerosols [72]. The strategy seems promising and may pave the way for effective treatment for TB.

Kaul et al. reported RIF and ofloxacin-loaded pegylated liposomes for the targeted therapy of mycobacterial infections. The size of the liposomes was 160.6 nm. They reported significant colloidal stability of the formulation up to 120 days. The pharmacokinetic analysis demonstrated a slow biphasic pattern together with a much longer terminal half-life of 19.13 h. High systemic concentration even up to 24 h post injection was reported. Maximum organ localization was also observed in the spleen, liver, and kidneys one hour post injection. *In vivo*, scintigraphic studies were performed in the murine model. The enhanced cellular uptake at infected lesions was recorded. The competitive inhibition and blocking imaging studies demonstrated reduced uptake, confirming specific and active targeting [73].

Another well-known approach is the prodrug design or lipid drug conjugates (LDCs)-based formulations. Lipid drug conjugates (LDCs) are also known as lipoidal prodrugs. Several merits of LDCs include increased bioactive(s) payload, enhanced retention of bioactive(s), and improved tumor and lymphatic targeting [74]. Pandit et al. reported Isoniazid-loaded lipid drug conjugates-based nanoparticles to improve its intracellular delivery to alveolar macrophages. Isonaizid has a lot of drawbacks because of its hydrophilic character involving poor gut permeability and bioavailability, failure to cross blood-brain barrier, etc. To address these limitations, Isoniazid was conjugated with a short lipid chain of stearoyl chloride to form a stable covalently linked lipid drug conjugate. The lipid drug conjugates-based nanoparticles were fabricated by using the cold-high pressure homogenization technique and extensively characterized. The results showed sustained drug release, improved cellular uptake by THP-1 cells, and increased lipophilicity [75]. Other examples of passive targeting of liposomes as drug delivery systems for the treatment of TB are enlisted in Table 1.

6.2 *Active Targeting of Liposomes as Drug Delivery Systems in TB Treatment*

The passive and active drug targeting strategies are equivocally involved in macrophage targeting in antitubercular therapy (Fig. 4). The normal bio-distribution pattern of nanoconstructs is modulated through modifications in their structure and composition; it is referred to as active targeting [63, 85]. These alterations involve surface modifications, the use of charged lipids, or attachment/conjugation/coating of a targeting moiety (monoclonal antibodies, glycoproteins, polysaccharides, proteins, glycolipids, peptides, aptamers, etc.). Active targeting entails the delivery of the bioactive(s) to the pathologic sites through the molecular recognition process [72, 86]. It depends on the cross talks/interaction between the receptors on the target cell and the specific targeting moieties on the surface of the nanoplatforms, resulting in receptor-mediated endocytosis or phagocytosis. Consequently, internalization of the nanoconstructs occurs, which further increases the intracellular bioactive(s) accumulation in the target cells/tissues. Moreover, the previously mentioned pegylation process is a part of the design of an active targeting module. It prolongs or extends the circulation half-life of the developed liposomal formulations. The incorporation of negatively charged lipids (e.g., dicetylphosphate) into the liposomal surface appears to improve the affinity of the alveolar macrophages to anionic liposomes through scavenger receptors. These receptors do not seem to have mechanisms of ligand-induced downregulation. The ligand-bound drug is rapidly cleared from the extracellular milieu to the macrophage phagolysosome. The ligand-anchored or surface-decorated liposomes represent the most successful and efficient approach for the delivery of bioactive(s) at a high concentration to the lungs.

In the case of TB, the receptors expressed on the surface of macrophages could be advantageous for targeting antitubercular bioactive(s) to these cellular tropics of

Table 1 Passive targeting of single or multiple drug-loaded liposomes as drug delivery systems in TB treatment

Formulations	Loaded bioactive(s)	Therapeutic benefits	References
Targeted pegylated liposomes	Rifampicin and Ofloxacin	Excellent antimycobacterial activity, higher cellular uptake at infected lesions was observed.	[73]
Rifabutin-encapsulated liposomes	Rifabutin	Higher concentration of drug was attained in the lungs, liver, and spleen at 24 h; lower bacterial burden was observed in the spleen; promising approach for extrapulmonary TB therapy.	[59]
Pyrazinamide liposomes	Pyrazinamide	Highly significant reduction in the bacterial count was recorded	[70]
Rifampicin- and Isoniazid-loaded liposomes	Rifampicin and Isoniazid	Improved therapeutic drug levels in the lungs; reduction in frequency of administration were observed	[76]
Amikacin-loaded liposomes	Amikacin	Improved anti-TB activity; the half-life of the drug-encapsulated liposomes was increased.	[77]
Rifampicin-loaded liposomes in dry powder	Rifampicin	Liposomes released the drug in a controlled manner over a protracted period of time.	[78]
Liposomes in dry powder	Isoniazid, Rifampicin, Pyrazinamide, Ethionamide, and Streptomycin	No excessive drug leakage or changes in mean vesicle diameter occurred during 3 weeks. This could be attributed to the stability of liposomes.	[79]
Inhalable Isoniazid-loaded surfactant liposomes	Isoniazid	Higher antimycobacterial activity and preferential accumulation of drug were measured in the lungs.	[80]
Isoniazid proliposome powders	Isoniazid	They were nontoxic to respiratory-associated cells. Isoniazid-proliposomes showed greater antimycobacterial activity than free drug.	[81]
Levofloxacin-proliposomes	Levofloxacin	The efficacy of developed drug-loaded proliposomes was significantly greater than free drug against *M. bovis*.	[81]
Isoniazid-loaded elastic liposomes	Isoniazid	The developed formulations were hemocompatible and demonstrated sustained drug release.	[82]
pH-responsive liposomes	Isoniazid	The developed formulation showed pH-dependent release behavior	[83]
Isoniazid-Phthalocyanine-cyclodextrin complex loaded liposomes	Isoniazid	Controlled release of drug was observed.	[84]

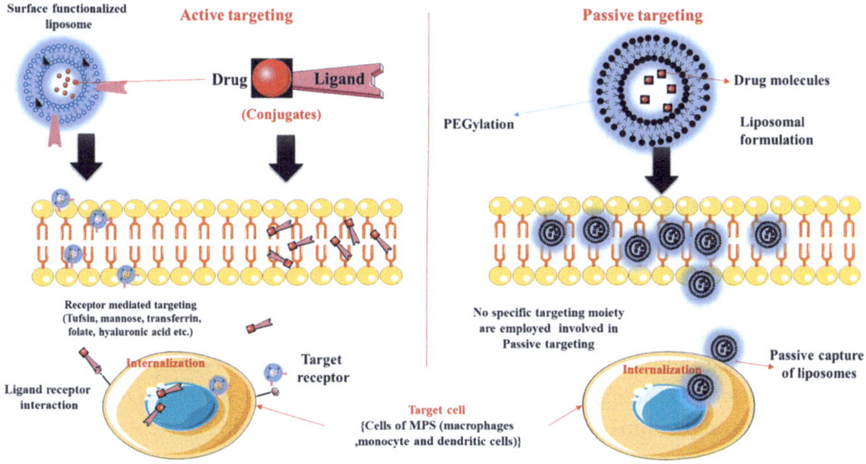

Fig. 4 Schematic representation of the principle of active and passive drug targeting

infection. Thus, the PRRs present on the macrophages are prime receptors for receptor-mediated targeting. These receptors play a significant role in recognizing pathogen-associated molecular patterns (PAMPs). The PRRs are of paramount significance for receptor recognition and targeting. The Toll-like receptors are primarily capable of recognizing heat shock proteins, lipoproteins, lipopolysaccharides, and flagellar proteins, whereas the C-type lectin receptors can recognize carbohydrate derivatives including fucose, mannose, β-glucans, and glycolipids [87–89]. These carbohydrate derivatives are expressed on the surface of the microbes. Apart from this, another receptor that also constitutes a subset of membrane-bound PRRs are referred to as scavenger receptors [90]. They could be advantageous candidates for site-specific delivery of the bioactive(s) due to their overexpression on the membrane of phagocytes. These receptors possess specificity for a wide range of ligands including proteoglycans, lipoproteins, polysaccharides, phospholipids, and bacterial components [91]. Apart from the PRRs, other types of receptors such as Fc receptors and complement receptors are also expressed on the surface of the phagocytes. They are involved in the opsonin-dependent phagocytosis and several other immunological responses as a result of which clearance of the opsonized microbe does occur [92, 93].

A diverse range of ligands has been investigated and reported in the literature that could be used to facilitate the active targeting of drugs to the macrophages. The specific targeting moieties for alveolar macrophages include O-steroyl amylopectin (O-SAP) and maleylated bovine serum albumin (MBSA). The O-SAP possesses a greater affinity toward alveolar macrophages, whereas the MBSA possesses a specific affinity for scavenger receptors that exist on the surface of alveolar macrophages [86]. One similar study was reported by Vyas and coworkers. They worked on O-SAP, MBSA-coated liposomes and reported their preferential accumulation in the lung macrophages [72].

The cells of the immune system such as macrophages excessively express mannose receptors (MR). The mannose receptor is also known as CD206, which belongs to a C-type lectin receptor family. The processes that are mediated by mannose receptors are endocytosis, phagocytosis, and other inflammatory consequences as well as intracellular trafficking pathways. The incorporation of mannose as a ligand onto the surface of the liposomes, producing so-called mannosylated liposomes, could be beneficial in the therapy of pulmonary TB. One similar study was reported by Shrivastava et al. The author developed dual drug-loaded mannosylated liposomes. The results demonstrated maximum anti-TB activity in Balb/C mice and preferential accumulation of mannosylated liposomes in the alveolar macrophages [94]. The strategy seems promising. Thus, mannose receptor-mediated macrophage targeting through liposomes-based delivery modules holds potential for TB chemotherapy.

Taciana and coworkers worked on usnic acid-loaded fucoidan-coated liposomes for Mtb-infected macrophage targeting. Fucoidan is a negatively charged sulfated polysaccharide that was used here as the coating material for liposomes. Fucoidan-coated liposomes demonstrated a lower IC_{50} value (8.26 ± 1.1 μM) in comparison to plain liposomes (18.37 ± 3.34 μM). The ligand-coated liposomes were avidly internalized through the C-type carbohydrate recognition domain and showed higher cell uptake than plain liposomes [95].

Tuftsin is a well-known ligand for macrophage targeting; it is a tetrapeptide, a natural macrophage activator. The peptide is found in leukophilic IgG. It specifically interacts and binds to phagocytic cells (polymorphonuclear leukocytes, monocytes, macrophages). Tuftsin demonstrates a wide array of activities including the activation of the immune system, potentiation of phagocytosis, immunological responses, and bactericidal activity [96]. Tuftsin grafting on the surface of liposomes facilitates their specific and selective transport to the macrophages. Moreover, it is also an activator of nonspecific cells to fight against infections.

Transferrin is another well-known PRR that could be used for macrophage targeting in the case of TB. It is an 80-kDa freely circulating iron-binding protein whereas transferrin receptor (Tfr) is a transmembrane-bound glycoprotein. It is involved in the transportation of ferric ions into the cells. It is internalized into the cell via receptor-mediated endocytosis after binding to its receptor. The transferrin receptor is functionally linked with the transportation of divalent metal ions (DMT-1) in endosomes [97, 98]. Regarding TB, Horwitz et al. [99] reported that exogenously administered Tf is transported to the Mtb phagosome, thus manifesting directly that the Mtb phagosome interacts with the early endosome. Tf and Tfr do not traffic via the lysosomal compartment [99]. The receptor holds the potential and constitutes the basis of the design and development of novel strategies for macrophage targeting of antitubercular bioactive(s). Other examples of active drug targeting of liposomes as bioactive(s) delivery modules for the treatment of TB are enlisted in Table 2.

Table 2 Active targeting of liposomes as drug delivery systems in TB treatment

Formulations	Loaded bioactive(s)	Targeting moiety/ligand used	Therapeutic benefits	References
Fucoidan-coated liposomes	Usnic acid	Fucoidan (A negative sulfated polysaccharide)	The developed formulations demonstrated higher cellular uptake and lower IC_{50} values than plain liposomes.	[95]
Mannosylated liposomes	Isoniazid and Rifampicin	Cholesten-5-yloxy-*N*-(4-((1-imino-2-D thiomannosylethyl) amino)alkyl)formamide (Man-C4-Chol)	Maximum anti-TB activity and preferential accumulation of drugs in lungs were recorded.	[94]
Ciprofloxacin-loaded mannosylated liposomes	Ciprofloxacin	Mannose	The developed mannosylated liposomes showed potent antibacterial effects against bacterial. However, plain drug-loaded liposomes were ineffective against many bacteria.	[100]
Covalent conjugation of ID93 antigen to liposomes	–	ID93 antigen	The vaccine formulation was stable. Significant secretion of IL-2, IF-γ, and TNF-α was observed.	[101]
Mannosyl phospholipid liposome system	–	Mannosyl phospholipid	Cellular uptake was enhanced.	[102]
Lung-specific stealth liposomes	Isoniazid and Rifampicin	O-stearylamylopectin	Preferential accumulation of drugs in the lungs was observed. The developed liposomal formulation was nontoxic to peritoneal macrophages than free drugs.	[60]
Lungs-specific stealth liposomes	Isoniazid and Rifampicin	O-stearylamylopectin	Mycobacterial burden was significantly reduced in the lungs, liver, and spleen of infected mice.	[103]

(continued)

Table 2 (continued)

Formulations	Loaded bioactive(s)	Targeting moiety/ligand used	Therapeutic benefits	References
Liposomal aerosols	Rifampicin	Maleylated bovine serum albumin (MBSA), and O-stearylamylopectin (O-SAP)	Preferential accumulation of drug in lungs at high concentration was recorded.	[72]
Tuftsin-bearing liposomes	Rifampin	Tuftsin	Rifampicin was delivered twice weekly for 2 weeks in infected mice. The results revealed that the developed formulation was 2000 times more effective compared to free drug. Mycobacterial burden in the lungs was also reduced.	[55]
Aerosolized liposomes	–	Mannose	Higher cellular uptake, efficiently delivered to alveolar macrophages upon pulmonary aerosolization to rats.	[104]
Mannosylated liposomes	–	Cholesten-5-yloxy-*N*-(4-((1-imino-2-D thiomannosylethyl) amino)alkyl)formamide (Man-C4-Chol)	Improved cellular uptake was observed.	[105]

Several other receptors are also present on the alveolar or pulmonary macrophages including CD200R, CD115 (colony-stimulating factor 1 receptor), CD14 (lipopolysaccharide (LPS) receptor), scavenger receptors (SR-A, CD36 (SR-B), Toll-like receptors (TLR4, TLR9), Dectin-1/2 (β-glucan receptor), Fc receptors (CD64, CD32, CD16), and macrophage receptor with collagenous structure (MARCO), CD163, CD204, CD68) [36, 98, 106]. These receptors can be used for macrophage targeting. They may provide new insights for the design and development of novel therapeutic options/approaches for macrophage targeting.

7 Recent Patents/Breakthroughs on Liposomes in TB

Some recent breakthroughs on liposomes-based systems for the treatment of TB are discussed in Table 3.

Table 3 Recent patents/breakthroughs on liposomes in TB

Patent number	Type of patent	Description of the patent	References
10,228,371 B2	US patent	US Patent 10,228,371 B2 shows the preparation of liposomes-based systems for the treatment of TB. The developed systems consist of a sterol-modified lipid and a pure mycobacterial lipid cell wall component. The mycobacterial lipid component is employed here as an antigen presentation determinant. It was used for the detection of antigen-specific biomarker antibodies in order to diagnose active TB.	[107]
2012/0244212 A1	US patent	US patent 2012/0244212 A1 revealed enhanced activity for the treatment of TB and HIV co-infected patients. The study described a unique approach in which liposomal system was used to deliver reduced glutathione to the sites of viral infection in combination with anti-TB drugs, and for HIV patients in combination with antiretroviral drugs. The formulation was administered orally. The patent claims the stability of liposomes and efficient delivery of contents to the infected macrophages.	[108]
8,795,719 B2	Us patent	US 8,795,719 B2 worked on immunotherapeutic agent used for the treatment and primary prophylaxis of TB. The immunotherapeutic agent was encapsulated in the liposomes-based system. The patent claims upon administration in Mtb-infected mice, the immunotherapeutic agent in the form of liposomes significantly reduced the bacterial burden.	[109]
2,661,253 B1	European patent	EP 2661253 B1 reported liposomal formulation for the treatment of TB. The developed formulation showed a lower average particle size in comparison to conventional liposome. Consequently, higher bioavailability and therapeutic effectiveness of anti-TB drug(s) were achieved.	[110]

8 Concluding Remarks and Future Prognosis

TB is one of the major public health scourges, with the number of cases adding every year. The conventional approach employed in TB chemotherapy requires high doses of multiple bioactive(s) for a prolonged period. This usually leads to drug-associated toxic effects and the development of multidrug resistance (MDR). Furthermore, the dosage regimens or bioactive(s) that are currently being employed in TB chemotherapy are also suffering some limitations in terms of permeability, solubility, and stability. Apart from this, it may develop resistance over time, consequently causing relapse. The infection may spread to the other peripheral areas of the body to cause secondary TB. The designing of smart nanoconstructs for the effective delivery of therapeutics has been extensively studied and has gained significant attention over the past few years. The nanomedicines-based systems presented an effective and versatile platform for not only alleviating the pharmacological constraints but also the drawbacks associated with conventional TB chemotherapy. In view of this, a significant step forward would be the design and development of targeted bioactive(s) delivery modules that possess the ability to improve the effectiveness and efficiency of conventional bioactive(s). Therefore, the therapy of TB could be revolutionized by developing new therapeutic approaches, including the design of anti-TB nanoconstructs such as liposomes. These are the most extensively studied and successful bioactive(s) delivery modules being explored to improve the efficacy of anti-TB bioactive(s). There is already a beginning to improve anti-TB chemotherapy using nanomedicines-based modules (liposomes) which are biomimetic, biocompatible, and biodegradable. The toxicity of bioactive(s) encapsulated in these nanoconstructs is decreased owing to their modified and preferential accumulation at the infected/desired sites. In addition, they provide higher flexibility for chemical amelioration to be specific against opportunistic intracellular infections, especially in the case of TB. The incorporation of targeting moieties at the surface of the liposomes would improve the targeting of the major and infected organ of this infection, i.e., the lungs. The composition/constitution of these carrier systems could also be modulated to select the most advantageous administration route. The anti-TB drugs in liposomes delivered via inhalational route have many discernible merits such as preferential accumulation of drugs in alveolar macrophages and reduction in the frequency of administration and drug-associated side effects. Additionally, it is a noninvasive route of administration that does not require healthcare assistance. Conclusively, using liposomes as anti-TB bioactive(s) delivery modules hold great potential for TB chemotherapy. The liposomal technology needs to be further explored to attain high therapeutic effectiveness via alternative routes such as inhalation.

Conflict of Interest The authors declare no competing financial/personal interest whatsoever.

References

1. MacNeil A, Glaziou P, Sismanidis C, Maloney S, Floyd K. Global epidemiology of tuberculosis and progress toward achieving global targets—2017. Morb Mortal Wkly Rep. 2019 Mar 22;68(11):263.
2. Shrivastava P, Gautam L, Jain A, Vishwakarma N, Vyas S, Vyas SP. Nanomedicine and its applications in combating resistance relating to opportunistic pathogens with special reference to tuberculosis. In: Handbook of research on nano-strategies for combatting antimicrobial resistance and cancer. IGI Global; 2021. p. 87–113.
3. Bhise SB, More AB, Malayandi R. Formulation and in vitro evaluation of rifampicin loaded porous microspheres. Sci Pharm. 2010 Jun;78(2):291–302.
4. Lakshminarayana SB, Huat TB, Ho PC, Manjunatha UH, Dartois V, Dick T, Rao SP. Comprehensive physicochemical, pharmacokinetic and activity profiling of anti-TB agents. J Antimicrob Chemother. 2015 Mar 1;70(3):857–67.
5. Hoagland DT, Liu J, Lee RB, Lee RE. New agents for the treatment of drug-resistant Mycobacterium tuberculosis. Adv Drug Deliv Rev. 2016 July 1;102:55–72.

6. Cardona PJ, Ruiz-Manzano J. On the nature of Mycobacterium tuberculosis-latent bacilli. Eur Respir J. 2004 Dec 1;24(6):1044–51.
7. Maison DP. Tuberculosis Pathophysiology and Pharmaceutical Intervention with Cancer Treatments. J Clin Tuberc Other Mycobact Diseases. 2022 Jan;19:100300.
8. Kashyap A, Singh PK, Silakari O. Mechanistic investigation of resistance via drug-inactivating enzymes in Mycobacterium tuberculosis. Drug Metab Rev. 2018 Oct 2;50(4):448–65.
9. Shrivastava P, Vyas S, Sharma R, Mody N, Gautam L, Jain A, Vishwakarma N, Vyas SP. Nanotechnology for oral drug delivery and targeting. In: Nanoengineered biomaterials for advanced drug delivery. Elsevier; 2020 Jan 1. p. 473–98.
10. Nasiruddin M, Neyaz M, Das S. Nanotechnology-based approach in tuberculosis treatment. Tuberculosis research and treatment. 2017 Jan 22;2017
11. Chaudhary KR, Puri V, Singh A, Singh C. A review on recent advances in nanomedicines for the treatment of pulmonary tuberculosis. J Drug Del Sci Technol. 2022 Feb 5:103069.
12. Kenchappa V, Cao R, Venketaraman V, Betageri GV. Liposomes as carriers for the delivery of efavirenz in combination with glutathione—an approach to combat opportunistic infections. Appl Sci. 2022 Jan;12(3):1468.
13. Moinuddin SM, Sajid I, Arif MH. Liposomal drug delivery system-A concise review. Acta Scientific Pharm Sci. 2022 Jan;6(1) (ISSN: 2581-5423)
14. Gaul R, Ramsey JM, Heise A, Cryan SA, Greene CM. Nanotechnology approaches to pulmonary drug delivery: targeted delivery of small molecule and gene-based therapeutics to the lung. InDesign of nanostructures for versatile therapeutic applications. William Andrew Publishing; 2018 Jan 1. p. 221–53.
15. Sharma AK, Sharma R, Jhorar R, Kumar R. Nanomedicine in therapeutic intervention of tuberculosis meningitis. Curr Nanosci. 2015 Feb 1;11(1):15–22.
16. Pandey R, Ahmad Z. Nanomedicine and experimental tuberculosis: facts, flaws, and future. Nanomedicine. 2011 Jun 1;7(3):259–72.
17. Kumar M, Jha A, Dr M, Mishra B. Targeted drug nanocrystals for pulmonary delivery: a potential strategy for lung cancer therapy. Expert Opin Drug Deliv. 2020 Oct 2;17(10):1459–72.
18. van Leent MM, Priem B, Schrijver DP, de Dreu A, Hofstraat SR, Zwolsman R, Beldman TJ, Netea MG, Mulder WJ. Regulating trained immunity with nanomedicine. Nat Rev Mater. 2022 Feb 7:1–7.
19. Karakousis PC, Bishai WR, Dorman SE. Mycobacterium tuberculosis cell envelope lipids and the host immune response. Cell Microbiol. 2004 Feb;6(2):105–16.
20. Ishikawa E, Mori D, Yamasaki S. Recognition of mycobacterial lipids by immune receptors. Trends Immunol. 2017 Jan 1;38(1):66–76.
21. Vieira AC, Chaves LL, Pinheiro S, Pinto S, Pinheiro M, Lima SC, Ferreira D, Sarmento B, Reis S. Mucoadhesive chitosan-coated solid lipid nanoparticles for better management of tuberculosis. Int J Pharm. 2018 Jan 30;536(1):478–85.
22. Gaspar DP, Faria V, Gonçalves LM, Taboada P, Remuñán-López C, Almeida AJ. Rifabutin-loaded solid lipid nanoparticles for inhaled antitubercular therapy: Physicochemical and in vitro studies. Int J Pharm. 2016 Jan 30;497(1–2):199–209.
23. Khatak S, Mehta M, Awasthi R, Paudel KR, Singh SK, Gulati M, Hansbro NG, Hansbro PM, Dua K, Dureja H. Solid lipid nanoparticles containing anti-tubercular drugs attenuate the Mycobacterium marinum infection. Tuberculosis. 2020 Dec 1;125:102008.
24. Laws M, Jin P, Rahman KM. Efflux pumps in Mycobacterium tuberculosis and their inhibition to tackle antimicrobial resistance. Trends Microbiol. 2022 Jan 1;30(1):57–68.
25. Kapp E, Malan SF, Joubert J, Sampson SL. Small molecule efflux pump inhibitors in Mycobacterium tuberculosis: a rational drug design perspective. Mini Rev Med Chem. 2018 Jan 1;18(1):72–86.
26. Pule CM, Sampson SL, Warren RM, Black PA, van Helden PD, Victor TC, Louw GE. Efflux pump inhibitors: targeting mycobacterial efflux systems to enhance TB therapy. J Antimicrob Chemother. 2016 Jan 1;71(1):17–26.

27. Demitto FD, do Amaral RC, Maltempe FG, Siqueira VL, Scodro RB, Lopes MA, Caleffi-Ferracioli KR, Canezin PH, Cardoso RF. In vitro activity of rifampicin and verapamil combination in multidrug-resistant Mycobacterium tuberculosis. PloS One. 2015 Feb 17;10(2):e0116545.
28. Choi JI, Kim HJ. Preparation of low molecular weight fucoidan by gamma-irradiation and its anticancer activity. Carbohydr Polym. 2013 Sep 12;97(2):358–62.
29. Gautam L, Sharma R, Shrivastava P, Vyas S, Vyas SP. Development and characterization of biocompatible mannose functionalized mesospheres: an effective chemotherapeutic approach for lung cancer targeting. AAPS PharmSciTech. 2020 Jul;21(5):1–3.
30. Rajan M, Raj V. Encapsulation, characterisation and in-vitro release of anti-tuberculosis drug using chitosan-poly ethylene glycol nanoparticles. Int J Pharm Pharm Sci. 2012;4(4):255–9.
31. Andrade F, Rafael D, Videira M, Ferreira D, Sosnik A, Sarmento B. Nanotechnology and pulmonary delivery to overcome resistance in infectious diseases. Adv Drug Deliv Rev. 2013 Nov 30;65(13–14):1816–27.
32. Zaru M, Sinico C, De Logu A, Caddeo C, Lai F, Manca ML, Fadda AM. Rifampicin-loaded liposomes for the passive targeting to alveolar macrophages: in vitro and in vivo evaluation. J Liposome Res. 2009 Mar 1;19(1):68–76.
33. Rawal T, Parmar R, Tyagi RK, Butani S. Rifampicin loaded chitosan nanoparticle dry powder presents an improved therapeutic approach for alveolar tuberculosis. Colloids Surf B: Biointerfaces. 2017 Jun 1;154:321–30.
34. Shah A, Aftab S, Nisar J, Ashiq MN, Iftikhar FJ. Nanocarriers for targeted drug delivery. J Drug Del Sci Technol. 2021 Apr 1;62:102426.
35. Nisini R, Poerio N, Mariotti S, De Santis F, Fraziano M. The multirole of liposomes in therapy and prevention of infectious diseases. Front Immunol. 2018 Feb 5;9:155.
36. Basha RY, TS SK, Doble M. Dual delivery of tuberculosis drugs via cyclodextrin conjugated curdlan nanoparticles to infected macrophages. Carbohydr Polym. 2019 Aug 15;218:53–62.
37. Chono S, Tanino T, Seki T, Morimoto K. Uptake characteristics of liposomes by rat alveolar macrophages: influence of particle size and surface mannose modification. J Pharm Pharmacol. 2007 Jan;59(1):75–80.
38. Lila AS, Ishida T. Liposomal delivery systems: design optimization and current applications. Biol Pharm Bull. 2017 Jan 1;40(1):1–10.
39. Gautam L, Shrivastava P, Yadav B, Jain A, Sharma R, Vyas S, Vyas SP. Multicompartment systems: a putative carrier for combined drug delivery and targeting. Drug Discov Today. 2021 Dec 11;
40. Bozzuto G, Molinari A. Liposomes as nanomedical devices. Int J Nanomedicine. 2015;10:975.
41. Bulbake U, Doppalapudi S, Kommineni N, Khan W. Pharmaceutics. 2017;9:E12.
42. Mirgorodskaya AB, Kushnazarova RA, Lukashenko SS, Zakharova LY. Self-assembly of mixed systems based on nonionic and carbamate-bearing cationic surfactants as a tool for fabrication of biocompatible nanocontainers. J Mol Liq. 2019 Oct 15;292:111407.
43. Nkanga CI, Bapolisi AM, Okafor NI, Krause RW. General perception of liposomes: formation, manufacturing and applications. Liposomes-advances and perspectives. 2019 Mar 26;
44. Malam Y, Loizidou M, Seifalian AM. Liposomes and nanoparticles: nanosized vehicles for drug delivery in cancer. Trends Pharmacol Sci. 2009 Nov 1;30(11):592–9.
45. Lindner LH, Hossann M. Factors affecting drug release from liposomes. Curr Opin Drug Discov Devel. 2010 Jan 1;13(1):111–23.
46. Kamboj S, Saini V, Magon N, Bala S, Jhawat V. Vesicular drug delivery systems: a novel approach for drug targeting. Int J Drug Del. 2013 Apr 1;5(2):121–30.
47. Bangham AD, Standish MM, Watkins JC. Diffusion of univalent ions across the lamellae of swollen phospholipids. J Mol Biol. 1965 Aug 1;13(1):238–IN27.
48. Owen RL, Strasters JK, Breyer ED. Lipid vesicles in capillary electrophoretic techniques: Characterization of structural properties and associated membrane-molecule interactions. Electrophoresis. 2005 Feb;26(4–5):735–51.

49. Bawarski WE, Chidlowsky E, Bharali DJ, Mousa SA. Emerging nanopharmaceuticals. Nanomedicine. 2008 Dec 1;4(4):273–82.
50. Farokhzad OC, Langer R. Nanomedicine: developing smarter therapeutic and diagnostic modalities. Adv Drug Deliv Rev. 2006 Dec 1;58(14):1456–9.
51. Koo OM, Rubinstein I, Onyuksel H. Role of nanotechnology in targeted drug delivery and imaging: a concise review. Nanomedicine. 2005 Sep 1;1(3):193–212.
52. Immordino ML, Dosio F, Cattel L. Stealth liposomes: review of the basic science, rationale, and clinical applications, existing and potential. Int J Nanomedicine. 2006 Sep;1(3):297.
53. Williams DF. On the nature of biomaterials. Biomaterials. 2009 Oct 1;30(30):5897–909.
54. Barratt G, Bretagne S. Optimizing efficacy of Amphotericin B through nanomodification. Int J Nanomedicine. 2007 Sep;2(3):301.
55. Agarwal AN, Kandpal HE, Gupta HP, Singh NB, Gupta CM. Tuftsin-bearing liposomes as rifampin vehicles in treatment of tuberculosis in mice. Antimicrob Agents Chemother. 1994 Mar;38(3):588–93.
56. Vyas SP, Khatri K. Liposome-based drug delivery to alveolar macrophages. Expert Opin Drug Deliv. 2007 Mar 1;4(2):95–9.
57. Hoet P, Legiest B, Geys J, Nemery B. Do nanomedicines require novel safety assessments to ensure their safety for long-term human use? Drug Saf. 2009 Aug;32(8):625–36.
58. Pandey R, Khuller GK. Antitubercular inhaled therapy: opportunities, progress and challenges. J Antimicrob Chemother. 2005 Apr 1;55(4):430–5.
59. Gaspar MM, Cruz A, Penha AF, Reymão J, Sousa AC, Eleutério CV, Domingues SA, Fraga AG, Longatto Filho A, Cruz ME, Pedrosa J. Rifabutin encapsulated in liposomes exhibits increased therapeutic activity in a model of disseminated tuberculosis. Int J Antimicrob Agents. 2008 Jan 1;31(1):37–45.
60. Deol P, Khuller GK. Lung specific stealth liposomes: stability, biodistribution and toxicity of liposomal antitubercular drugs in mice. Biochimica et Biophysica Acta (BBA)-General Subjects. 1997 Mar 15;1334(2–3):161–72.
61. Pison U, Welte T, Giersig M, Groneberg DA. Nanomedicine for respiratory diseases. Eur J Pharmacol. 2006 Mar 8;533(1–3):341–50.
62. Gill S, Löbenberg R, Ku T, Azarmi S, Roa W, Prenner EJ. Nanoparticles: characteristics, mechanisms of action, and toxicity in pulmonary drug delivery—a review. J Biomed Nanotechnol. 2007 Jun 1;3(2):107–19.
63. Shrivastava P, Gautam L, Vyas S, Vyas SP. Macrophage-targeted chemotherapy for tuberculosis. In: Gupta S, Pathak YV, editors. Macrophage targeted delivery systems. Cham: Springer; 2022. https://doi.org/10.1007/978-3-030-84164-5_17.
64. Attia MF, Anton N, Wallyn J, Omran Z, Vandamme TF. An overview of active and passive targeting strategies to improve the nanocarriers efficiency to tumour sites. J Pharm Pharmacol. 2019 Aug;71(8):1185–98.
65. Fang J, Nakamura H, Maeda H. The EPR effect: unique features of tumor blood vessels for drug delivery, factors involved, and limitations and augmentation of the effect. Adv Drug Deliv Rev. 2011 Mar 18;63(3):136–51.
66. Kobayashi H, Watanabe R, Choyke PL. Improving conventional enhanced permeability and retention (EPR) effects; what is the appropriate target? Theranostics. 2014;4(1):81.
67. Omidi Y, Barar J. Targeting tumor microenvironment: crossing tumor interstitial fluid by multifunctional nanomedicines. Bioimpacts. 2014;4(2):55.
68. Torchilin V. Tumor delivery of macromolecular drugs based on the EPR effect. Adv Drug Deliv Rev. 2011 Mar 18;63(3):131–5.
69. Noble GT, Stefanick JF, Ashley JD, Kiziltepe T, Bilgicer B. Ligand-targeted liposome design: challenges and fundamental considerations. Trends Biotechnol. 2014 Jan 1;32(1):32–45.
70. El-Ridy MS, Mostafa DM, Shehab A, Nasr EA, Abd E-AS. Biological evaluation of pyrazinamide liposomes for treatment of Mycobacterium tuberculosis. Int J Pharm. 2007 Feb 7;330(1–2):82–8.
71. Schreier H. Drug targeting technology. Marcel Dekker; 2001 Aug.

72. Vyas SP, Kannan ME, Jain S, Mishra V, Singh P. Design of liposomal aerosols for improved delivery of rifampicin to alveolar macrophages. Int J Pharm. 2004 Jan 9;269(1):37–49.
73. Kaul A, Chaturvedi S, Attri A, Kalra M, Mishra AK. Targeted theranostic liposomes: rifampicin and ofloxacin loaded pegylated liposomes for theranostic application in mycobacterial infections. RSC Adv. 2016;6(34):28919–26.
74. Shrivastava P, Gautam L, Jain A, Vishwakarma N, Vyas S, Vyas SP. Lipid drug conjugates for improved therapeutic benefits. Curr Pharm Des. 2020 Aug 1;26(27):3187–202.
75. Pandit S, Roy S, Pillai J, Banerjee S. Formulation and intracellular trafficking of lipid–drug conjugate nanoparticles containing a hydrophilic antitubercular drug for improved intracellular delivery to human macrophages. ACS Omega. 2020 Feb 26;5(9):4433–48.
76. Pandey R, Sharma S, Khuller GK. Nebulization of liposome encapsulated antitubercular drugs in guinea pigs. Int J Antimicrob Agents. 2004 July 1;24(1):93–4.
77. Dhillon J, Fielding R, Adler-Moore J, Goodall RL, Mitchison D. The activity of low-clearance liposomal amikacin in experimental murine tuberculosis. J Antimicrob Chemother. 2001 Dec 1;48(6):869–76.
78. Patil JS, Devi VK, Devi K, Sarasija S. A novel approach for lung delivery of rifampicin-loaded liposomes in dry powder form for the treatment of tuberculosis. Lung India: Official Organ of Indian Chest Society. 2015 Jul;32(4):331.
79. Justo OR, Moraes ÂM. Incorporation of antibiotics in liposomes designed for tuberculosis therapy by inhalation. Drug Deliv. 2003 Jan 1;10(3):201–7.
80. Chimote G, Banerjee R. In vitro evaluation of inhalable isoniazid-loaded surfactant liposomes as an adjunct therapy in pulmonary tuberculosis. J Biomed Mater Res B Appl Biomater 2010 Jul;94(1):1–0.
81. Rojanarat W, Nakpheng T, Thawithong E, Yanyium N, Srichana T. Levofloxacin-proliposomes: opportunities for use in lung tuberculosis. Pharmaceutics. 2012 Sep;4(3):385–412.
82. Altamimi MA, Hussain A, Imam SS, Alshehri S, Singh SK, Webster TJ. Transdermal delivery of isoniazid loaded elastic liposomes to control cutaneous and systemic tuberculosis. J Drug Del Sci Technol. 2020 Oct 1;59:101848.
83. Nkanga CI, Walker RB, Krause RW. pH-Dependent release of isoniazid from isonicotinic acid (4-hydroxy-benzylidene)-hydrazide loaded liposomes. J Drug Del Sci Technol. 2018 Jun 1;45:264–71.
84. Nkanga CI, Krause RW. encapsulation of isoniazid-conjugated phthalocyanine-in-cyclodextrin-in-Liposomes Using Heating Method. Sci Rep. 2019 Aug 7;9(1):1–6.
85. Santos NC, Castanho MA. Liposomes: has the magic bullet hit the target? Química nova. 2002;25:1181–5.
86. Huynh NT, Passirani C, Saulnier P, Benoît JP. Lipid nanocapsules: a new platform for nanomedicine. Int J Pharm. 2009 Sep 11;379(2):201–9.
87. Areschoug T, Gordon S. Pattern recognition receptors and their role in innate immunity: focus on microbial protein ligands. Trends in Innate Immunity. 2008;15:45–60.
88. Takeuchi O, Akira S. Pattern recognition receptors and inflammation. Cell. 2010 Mar 19;140(6):805–20.
89. Kumar H. Metabolic pathways and metabolites shaping innate immunity. Int Rev Immunol. 2020 May 3;39(3):81–2.
90. Zani IA, Stephen SL, Mughal NA, Russell D, Homer-Vanniasinkam S, Wheatcroft SB, Ponnambalam S. Scavenger receptor structure and function in health and disease. Cell. 2015 Jun;4(2):178–201.
91. Postlethwait EM. Scavenger receptors clear the air. J Clin Invest. 2007 Mar 1;117(3):601–4.
92. Collins HL, Kaufmann SH. The many faces of host responses to tuberculosis. Immunology. 2001 May;103(1):1.
93. Costa A, Sarmento B, Seabra V. Targeted drug delivery systems for lung macrophages. Curr Drug Targets. 2015 Dec 1;16(14):1565–81.

94. Shrivastava P, Gautam L, Sharma R, Dube D, Vyas S, Vyas SP. Dual antitubercular drug loaded liposomes for macrophage targeting: development, characterisation, ex vivo and in vivo assessment. J Microencapsul. 2021 Feb 17;38(2):108–23.
95. Lima Salviano T, dos Santos M, Charles D, de Siqueira Ferraz Carvalho R, Pereira MA, de Arruda B, Santos V, dos Santos Aguiar J, Souto FO, da Silva C, da Paz M. Fucoidan-coated liposomes: a target system to deliver the antimicrobial drug usnic acid to macrophages infected with mycobacterium tuberculosis. J Biomed Nanotechnol. 2021 Aug 1;17(8):1699–710.
96. Agrawal AK, Gupta CM. Tuftsin-bearing liposomes in treatment of macrophage-based infections. Adv Drug Deliv Rev. 2000 Mar 30;41(2):135–46.
97. Talukder J. Role of transferrin: an iron-binding protein in health and diseases. In: Nutraceuticals. Academic; 2021 Jan 1. p. 1011–25.
98. Pati R, Sahu R, Panda J, Sonawane A. Encapsulation of zinc-rifampicin complex into transferrin-conjugated silver quantum-dots improves its antimycobacterial activity and stability and facilitates drug delivery into macrophages. Sci Rep. 2016 Apr 26;6(1):1–4.
99. Horwitz MA, Clemens DL. Antimicrobial for targeting intracellular pathogens. US6054133, 2000.
100. Chono S, Tanino T, Seki T, Morimoto K. Efficient drug targeting to rat alveolar macrophages by pulmonary administration of ciprofloxacin incorporated into mannosylated liposomes for treatment of respiratory intracellular parasitic infections. J Control Release. 2008 Apr 7;127(1):50–8.
101. Adeagbo BA, Akinlalu AO, Phan T, Guderian J, Boukes G, Willenburg E, Fenner C, Bolaji OO, Fox CB. Controlled Covalent Conjugation of a Tuberculosis Subunit Antigen (ID93) to Liposome Improved In Vitro Th1-Type Cytokine Recall Responses in Human Whole Blood. ACS omega. 2020 Nov 20;5(48):31306–13.
102. Shao J, Ma JK. Characterization of a mannosylphospholipid liposome system for drug targeting to alveolar macrophages. Drug Deliv. 1997 Jan 1;4(1):43–8.
103. Labana S, Pandey R, Sharma S, Khuller GK. Chemotherapeutic activity against murine tuberculosis of once weekly administered drugs (isoniazid and rifampicin) encapsulated in liposomes. Int J Antimicrob Agents. 2002 Oct 1;20(4):301–4.
104. Chono S, Kaneko K, Yamamoto E, Togami K, Morimoto K. Effect of surface-mannose modification on aerosolized liposomal delivery to alveolar macrophages. Drug Dev Ind Pharm. 2010 Jan 1;36(1):102–7.
105. Wijagkanalan W, Kawakami S, Takenaga M, Igarashi R, Yamashita F, Hashida M. Efficient targeting to alveolar macrophages by intratracheal administration of mannosylated liposomes in rats. J Control Release. 2008 Jan 22;125(2):121–30.
106. Hasnain SE, Ehtesham NZ, Grover S, editors. Mycobacterium tuberculosis: molecular infection biology, pathogenesis, diagnostics and new interventions. Springer; 2019 Nov 30.
107. Baumeister C, Shaw WA, Verschoor JA, inventors; University of Pretoria, assignee. Liposomal composition comprising a sterol-modified lipid and a purified mycobacterial lipid cell wall component and its use in the diagnosis of tuberculosis. United States patent US 10,228,371. 2019 Mar 12.
108. Guilford FT, inventor. Enhanced method and composition for the treatment of hiv+ tuberculosis patients with anti-retroviral drugs and liposomal encapsulation for delivery of reduced glutathione. United States patent application US 13/458,449. 2012 Sep 27.
109. Iglesias PJ, Riera IA, inventors; Archivel Farma SL, assignee. Immunotherapeutic agent suitable for the primary prophylaxis of tuberculosis. United States patent US 8,795,719. 2014 Aug 5.
110. Cardona PJ, Amat I, Reyes B, Selga A, Amat M. Inventors; Archivel Farma, Sl, assignee. Liposome formulation suitable for treating or preventing tuberculosis 2012 July 12.

Drug Delivery by Micro, Nanoemulsions in Tuberculosis

Dignesh Khunt, Bhupendra G. Prajapati, Mehul Prajapti, Manju Misra, Sagar Salave, Jayvadan K. Patel, and Ravish J. Patel

Abstract *Mycobacterium tuberculosis* is the single pathogen responsible for the spread of tuberculosis (TB) in humans. TB is present from ancient times, but is still one of the deadliest diseases due to the nature of bacteria, chronic treatment and also its spread into different organs such as lymph nodes, lungs, brain, eye, spleen and liver. The drawback of currently available drug treatment is the low oral bioavailability, toxicity of drug, and drug resistance. Recently, nano-based therapy is widely explored for drug targeting, enhancing bioavailability and improving the stability of drugs. From the different nanocarriers, micro/nanoemulsions (MEs/NEs) are the promising nanocarriers for the improvement of bioavailability and Pgp modulation. This chapter highlights the composition, formulation, characterization, and pharmacokinetics of NEs/MEs with reference to tuberculosis.

Keywords Nanoemulsion · Microemulsion · Tuberculosis · *Mycobacterium tuberculosis* · Cationic nanoemulsion

The original version of this chapter was revised. The correction to this chapter is available at https://doi.org/10.1007/978-3-031-14100-3_16

D. Khunt · M. Prajapati · M. Misra
Department of Pharmaceutics, Graduate School of Pharmacy, Gujarat Technological University, Gandhinagar, Gujarat, India

B. G. Prajapati (✉)
Shree S K Patel College of Pharmaceutical Education and Research, Ganpat University, Mahesana, Gujarat, India
e-mail: Email-bhupen27@gmail.com

S. Salave
Department of Pharmaceutics, NIPER-Ahmedabad, Gandhinagar, Gujarat, India

J. K. Patel
Nootan Pharmacy College, Sankalchand Patel University, Mehsana, Gujarat, India

R. J. Patel
Ramanbhai Patel College of Pharmacy, Charotar University of Science and Technology, Changa, Gujarat, India

R. Shegokar, Y. Pathak (eds.), *Tubercular Drug Delivery Systems*,
https://doi.org/10.1007/978-3-031-14100-3_9

Abbreviations

%DR	Percentage drug loading
AIDS	Acquired-Immuno Deficiency Syndrome
BBD	Box–Behnken design
DPI	Dry powder inhaler
DSC	Differential Scanning Calorimetry
GI	Gastrointestinal
HIV	Human immunodeficiency virus
HPH	High pressure homogenization
INH	Isoniazid
MDR	Multi-drug resistance
ME	Microemulsion
MT	Mtb
NE	Nanoemulsion
PBS	Phosphate buffer saline
PDI	Polydispersity index
PIT	Phase Inversion Temperature
PYZ	Pyrazinamide
RFB	Rifabutin
RIF	Rifampicin
SDR	Single-drug resistance
SEM	Scanning electron microscopy
STRS	Streptomycin sulphate
TB	Tuberculosis
TEM	Transmission electron microscopy
XDR	Extensive-drug resistance

1 Introduction

Mycobacterium tuberculosis (Mtb)(M. tuberculosis or Mtb) is a communicable bacteria from the Mycobacteriaceae family which is known for the infection of the chronic most deadly airborne disease of tuberculosis (TB) [1]. TB was the first single bacterial-infected disease before the COVID-19 pandemic, and its ranking is ahead of HIV/AIDS. As per World Health Organization (WHO) (define) 2021 report, there were around 1.3 million deaths due to TB in 2020 [2]. In developing countries, TB is declared as an alarming issue due to non-follow-up of medicine by patients up to the treatment plan (more than 6 months) [3]. Over the 50 years, effective pharmacotherapy has been available, but still date TB is the deadliest disease [4, 5]. The reason for the failure of the treatment is the multiple drug resistance (MDR) (Rifampicin and Isoniazid), concurrent infection of HIV, Spread of TB in different organs (brain, lymph nodes, lung, liver, eye), chronic toxicity of drugs, failure of patient compliance, drug-drug interaction and low oral bioavailability [6, 7]. Conventional drugs for TB are useful for the 6-month duration of treatment for fighting TB. But MDR and extensive drug resistance (XDS) are the bigger challenges for the healthcare industry [3]. Again, the development of new drugs is also challenging. So, there is a need to find an alternative solution to tackle the above situation. Recently, nanotechnology-based carriers are the best choice due to their smaller size, surface modification and absorption-enhancing properties. Different type of nanocarriers are lipid nanoparticles, microemulsions (MEs)/nanoemulsions (NEs), liposomes, etc. In this sense, nano/microemulsions are seen as one of the most promising options for increasing antitubercular medication absorption and thereby improving therapeutic efficacy [8].

In the early 1940s, Hoar and Schulman introduced the concept of ME, which is one-phase system made up of titration of milky emulsion with hexanol. Commonly, ME is an optically isotropic, and thermodynamically stable formulation of two non-miscible solvents (oil and water). Two non-miscible solvents are stabilized with the help of a surfactant or mixture of surfactant with co-surfactant. The free energy of ME is low in contrast to water and oil and that is why it is thermodynamically stable. Formulation of ME is by spontaneous method with minor heating and/or stirring, which create the nanodroplets of 1–100 nm size. Due to the smaller size of droplet, MEs are a transparent system. Furthermore, the smaller diameter of droplets increases the shelf life, hydrophobic nature, near to zero interfacial tension and minimum viscosity of ME, which benefit physicochemical properties of ME. There is indistinctness between NE and ME terminology and both are represented as an alternative. The first misunderstanding is related to the size of globules as ME represents the micron size while NE stands for nano size. But the size of ME droplets is below 100 nm. MEs are formulated with more than 10% surfactant, and due to this they form spontaneously. In comparison to MEs, NEs are made up of the below 10% surfactant using high-energy input with globule size ranges of 20–200 nm [9]. NEs and MEs are crucial drug delivery vehicles due to solubility and high entrapment of lyophilic and lyophobic drugs due to their composition of oil, aqueous, surfactant and co-surfactant. MEs/NEs are classified based on the composition of oil and water portion namely, (1) oil in water (O/W) (2), water in oil (W/O) and (3) bicontinuous ME/NE. Nowadays, cationic ME/NE are explored for drug delivery to improve absorption. This chapter emphasizes on the composition, formulation, characterization and application of MEs/NEs with regard to TB.

2 Formulation and Characterization of ME/NE

2.1 Composition of ME/NE

MEs/NEs are generally made of oil, water, surfactant and co-surfactant. Selection of ME/NE excipients are done by solubility studies, types and safety aspects. Table 1 illustrates the commonly used oils, surfactants and co-surfactants.

Oils can insert into a lipophilic portion (tail) of surfactant single layer resulting in swelling and affecting the curvature of ME/NE. Small-chain length oils have great capability to pierce the lipophilic part of surfactant in comparison to a long chain. Oil affects the drug loading, release of drug and absorption of drug, and that is the reason selection of oil is a key part [9]. Commonly used oils are saturated fatty acid, unsaturated fatty acid, vegetable oils and glycerides (medium-chain mono, di and tri) owing to permeation-enhancing properties and many of them have Pgp modulation properties [10]. One of the well-known example is the Capmul® MCM which blocks the MDR and by that modulated the Pgp ([11]. Emulsification of oil is affected by the molecular volume and chain length of oils. Oil composed of long-chain fatty acids such as soybean oil has difficulties with emulsification. Emulsification of oil increases with decreasing lyophobicity of oil but it impacts drug solubility. Water-insoluble drug solubility is higher in medium-chain mono and diglycerides in comparison to medium-chain triglycerides and vegetable oils. Solubility and emulsification are important for the selection of oil. Higher viscosity oil (olive oil, corn oil, linseed oil, etc.,) is needed for the high-energy method for the preparation of NE [12–14].

Surfactant is the heart of the ME/NE formulation which reduces energy difference of two immiscible liquids and improves the thermodynamic stability. Surfactant is composed of hydrophilic and lyophobic parts which make its amphiphilic nature. An example of surfactant is reported in Table 2. Tween and Labrasol® are the most used surfactants due to modulation of Pgp and improvement of permeation of drugs [15]. Other than these poloxamers, sorbitans, lecithin and phospholipids are widely used. Surfactant also plays a role in the solubilization of drugs. The non-ionic surfactant is more preferable over cationic or anionic surfactants due to its safety [16]. The ideal characteristics of surfactant are (a) reduction of surface tension below 10 dynes/cm, (b) absorption of surfactant above dispersed phased globules are fast so that made full and coherent film and protect against coalesce, (c) obtain optimized

Table 1 Composition of ME/NE

Oils	Surfactants	Co-surfactants
Capmul® MCM, IMWITOR® 988, IMWITOR®742, Miglyol, Capryol®90, Captex®200P, Captex®355, Lauroglycol™ 90, Olive oil, Labrafil®M2125, Labrafil® M 1944, Peceol™, Soyabean oil, Maisine® 35–1, Oleic acid, Ethyl oleate, castor oils	Tween 80, Tween 20, Brij® 93, Brij® 92 V, Cremophore® RH 40, Span20, Tween 60, Span 80, Solutol® HS 15, Labrasol®	PEG 400, Transcutol® P Transcutol® HP, Propylene glycol, PEG 200, Cremophore® EL

Table 2 Different types of NE/ME preparation with their characterization

Drugs name	Type of method	Subtype of ME/NE	Characterization			References
			Globule size (nm)	PDI	Zeta potential (mV)	
Pretomanid	Ultrasonication	Pretomanid NE	186.46 ± 0.38	0.058 ± 0.007	0.086 ± 0.13	[25]
Rifampicin	Spontaneous emulsification	RIF NE	43.89 ± 0.36	0.160 ± 0.03	−2.50 ± 1.06	[26]
		RIF Chitosan NE	52.12 ± 0.36	0.250 ± 0.03	4.18 ± 0.13	
		RIF Chitosan-folate NE	59.69 ± 0.26	0.230 ± 0.01	0.70 ± 0.24	
	High-pressure homogenization	RIF NE	133.1 ± 1.1	0.200 ± 0.01	−32.7 ± 1.50	[23]
		RIF Chitosan NE	136.8 ± 1.9	0.170 ± 0.02	+34.1 ± 1.20	
		RIF Polymyxin B ME	134.2 ± 1.6	0.210 ± 0.01	+4.70 ± 0.30	
	Spontaneous emulsification	RIF Oleylamine NE	86.1 ± 0.9	0.150	+36.9 ± 0.5	[3]
	Spontaneous emulsification	RIF NE	47.41 ± 4.36	0.092	+15.11 ± 1.51	[27]
	Phase inversion temperature	RIF NE	25.75 ± 0.01	0.18 ± 0.01	−8.22 ± 4.59	[28]
	Spontaneous emulsification	RIF Cationic NE	97.25	Not reported	+22.21	[29]
	Spontaneous emulsification	RIF Cationic NE	88.4 ± 5.3	0.12	+27.2 ± 0.6	[3]

zeta potential for charge surfactant and viscosity, and (d) minimum concentration of surfactant are effective [17]. Surfactants should be solubilized with higher amount of drug without changing their emulsification capacity [12].

Alone, surfactant does not reduce a sufficient amount of interfacial tension and also form rigid films. To overcome this problem, there is a need for a high concentration of surfactant or co-surfactant. Co-surfactants are used to decrease interfacial tension (achieve negative) and also reduce the stress of two immiscible liquids. Short-chain alcohols are the first choice of co-surfactants. Co-surfactant is penetrating the layer of surfactant at the boundary of two immiscible liquids. It also alters the packing, fluidity and interfacial films. It also improves the permeation of drug, solubility and viscosity [18].

Selection of oils, surfactants and co-surfactants is done based on the solubility studies and pseudo-ternary diagram [19]. Artificial neural network (ANN)-based techniques are used. Agatonovic-Kustrin et al. [20] used the ANN for the prediction of stable ME formulation of RIF and INH. MS-Windows–based ANNs simulator software, Statistica Neural Networks 0.0F are used for the prediction of the model. The software was predicting the many ME formula. ANN suggested the formula with target concentration for the treatment of TB [20]. A similar set of experiments was formed by B. D., S. Agatonovic-Kustrin and M. H. Wisch [21] to optimize the combination of Rifampicin (RIF), Isoniazid (INH) and Pyrazinamide (PYZ) ME. The results suggest that ANN successfully predicts the stability and composition of ME [21].

2.2 *Different Formulation Approaches of Micro and Nanoemulsions*

NEs/MEs are developed by two broad methods, namely, (a) high-energy and (b) low-energy method. Graphical representation of different formulation methods is shown in Fig. 1 and Table 2.

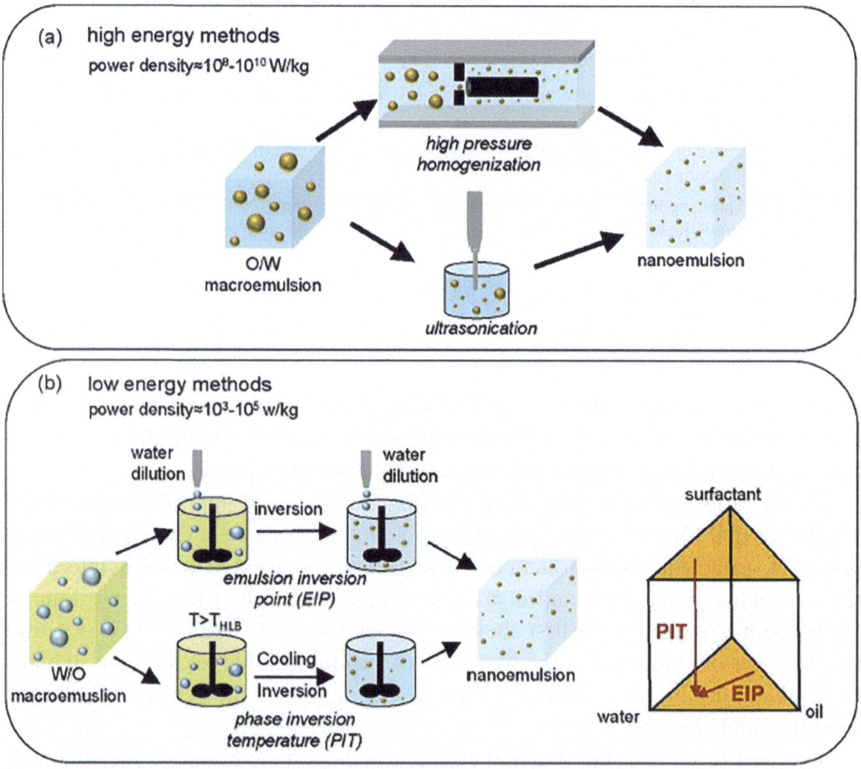

Fig. 1 Graphical illustration of high- and low-energy methods of ME/NE. (Reproduced from Gupta et al. [22], an open access article distributed under the Creative Commons Attribution License that permits unrestricted use, distribution, and reproduction in any medium)

(a) *High-energy method*

High energy is applied to break the droplets into nanoscales. Generally used instruments for the formulation of NEs/MEs are the high-pressure homogenizer (HPH), high-speed homogenized (HSH), small pore membrane and ultrasonication. Different process variables are time and RPM. HPH or microfluidizations are commonly employed industrial methods due to scalability features.

High-pressure Homogenization/Microfluidization

Premix of oil, surfactant, water and co-surfactant is homogenized using high share force. After that developed ME/NE are passed through the HPH for the reduction of size. In HPH, fluids are passed through the small orifice with pressure (500–5000 PSI) which reduces globule size. Globule size is broken into smaller globules by elongation and shear stress. The process of homogenization is repeated several times, which is known as a number of cycles or passes until the globule size is constant. One of the proven methods for the NE formulation, however high energy consumption increases process temperature [22].

Henostroza et al. [23] developed the cationic RIF NE for the ocular TB. For the formulation of RIF NE, the HPH method was used. Oleic acid as oil, Tween-80 and Poloxamer 188 as a surfactant was taken for NE. Firstly pre-mixture was homogenized using Ultra-Turrax at 10,000 RPM for 5 min. The resultant NE was passed through HPH at 10,000 PSI and five cycles. Surface modification was done using Chitosan chloride and polymaxin B sulphate. Optimized RIF NE without surface modification had 133.1 nm globule size with 0.20 PDI and − 38.3 mV zeta potential. Chitosan surface-modified NEs were shown to be 136.0–202.6 nm in size with PdI 0.17 to 0.27 and zeta potential values from +26.1 to +57.7 mV. It was predicted that the cationic charge of Chitosan increases the permeation across the corneal epithelial. Positive-charge Chitosan reacts with the negative charge of the cell wall which increases the efficacy of RIF. Polymyxin B-modified NE showed the size range of 126.4 to 135.1 nm; PDI between 0.18 to 0.21 and Zeta potential from +2.7 to +5.7 mV. Polymyxin B is a positive antibiotic for the killing of most of the gram-negative bacteria (Table 2) [23].

Microfluidization is done using the microfluidizer device which used high-pressure positive displacement pump (500–20,000 psi) to form small droplets. It is a similar technique to the HPH and it is used interchangeably. The mixture is repeatedly circulated through the microfluidizer until the required particle size is achieved. The resultant product is also passed through the filter to separate smaller droplets from larger ones and to obtain a uniform nanoemulsion. Orr et al. [24] used the microfluidization technique for the development of NE formulation of vaccine for TB. Premixture of oil (DMPC and GLA added into squalene) and ammonium phosphate buffer water phase (Poloxamer 188 and glycerol) was made. Premix was passed through the Microfluidics M110P at 30,000 PSI (12 cycles). Size obtained was obtained below the 100 nm [24].

In ultrasonication method, ultrasonic frequency is used for the reduction of globule size of prepared premixed ME/NE. Commonly, probe sonicators are used with and without a water jacket. In this, sonicator probes are used as a source of sonic waves. The sonicator probe is made of piezoelectric quartz crystal and coated with zirconium. On the application of voltage, probe generates sonic waves which reduce size using a cavitation mechanism [22]. Shobo et al. [25] formulated the Pretomanid NE using solvent evaporation technique and further size reduction was done using probe sonicator at 40% amplitude for the 20 min. They found that the developed NE was below 200 nm and suitable for brain delivery via the intranasal route [25]. Other NE/ME formulation methods are shown in Fig. 1 namely, evaporative ripening and bubble bursting at the interface.

(b) *Low-energy method (LEM)*

In this method, smaller globules are developed by phase inversion using the composition or temperature. LEM required low energy input which can be applied by simple stirring. Emulsion inversion point (phase inversion point) and phase inversion temperature are commonly used in LEM (Fig. 2). In emulsion inversion point,

Fig. 2 Difference characterization parameter of micro/nanoemulsion

the W/O ME is first prepared at room temperature and then water is added slowly. The addition of water leads to the transformation of Water-in-oil ME to Oil-in-Water ME. In phase inversion temperature, Water-in-oil ME was formulated at higher than the phase inversion temperature of the mixture. After that, the mixture is cooled at room temperature which leads to inversion of ME. At the inversion point, interfacial tension of water and oil interface is very low, which induces small-size globules [22]. Halicki et al. [28] prepared the Rifampicin (RIF) NE using the hot solvent diffusion with phase inversion temperature method. The developed NE was 25 nm size with 0.18 PDI [28].

Another method is spontaneous emulsification. In this method, oil, surfactant and co-surfactant are mixed using simple stirring. After that, water is added to the above mixture with continuous stirring [9, 22, 30]. Mehta et al. [30] prepared ME using spontaneous methods. For the preparation of ME, oleic acid as oil, tween-80 as surfactant and ethanol as co-surfactant are used. They focused on the evolution of the microstructure of ME/NEs. It was found that INH-loaded ME is stable and optically clear without any phase separation [30]. Hussain et al. [29] formulated the RIF cationic NE using the slow spontaneous titration method. Oleylamine was added for the cationic charge. The developed NE was below 100 nm of globules size [29].

2.3 Different Characterization Techniques of Micro and Nanoemulsion

Different characterization parameters of ME/NEs are shown in Fig. 2. Critical quality attributes of NE/MEs are the size, PDI and zeta potential which prove that NE/ME are formed [31]. Additional, refractive index, dye test and conductive test give the idea of which type of NE/ME are formed [19]. The stability of ME/NEs is proven by the interfacial tension determination and phase separation studies. Drug compatibility with oils, surfactants, and co-surfactant is done by the measurement of drug content through HPLC assay and visual changes. ME/NE pourability is the prime characteristic which can be determined using the viscosity test. Viscosity of ME/NEs is also impacted on the drug release. Drug release determines the efficacy of the developed formulation. Safety parameter of NE/MEs is determined using cell line studies and also animal toxicity study.

Globule size, PDI and zeta potential are commonly measured using the photon correlation spectroscopy [32]. Globule size is dependent on the concentration and chain length of surfactant, and applied energy. Kifayatullah et al. (2017) reported that increasing the surfactant concentration leads to the decrease of the globule size (from 250 nm to 50 nm) due to reduction of interfacial tension. Non-ionic surfactant stabilized globules by steric stabilization [26]. A similar result was also found by Hussain et al. (2020) [29] and Orr et al. [24], It was reported that the globules size impact on the vaccine efficacy and sizes below 200 nm directly target the lymph nodes [24]. PDI gives the idea of the homogeneity of NE/ME. Near to zero value of

PDI shows the higher homogeneity of NE/ME, which is desirable. The stability of NE/ME is also evaluated using zeta potential. Zeta potential is the surface charge of globules which suggests the attraction or repulsion in between dispersed phase and globules. The lower charge of globules is prone to coalescence owing to the attraction of droplets for charged surfactants. Idea value of zeta potential is the ±30 mV for better stability of ME/NE [33].

Viscosity is measured by the different types of viscometer such as capillary tube viscometer. pH and osmolarity are important parameters for the parenteral route NE.

Scanning and transmission electron microscopy are utilized for the evaluation of surface properties [34]. Aerosolization studies were also performed for the pulmonary delivery of anti-TB drugs. Kifayatullah et al. (2017) measured the aerodynamic and inhalation properties using Anderson cascade impactor [26]. From the result of total aerosol input and output rate, it was concluded that fine particle fraction was deposited into respiratory bronchioles and alveoli. Lower globule size leads to decrease in the surface tension and ultimately improves the aerosol output and their rate [26].

Refractive index studies are performed for the evaluation of optical properties of NE/ME [26]. Phase separation studies are done by centrifugation and dilution test [25]. In the centrifugation test, NE/ME are filled into a centrifuge tube. If ME/NEs are formed and stable, then no phase separation is observed [35]. In dilution studies, NE/ME diluted with continuous phase to check phase separation [15]. NMR (Nuclear Magnetic Resonance) studies are done to evaluate where the drug is located. Mehta et al. [15] performed ^{1}H NMR to find INH location in ME. NE/ME is composed of long hydrocarbon with oxyethylene group which hinders the resolution of the peak. So, to avoid that problem deuterium-based NME was performed. Results revealed the INH is located in the dispersion phase [30].

Surface tension is measured by a tensiometer. It is a contractive parameter of liquid surface that resists external force. It is present at the interface of two immiscible liquid [26].

Kifayatullah et al. (2017) measured the surface tension of RIF NE, Chitosan NE and Chitosan folate NE. The lower surface tension of NE leads to interaction with the alveolar surface and leads to phagocytosis. It is an important parameter for the nebulizer and pulmonary delivery. From the result, it is concluded that surface tension has impact on the nebulisation time and ultimate effect dispensability of NE [26].

NE is kinetic stable which provides time to phase separation of different phases. Ostwald ripening, flocculation, coalescence and creaming or sedimentation are the main mechanisms of instability of NE. Globules of NE are joined to each other due to attractive force and make single globules without merging. If globules merge with each other and make single bigger globules, then it is called coalescence. Different mechanisms of instability are shown in Fig. 3.

Coalescence and flocculation are avoided by increasing surface concentration. If the chemical composition of smaller and larger globules is varying, then the difference of these leads to the Ostwald ripening. Dispersed phase chemical potential is greater than smaller globules which stimulate the mass transfer from smaller to

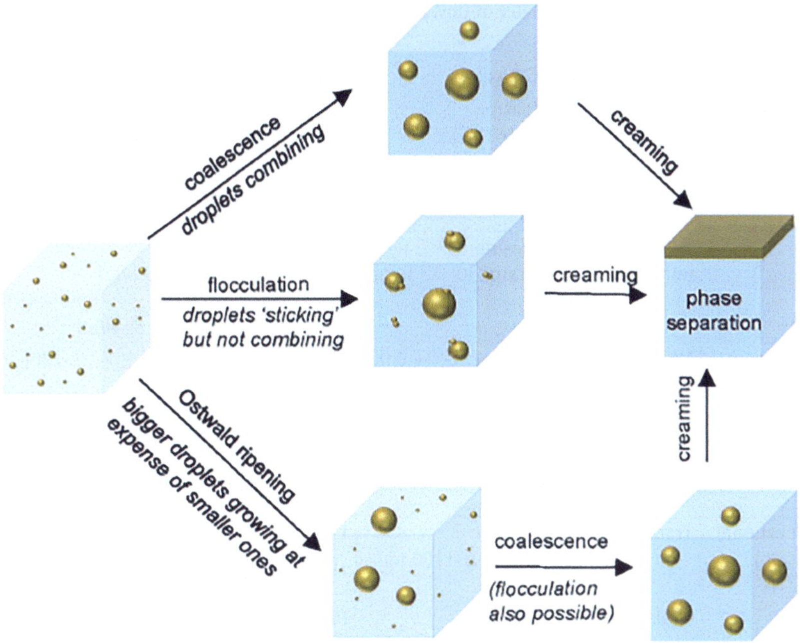

Fig. 3 Pictorial representation of different mechanisms for instability of NE. (Reproduced from Gupta et al. [22] an open access article distributed under the Creative Commons Attribution License that permits unrestricted use, distribution, and reproduction in any medium)

bigger globules. Due to these smaller globules, small and larger globules become larger (Fig. 3) [22].

Drug release from ME/NE or in vitro drug release is used to predict in vivo performance of drugs. It gives the idea of how drugs are partitioned out from the NE to the media. Hussain et al. [29] was shown the RIF released from RIF NE a faster rate from dialysis membrane in comparison to solution due to smaller globule size of NE. While NE loaded into gels was shown the slow release which proves retain of drug. For further confirmation, perform the *ex vivo* release studies using rat skin. It was found that RIF NE penetrates skin more in comparison to RIF solution [29].

2.4 Pharmacokinetics and Pharmacodynamic of Developed NE/ME for TB

Pharmacokinetics and dynamic parameters are the most essential features of NE/ME to prove their efficacy and safety. Table 3 shows different pharmacokinetics parameters of developed ME/NE.

Pretomanid O/W NE was formulated by Shobo et al. [25] for the treatment of brain TB. They have developed the LC-MS/MS method for analysing the drug

Table 3 Pharmacokinetics parameter of developed different NEs/MEs

Drug with NE/ME type	Route of administration	Cmax	Tmax (H)	AUC	Reference
Pretomanid	NE intranasal	Plasma: 3616.9 ng/g Brain: 120262.3 ng/g	Plasma: 8 Brain: 8	Plasma: 47908.1 ng-h/ml Brain: 183465.8 ng-h/ml	[25]
	Solution intranasal	Plasma: 109.3 ng/g Brain: 3060.3 ng/g	Plasma: 4 Brain: 4	Plasma: 679.2 ng-h/ml Brain: 36589.0 ng-h/ml	
Rifampicin	NE lung	Plasma: 0.62+ 0.06	Plasma: 1.00 + 0.00	Plasma: 5.41 + 0.58	[26]
	Chitosan NE lung	Plasma: 0.47 + 0.04	Plasma: 2.75 + 1.29	Plasma: 8.32 + 0.49	
	Chitosan folate NE lung	Plasma: 0.69 + 0.07	Plasma: 1.33 + 0.47	Plasma: 6.34 + 0.71	
RIF cationic NE	Transdermal	Plasma: 27,900 ± 1106 ng/mL	Plasma: 6.0	Plasma: 351.9 ± 17.2 μg. hr. mL^{-1}	[29]
	Oral	Plasma: 5890 ± 112.9	Plasma: 2.0	Plasma: 81.09 ± 4.9	

concentration in body fluid. It was found that a higher concentration of drug in the brain and plasma showing the potential NE given via intranasal route. MALDI-MSI was performed for the evaluation of brain distribution studies. From the result, it was concluded that NE transports the drug through nasal route to the cortical region and BBB to the cerebral capillary network [25].

Kifayatullah et al. (2017) prepared the RIF NE (first generation), RIF Chitosan NE (second generation) and RIF Chitosan Folate conjugate NE (third generation) for treatment of TB via the intratracheal route (Fig. 4). Results suggested that Chitosan and Chitosan folate NE showed a higher plasma concentration in comparison to NE. In a later phase, plasma drug concentration of RIF did not increase when Chitosan folate NE was given due to the accumulation into the alveolar macrophage using receptor targeting. Cell viability gives assurance safety of formulation [26].

Hussain et al. [29] prepared the cationic RIF for treatment of TB via transdermal route. Developed NE was again loaded into 0.5% carbopol 934 gel. The toxicity of developed cationic NE gel was evaluated using a haemolysis assay. Results suggest that developed formulations were not shown in haemolysis which revealed the safety of NE. Cationic NE-loaded gel showed reversible changes in the skin which were evaluated using scanning electron microscopy. Pharmacokinetics studies of cationic NE-loaded gel revealed 4.74-times higher AUC in comparison to oral RIF suspension. Skin irritation study proved the formulation was not irritant to skin with no sign of erythema for edema which suggested the safety of Cationic NE [29].

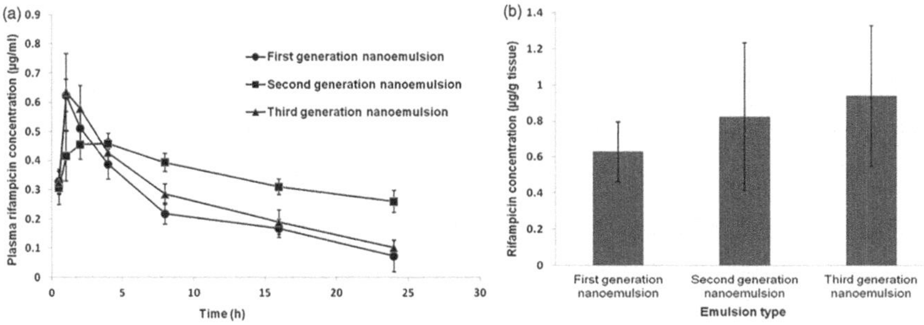

(c)

Parameters	First generation nanoemulsion	Second generation nanoemulsion	Third generation nanoemulsion
C_{max} (μg/ml)	0.62 ± 0.06	0.47 ± 0.04	0.69 ± 0.07
T_{max} (h)	1.00 ± 0.00	2.75 ± 1.29	1.33 ± 0.47
K (h^{-1})	0.11 ± 0.04	0.03 ± 0.01	0.08 ± 0.03
$t_{1/2}$ (h)	10.38 ± 5.91	27.76 ± 10.87	10.37 ± 2.09
MRT	8.43 ± 0.64	10.89 ± 0.44	8.67 ± 0.30
AUC_{0-t} (μg/ml.h)	5.41 ± 0.58	8.32 ± 0.49	6.34 ± 0.71

C_{max} maximum drug concentration, T_{max} time at which C_{max} occurs, K elimination rate constant, $t_{1/2}$ half life, *MRT* mean residence time, AUC_{0-t} is area under the curve from t = 0 to t = 24 h

Fig. 4 Graphical representation of drug reaching to plasma from NE, Chitosan NE, and Chitosan Folate NE via pulmonary routes. (Reproduced from Shah et al. [26] an open access article distributed under the Creative Commons Attribution License that permits unrestricted use, distribution, and reproduction in any medium)

Halicki et al. [28] developed RIF NE which were evaluated for antimycobacterial activity. RIF NE was effective against MT strains. The minimum effective concentration (MIC) of RIF NF and RIF was 7.8 μg/mL and 1024 μg/mL. From the result, it was concluded that RIF NE is highly effective in comparison to RIF [28].

Alshehri et al. [3] fabricated cationic NE using oleylamine as a positive charge. Confocal Laser Scanning Microscopy studies showed the intense intensity of fluorescence in the jejunum, ileum and duodenum in comparison to drug solution. It was concluded that present components of NE enhance the permeation of rhodamine dye and also inhibit the Pgp efflux [3].

Okonkwo Sylvia et al. [36] prepared the NE of Ocimum Gratissimum Leaf Extract and evaluated antimycobacterial activity (*Mycobacterium bovis* BCG strain (ATCC 35737) and *Mycobacterium smegmatis* (650)). NE-loaded saponin sample increased the activity 2 and 4 times against *M. bovis* BCG (MIC: 312.5 μg/mL) and *M. smegmatis* (MIC: 39 μg/mL) respectively [36].

3 Challenges, Opportunities and Future Perspectives

NE/ME are the oil-based drug vehicle systems that stabilize using the surfactant. One of the problems of NE/ME is the rancidity of oil which needs to be considered. Rancidity can be delayed by the addition of antioxidants or kept in control temperature. Toxicity of surfactant is necessary for anti-TB ME/NE. One of the examples of ethylene glycol impurities is present in Transcutol, which leads to kidney and brain damages in chronic use [37]. These problems are overcome by reducing concentration or avoiding of Transcutol due to chronic treatment of TB. Another parameter is the loading of the drug, which is decided based on the solubility studies. The development of self-nano-emulsifying/micro-emulsifying delivery (SNEDS/SMEDS) of anti-TB is increase the acceptance by patients. SNEDS/SMEDS is a simple mixture of oil, surfactant and co-surfactant. It is solidified using a spray dryer or adsorption on an adsorbent. Solidified products can be delivered through tablets or capsules which makes NE/ME in the gastrointestinal tract with a dilution of fluid.

4 Conclusion

Anti-TB drug-loaded NE/ME improve the permeation, drug loading, bioavailability and also targeting. ME/NE formulation is easy to scale up and has no critical factors. But still date, in market ME/NE-loaded anti-TB products are available so it is a great opportunity. Also, lack of pharmacokinetics and dynamic-based research article are available which are needed before going to clinical studies.

References

1. Jamison DT, Alwan A, Mock CN, et al. Universal health coverage and intersectoral action for health. Dis Control Priorities Improv Heal Reducing Poverty 3rd Ed; 2017.
2. (WHO) WHO. Global tuberculosis report 2021. Geneva: WHO; 2021.
3. Alshehri S, Altamimi MA, Hussain A, Imam SS, Singh SK, Faruk A. Morphological transition of M. tuberculosis and modulation of intestinal permeation by food grade cationic nanoemulsion: In vitro-ex vivo-in silico GastroPlus™ studies. J Drug Deliv Sci Technol. 2020;60:101971.
4. Kwan CK, Ernst JD. HIV and tuberculosis: a deadly human syndemic. Clin Microbiol Rev. 2011;24(2):351–76.
5. Kaur IP, Singh H. Nanostructured drug delivery for better management of tuberculosis. J Control Release. 2014;184:36–50.
6. Smith I. Mycobacterium tuberculosis pathogenesis and molecular determinants of virulence. Clin Microbiol Rev. 2003;16(3):463–96.
7. Sundell J, Bienvenu E, Janzén D, Birgersson S, Äbelö A, Ashton M. Model-based assessment of variability in isoniazid pharmacokinetics and metabolism in patients co-infected with tuberculosis and HIV: implications for a novel dosing strategy. Clin Pharmacol Ther. 2020;108(1):73–80.

8. Beg S, Saini S, Imam SS, Rahman M, Swain S, Hasnain MS. Nanoemulsion for the effective treatment and management of anti-tubercular drug therapy. Recent Pat Antiinfect Drug Discov. 2017;12(2):85–94.
9. Shah B. Microemulsion as a promising carrier for nose to brain delivery: journey since last decade. J Pharm Investig. 2021:1–24.
10. Bhattacharya S, Prajapati BG. Self Emulsifying Drug Delivery System (SEDDS): a novel approach.
11. Negi LM, Tariq M, Talegaonkar S. Nano scale self-emulsifying oil based carrier system for improved oral bioavailability of camptothecin derivative by P-Glycoprotein modulation. Colloids Surfaces B Biointerfaces. 2013;111:346–53.
12. Sapra B, Thatai P, Bhandari S, Sood J, Jindal M, Tiwary A. A critical appraisal of microemulsions for drug delivery: part I. Ther Deliv. 2013;4(12):1547–64.
13. Choudhury H, Gorain B, Chatterjee B, Mandal UK, Sengupta P, Tekade RK. Pharmacokinetic and pharmacodynamic features of nanoemulsion following oral, intravenous, topical and nasal route. Curr Pharm Des. 2017;23(17):2504–31.
14. Warisnoicharoen W, Lansley AB, Lawrence MJ. Nonionic oil-in-water microemulsions: the effect of oil type on phase behaviour. Int J Pharm. 2000;198(1):7–27.
15. Mehta SK, Kaur G, Bhasin KK. Tween-embedded microemulsions--physicochemical and spectroscopic analysis for antitubercular drugs. AAPS PharmSciTech. 2010;11(1):143–53.
16. Ge X, Wei M, He S, Yuan W-E. Advances of non-ionic surfactant vesicles (niosomes) and their application in drug delivery. Pharmaceutics. 2019;11(2):55.
17. Gupta R, Shea J, Scaife C, Shurlygina A, Rapoport N. Polymeric micelles and nanoemulsions as drug carriers: Therapeutic efficacy, toxicity, and drug resistance. J Control Release. 2015;212:70–7.
18. Khunt D, Shah B, Misra M. Role of butter oil in brain targeted delivery of Quetiapine fumarate microemulsion via intranasal route. J Drug Deliv Sci Technol. 2017;40:11–20.
19. Prajapati BG, Patel AG, Paliwal H. Fabrication of nanoemulsion-based in situ gel using moxifloxacin hydrochloride as model drug for the treatment of conjunctivitis. Food Hydrocoll Heal. 2021:100045.
20. Agatonovic-Kustrin S, Glass BD, Wisch MH, Alany RG. Prediction of a stable microemulsion formulation for the oral delivery of a combination of antitubercular drugs using ANN methodology. Pharm Res. 2003;20(11):1760–5.
21. Glass BD, Agatonovic-Kustrin S, Wisch MH. Artificial neural networks to optimize formulation components of a fixed-dose combination of rifampicin, isoniazid and pyrazinamide in a microemulsion. Curr Drug Discov Technol. 2005;2(3):195–201.
22. Gupta A, Eral HB, Hatton TA, Doyle PS. Nanoemulsions: formation, properties and applications. Soft Matter. 2016;12(11):2826–41.
23. Henostroza MAB, Melo KJC, Yukuyama MN, Löbenberg R, Bou-Chacra NA. Cationic rifampicin nanoemulsion for the treatment of ocular tuberculosis. Colloids Surfaces A Physicochem Eng Asp. 2020;597:124755.
24. Orr MT, Kramer RM, Barnes L, et al. Elimination of the cold-chain dependence of a nanoemulsion adjuvanted vaccine against tuberculosis by lyophilization. J Control Release. 2014;177:20–6.
25. Shobo A, Pamreddy A, Kruger HG, et al. Enhanced brain penetration of pretomanid by intranasal administration of an oil-in-water nanoemulsion. Nanomedicine. 2018;13(9):997–1008.
26. Shah K, Chan LW, Wong TW. Critical physicochemical and biological attributes of nanoemulsions for pulmonary delivery of rifampicin by nebulization technique in tuberculosis treatment. Drug Deliv. 2017;24(1):1631–47.
27. Ahmed M, Ramadan W, Rambhu D, Shakeel F. Potential of nanoemulsions for intravenous delivery of rifampicin. Die Pharm Int J Pharm Sci. 2008;63(11):806–11.
28. Halicki PCB, Hädrich G, Boschero R, et al. Alternative pharmaceutical formulation for oral administration of rifampicin. Assay Drug Dev Technol. 2018;16(8):456–61.

29. Hussain A, Altamimi MA, Alshehri S, Imam SS, Shakeel F, Singh SK. Novel approach for transdermal delivery of rifampicin to induce synergistic antimycobacterial effects against cutaneous and systemic tuberculosis using a cationic nanoemulsion gel. Int J Nanomedicine. 2020;15:1073.
30. Mehta SK, Kaur G, Bhasin KK. Incorporation of antitubercular drug isoniazid in pharmaceutically accepted microemulsion: effect on microstructure and physical parameters. Pharm Res. 2008;25(1):227–36.
31. Malay NJ, Chandresh PP, Bhupendra GP. Nanoemulgel innovative approach for topical gel based formulation. Res Rev Heal Care Open Acc J. 2018;1(2)
32. Prajapati BG, Jivani M, Paliwal H. Formulation and optimization of topical nanoemulsion based gel of mometasone furoate using 32 full factorial design. Indian Drugs. 2021;58(06):19.
33. Ding Z, Jiang Y, Liu X. Chapter 12 – Nanoemulsions-based drug delivery for brain tumors. In: Kesharwani P, Gupta UBT-N-BTDDS for BT, editors. Academic Press, 2018:327–358.
34. Bhattacharya S, Prajapati BG. Formulation and optimization of celecoxib nanoemulgel. Asian J Pharm Clin Res. 2017;10(8):355–6.
35. Shah B, Khunt D, Misra M, Padh H. Non-invasive intranasal delivery of quetiapine fumarate loaded microemulsion for brain targeting: formulation, physicochemical and pharmacokinetic consideration. Eur J Pharm Sci. 2016;91:196–207.
36. Okonkwo S, Emeje M, Peters O, Okhale S. Extraction and nanoencapsulation of ocimum gratissimum leaf extract and its anti-mycobacterial activities. Extraction. 2020;12(2)
37. Ether DGM. Opinion on diethylene glycol monoethyl ether (degee). 2013.

Nanosuspensions in Treatment of Tuberculosis

Atul Garkal, Priyanka Bangar, Khushboo Mathur, Dhaivat Parikh, and Tejal Mehta

Abstract Tuberculosis (TB) is one of the prime reason of death globally and is tough to treat by most antibiotics, together with tuberculosis (TB) multidrug therapy (MDT). Hence, TB rests as the foremost universal eminent health problem. Nanotechnology-based treatment has contributed an enormous development for the treatment by designing and developing advanced drug delivery systems, which can aim for phagocytic cells affected by intracellular pathogens, namely, mycobacteria. Advanced therapeutics can increase the therapeutic index of antimycobacterial treatments, reduce dosing recurrence, and also improve the efficacy. Nanosuspension is widely applicable for the management of TB because of its favourable characteristics. Therefore, nanosuspension is new ambition towards anti-TB drug development. In this chapter, the authors have tried to briefly summarize the pathogenesis of TB, the importance of nanosuspension, characterization and application of nanosuspension in the management of TB.

Keywords Tuberculosis · Nanosuspension · Nanotechnology · Drug delivery system · Nanoformulation · MDR-TB

1 Introduction

Tuberculosis (TB) is a granulomatous illness that causes persistent inflammation and is a serious health issue in underdeveloped and developing countries [1]. Therefore, controlling the tuberculosis pandemic is consequently a top concern for worldwide public health. According to the data from the survey in 2018, around 1.4 million people till that date, 2,51,000 of whom were also Human Immunodeficiency Virus (HIV) infected were killed due to the spread of *Mycobacterium tuberculosis*

A. Garkal · P. Bangar · K. Mathur · D. Parikh · T. Mehta (✉)
Department of Pharmaceutics, Institute of Pharmacy, Nirma University, Ahmedabad, Gujarat, India
e-mail: tjshah3@gmail.com

R. Shegokar, Y. Pathak (eds.), *Tubercular Drug Delivery Systems*,
https://doi.org/10.1007/978-3-031-14100-3_10

(M. tb) [2, 3]. Overpopulation, poverty, malnutrition, liquor abuse, HIV, chronic renal failure requiring dialysis, silicosis, fibro-apical radiographic abnormalities, diabetes, cigarette smoking, and immune-suppressive medication are all important risk factors for TB [4].

1.1 Pathogenesis of TB

Tuberculosis is transmitted when an infected patient coughs up the organism, which is then gasped hooked on the alveoli of a fresh crowd [4]. Close contacts of infectious TB patients are at risk of being infected and, if infected, develop tuberculosis, especially in the first year following exposure [5]. Coughing, shouting, singing, sneezing, other assertive expiratory manoeuvre that shears respiratory discharges from the airways is the most resourceful at causing infective aerosols in people with active pulmonary or laryngeal tuberculosis, with coughing being the maximum effectual at giving rise to infectious aerosols [5]. Only a fraction of MTB-containing droplet nuclei from infectious patients reach alveoli; the majority are confined in the upper airway and are thrown out by ciliated mucosal cells [6]. The human immune response can stop bacterial development and eliminate the germs in most *M. tuberculosis* infections, or it can establish latent tuberculosis infection (LTBI). However, 5–15% of latent tuberculosis infection patients develop lively tuberculosis with extrapulmonary involvement. Active tuberculosis usually appears soon after infection, but it might appear years later due to a weakened immune response, highlighting the need for both innate and adaptive immunity in *M. tuberculosis* management [7] (Fig. 1).

MTB bacilli spread hematogenously and lymphatically during initial infection, generating main Ghon's complex in the hilar and mediastinal lymph nodes [6, 8–11]. Bacilli eventually enter the bloodstream and then they make their way to various organs. Extrapulmonary TB occurs as a consequence of lympho-hematogenous feast during original contagion or later in life during disease reactivation [6]. The infection may spread to different organs. This is known as extra-pulmonary tuberculosis (EPTB).

2 Current Treatment for TB

Table 1 summarizes the details of anti-TB drugs with examples.

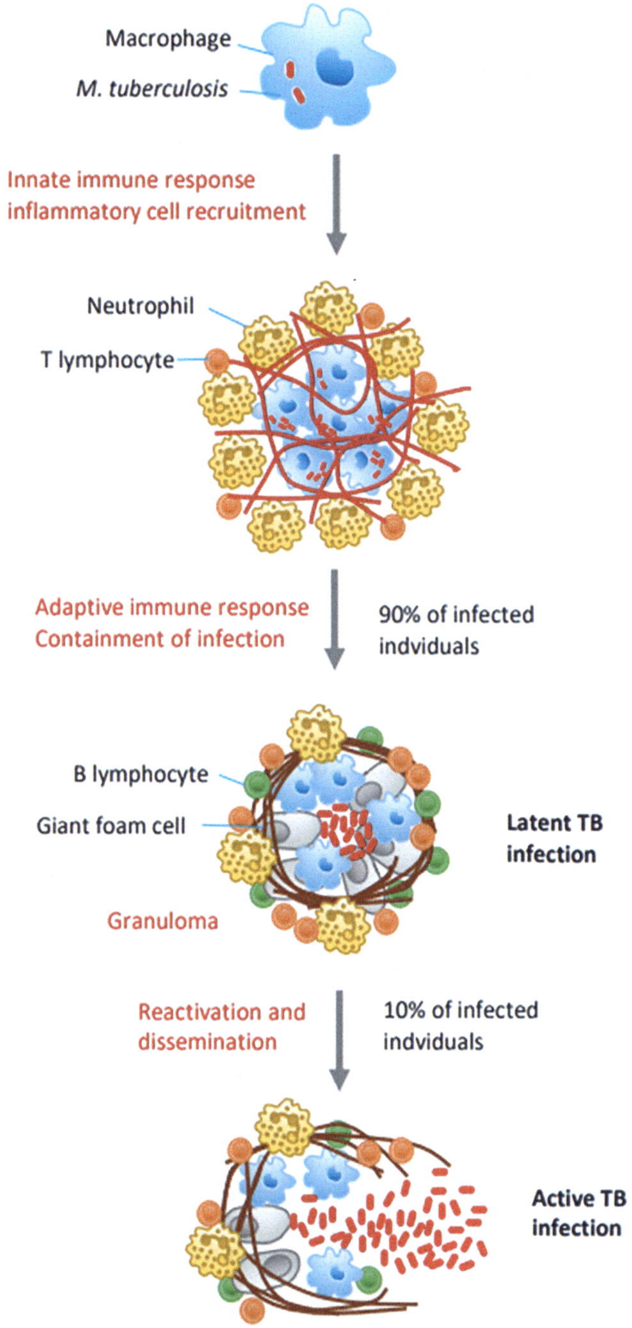

Fig. 1 Tuberculosis pathogenesis and disease progression [7]

Table 1 Details of anti-TB drugs [1]

Group	Description	Example of drugs
Group I	First-line oral anti-TB drugs	Isoniazid, Rifampin, Pyrazinamide, Ethambutol
Group II	Injectable anti-TB drugs	Streptomycin, Kanamycin, Amikacin, Capreomycin
Group III	Fluoroquinolones Well-tolerated bactericidal oral drugs	Ofloxacin, Moxifloxacin, Levofloxacin, Ciprofloxacin
Group VI	Oral anti-TB medicines used as a second line of protection mechanism	Terizidone, Ethionamide, Cycloserine, Para-aminosalicylic acid, Prothionamide
Group V	Drugs which show unclear efficacy	Clarithromycin, Thiacetazone, Clofazimine, Linezolid, Amoxicillin/clavulanate, Imipenem/cilastatin

2.1 Limitations of Current Treatment

The MDR-TB is even more complex and challenging; the human body's complicated immunological response to Mtb causes a wide range of clinical symptoms, making clinical and radiological diagnosis difficult [12]. Ever since the finding of the first anti-TB medicine, that is, streptomycin, the drug resistance has also been documented for *Mycobacterium tuberculosis* as a biological phenomenon [13].

3 Nanoformulations for Anti-tuberculosis Therapy

Targeted and customized drug delivery systems were created to address the drawbacks of traditional dosage forms. A sort of targeted medication delivery technology is nanoparticles [14]. Micronization is a frequently used method for increasing the bioavailability of medicines with low solubility [15]. Because, novel chemical entities (NCEs) are essentially insoluble in aqueous fluids and difficult to formulate using standard approaches, the nanosuspension approach is common and universal formulation nanotechnology [14, 16–22].

They exhibit several advantages such as providing a larger surface area for the surface interaction of the drugs. They can readily enter smaller capillaries and target the cell because of their tiny size. The other advantage is prolonged clearance period, which implies that a modest quantity of medicine is needed to achieve therapeutic efficacy and reduced toxicity. They are simple to customize and regulate and exhibit decreased patient-to-patient variability. The beginning of therapeutic activity is quicker. Various "smart" formulations can also be prepared, that is, they may be developed to be pH-responsive, temperature-controlled, enzyme-responsive, light-responsive, or even such that they act on magnetic stimuli. Different types of formulations consisting of nanoparticles are prepared as shown in Fig. 2 [23].

Nanosuspensions, which are liquid dispersions that show a nanoscale distribution of size, are being developed and becoming more used in pharmaceutical

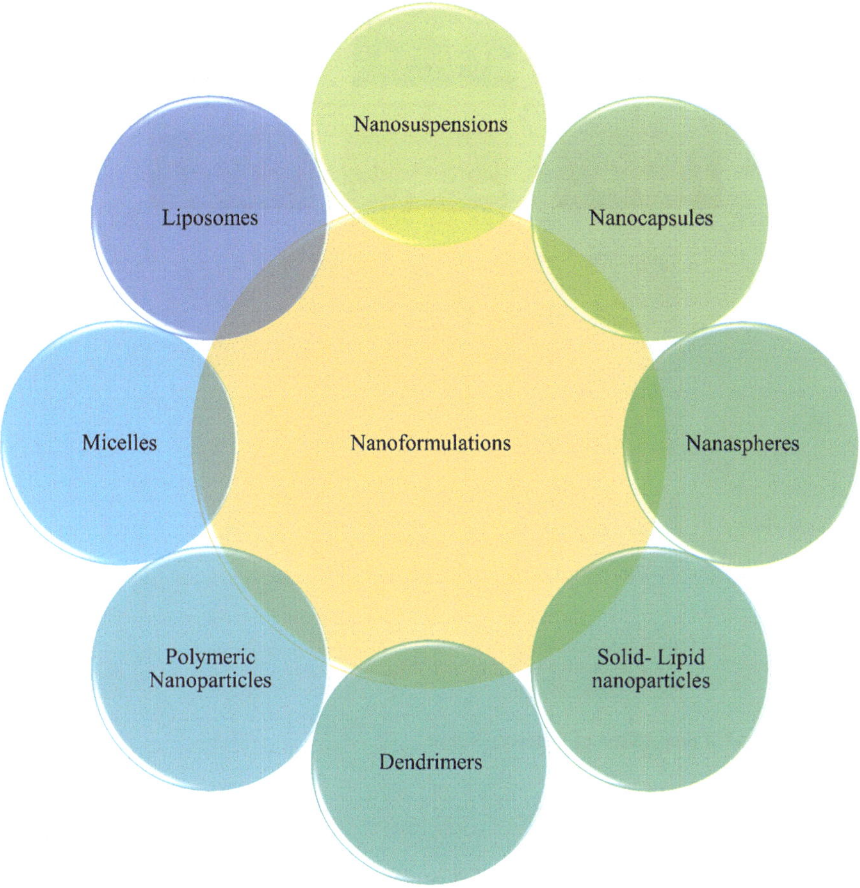

Fig. 2 Types of nanoformulation that can be used for treatment of TB

practice as a way to synthesize poorly water-soluble medicines and boost their bioavailability. The prepared nanosuspension must exhibit the desired properties and formulation aspects [24, 25]:

4 Methods for Preparation of Nanosuspensions

Different methods are employed for the preparation of nanosuspension. The schematic diagram is given in Figs. 3 and 4 [26].

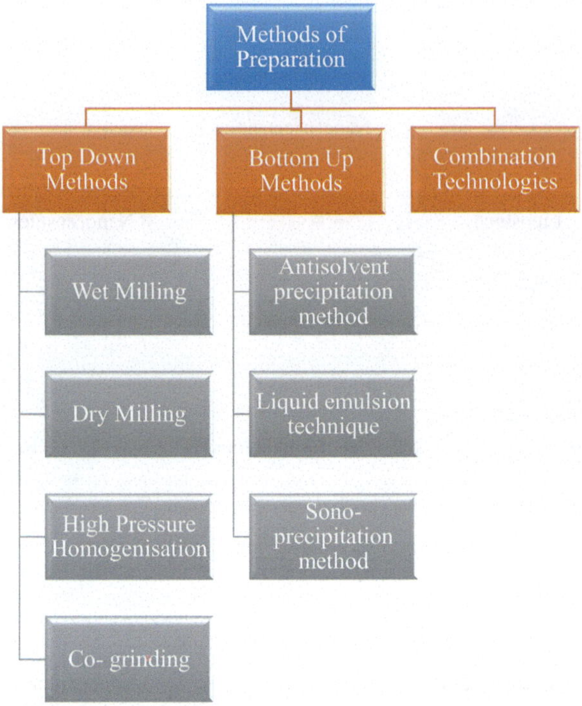

Fig. 3 Method of preparation of nanosuspension

4.1 Bottom-Up Methods

This technique begins at a molecular plane and works its way up to the production of a solid nanoparticle. The medication is diffused in an organic solvent, and then it is blended with a miscible antisolvent to induce quick precipitation to a dispersed product in bottom-up procedures [15].

4.1.1 Precipitation

The bottom-up precipitation approach has been successfully used to manufacture nanosuspensions. High supersaturation conditions are caused by a change in solvent, resulting in fast nucleation while preventing supersaturation near the nucleating crystals. According to the traditional Ostwald law of nucleation, high-speed nucleation in addition to moderate growth rate form the fundamental determinants for a successful thermodynamically stable crystal structure [26].

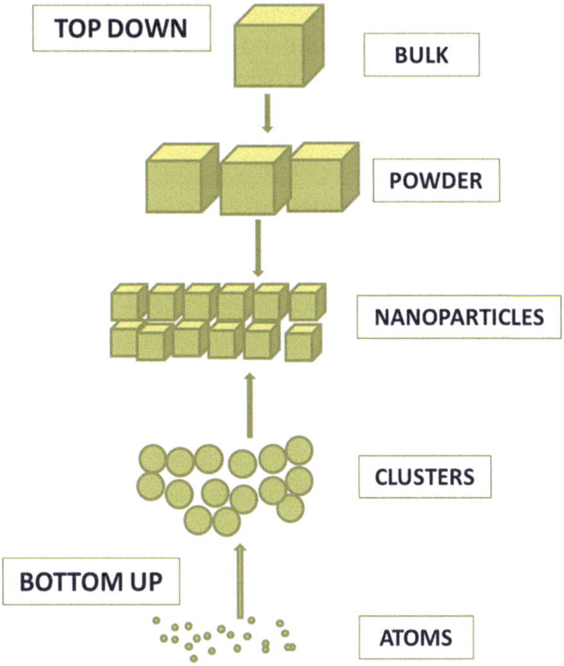

Fig. 4 Schematic representation of bottom-up and top-down technologies and for the composition of nanosuspensions

4.1.2 Hydrosols

The drug will dissolve in a organic solvent, which is then added into a aqueous solvent, which helps to start the fast precipitation settle of the product. In order to do this, the so-called Ostwald–Mier region must be passed swiftly, which implies that the solvent quantity must be reduced rapidly. This can be accomplished by combining a solvent with a nonsolvent [15, 27–31].

4.1.3 Nanomorph

This method reportedly creates amorphous particles. Solution of polymer is used to precipitate the amorphous form of the material. The amorphous nature of nanomorph gives it a distinct advantage in terms of dissolving velocity. It is increased significantly [15].

4.1.4 Supercritical Fluid Technology

Supercritical fluids may be used to make nanosized drug particles utilising a variety of methods, including the fast expansion of supercritical solution, the gas antisolvent techniques, and the supercritical antisolvent techniques [15]. The drug solution is subjected to an ultrasonic field created due to the vibrating surface at the core of the supercritical medium, which allows the shape and particle size of the prepared nanoparticles to be additionally regulated. Vibration frequency was modified to generate particles of various sizes and morphologies [26].

4.1.5 Other Precipitation Techniques

Some precipitation approaches for creating nanosized particles have recently been published [15] including the following.

4.2 Top-Down Technologies

Large drug crystals are reduced to the micrometre range in the top-down technique, then reduced to the nanodimension in a stabilizer solution [32].

4.2.1 High-Pressure Homogenization

Several researchers have successfully employed the high-pressure homogenization approach to generate nanosuspensions [26].

The second approach is microfluidization, which works on the notion of a jet surge. This causes the suspension to be accelerated and passed through a specifically built homogenization chamber at a high velocity, resulting in particle size reduction owing to particle collisions and shear forces created. Hydrosol, dissocubes, nanomorph, nanocrystals, nanopure, and NANOEDGE are all HPH-based techniques [15] (Fig. 5).

4.2.2 Media Milling

Media mills are used to make nanosuspension in this procedure. The API is broken down into nanoparticles during the shearing process. It also maintains continuous output due to its link to the recirculating chamber [32]. The milling contains an aqueous solution of the drug and stabilizer, and the milling medium, or pearls, rotate at a very high shear rate, generating drug nanoparticles by friction and collisions [32]. Shearing produces a lot of heat, yet it is an operation that may be done at a regulated temperature. However, this approach may result in pearl erosion, which might contaminate the drug nanoparticle product [32].

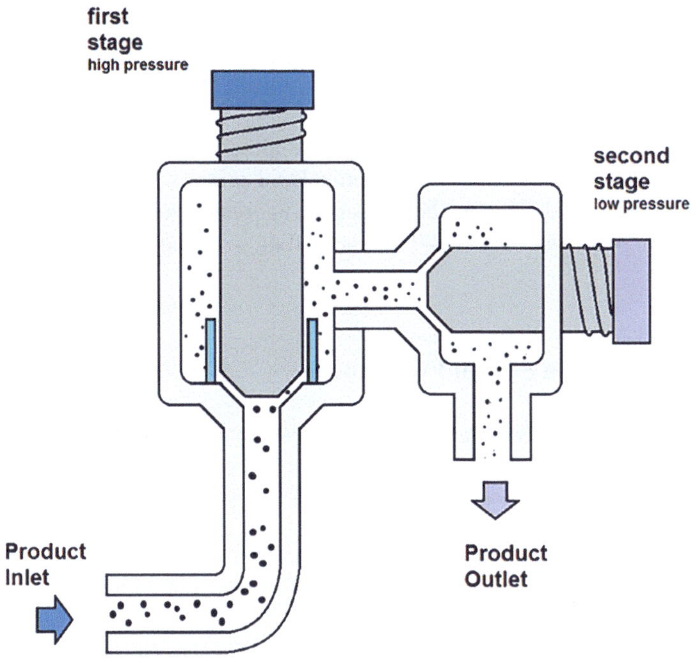

Fig. 5 High-pressure homogenizer [33]

4.3 Combination Technologies

In some circumstances, the preceding nanosuspension technologies are coupled to provide superior size reduction and system stability. As a result of combining these technologies, the advantages of nanosuspension technology have enhanced. NANOEDGE is a patented method that uses a mix of precipitation and homogenization to create a cutting-edge product. NANOEDGE is a technology that makes it possible for water-insoluble medicines to be used as pharmaceuticals. Particle size distribution and stability are frequently improved by combining precipitation and homogenization processes [32]. The use of NANOEDGE technique will help solve the disadvantages of both the homogenization and precipitation techniques. The precipitated suspension is then homogenized with the help of NANOEDGE technique, causing particle size to be reduced and crystal development is slowed. After the homogenization stage, particles which are in the nano range with enhanced thermodynamics are produced because the crystal development of nanoparticles is checked [32, 34].

4.4 Other Technologies

4.4.1 Supercritical Fluid Technology

The supercritical solution process (RESS), the supercritical antisolvent process, and the precipitation with the compressed antisolvent process (PCA) are some of the ways that may be used. This method was used to create cyclosporine nanoparticles with sizes ranging from 400 to 700 nm, according to a study done by Young et al., 2000 [32].

4.4.2 Emulsification Solvent Evaporation

Drug solution is initially created in the emulsification-solvent evaporation technique. The emulsification was then preformed in a liquid with low solubility of the medication. The drug nanocrystals precipitate out when all of the solvents have been evaporated [32].

4.4.3 Emulsion Diffusion

The emulsion diffusion technique is utilized for medications which are diffusible in volatile organic solvents, also partly water miscible organics. In an organic solvent or a combination of solvents, the medicine particles are completely dissolved. The organic solution is then stirred into an aqueous phase containing appropriate surfactants to generate an emulsion. The emulsion diffusion technique has the benefit of requiring no specific equipment. Controlling the emulsion droplet size allows for easy tuning of particle size. The formula is very simple to scale up. The main disadvantage of the emulsion-diffusion approach is that it cannot produce nanosuspensions of the medicines with low solubility in aqueous as well as organic mediums. In addition, toxic solvents are employed in the manufacturing process [32].

4.4.4 Melt Emulsification

The medicament is initially dispersed in an aqueous solution along with a stabilizer in the melt emulsification process. The solution was then homogenized and temperature is increased above the drug's melting point, resulting in an emulsion. While this procedure is going on, the emulsion is kept at a temperature greater than the drug's melting point. The drug emulsion is carefully chilled to room temperature or cooled using an ice bath [32].

4.4.5 Lipid Emulsion

One of the methods for making nanosuspensions is to emulsify the drug particles into a partly water-miscible solvent and then dilute the emulsion. The oil and water can produce thermodynamically stable and isotropically transparent microemulsions. The usage of organic solvents, which might induce toxicity, adds as a major disadvantage. In addition, a high number of surfactants and stabilizers must be added [32].

4.4.6 Nanojet

A process works in the other direction. A chamber is used in this method; here the suspension surge is split into two or more than that halves and the particles collide with amongst each other under high pressure. Therefore, the strong shear force produced by the high-pressure technique aids in particle size reduction. The main downside of this method is that it can produce a huge number of microparticles when a significant mass passes through the microfluidizer [32].

5 Characterization of Nanosuspension

The nanosuspension characterization usually includes physical, chemical and biological examinations. The size of particle and distribution of particle with different sizes, state of particles and morphology are the important evaluations for any nanosuspension. Unluckily, because of low power of detection, it becomes tough to detect nanosuspensions with high particle size. Hence, the laser diffractometric analysis is preferred. Laser diffractometry is a technique for detecting particles with sizes ranging from 0.05–80 μm. Furthermore the nanosuspensions are characterized for different parameters such as [17] pharmacokinetic behaviour of a formulation, which can be greatly affected by the size of the particles. In a nanocrystalline suspension, the particle size must be less than 5 μm. When long-term dissolution is desired, the diameter of the average particle must be in the nanometre range such as 800–1000 nm. The particle distribution is the critical parameter of nanosuspension because the particle size is directly correlated to physical stability, dissolution velocity, saturation solubility, along with their biological action [35]. Specifically, for nasal drug delivery, the specific requirement of nanosuspension is measured with different techniques; however, the particle size of suspended particles is measured with low-resolution dynamic light scattering (DLS) techniques. DLS is the one of the most preferred sizing technologies for measuring the particle size of formulations. The particle size detection techniques such as electron microscopy, field flow fractionation connected to online sizing detectors, centrifugal techniques, particle tracking analysis, and adjustable resistive pulse sensing are rarely employed. The depictions of molecular morphology and glasslike design comprehend the morphological or main changes that drugs may undergo when subjected to nano-measuring devices. Because shapeless pharmaceutical nanoparticles are likely to be supplied during nanosuspension planning, it is vital to pay attention to how much is in the nanosuspensions. The shapeless structure is thermodynamically shaky and will in general change into a glasslike structure during capacity. Such change over capacity ought to be thought about while planning or forming nanosuspensions. Thus, X-beam powder diffraction (XRD) and examining electron microscopy (SEM) are normally used to decide the glasslike design and molecule morphology of medications, individually. In addition, differential filtering calorimetry (DSC) is one more helpful strategy for portraying the glasslike structure and deciding the nebulous parts of medications.

5.1 *Redispersibility and Reconstitution Time*

The nanocrystals must redisperse readily with modest agitation and retain their integrity following reconstitution. The USP recommends certain tests for homogeneity of dosage units and water content, which may be used for nanosuspension as well.

5.2 *Stabilizers*

To moisten the API particles and inhibit Ostwald's ripening, stabilizers are utilized. Due to the high surface energy of drug nanocrystals, the absence of stabilizers might cause aggregation formation. They should be chosen based on how well they interact with the active pharmaceutical component and how well they can stabilize it. Ionic surfactants that display electrostatic repulsion or polymers that produce steric hindrance also inhibit particle accumulation are examples used to describe stabilizers. To offer greater stability, a mix of surfactants and polymers can also be utilized.

5.3 *Organic Solvents*

For producing the drug solution, organic solvents such as ethanol, isopropyl alcohol, methanol, acetone, and N-Methylpyrrolidone can be employed. Residual organic solvents, on the other hand, may provide a toxicity risk. Class III solvents can be used in the manufacture of nanocrystals, according to ICH rules, because they are considered less harmful.

5.4 pH/Buffers

Inflammatory responses and other negative effects may occur if a formulation has an extremely high pH. To modify the pH, buffering agents are utilized. Nanosuspensions can be administered through various routes [16].

6 Preferred Routes for Administration of Nanosuspension in Treatment of TB

Oral Delivery They show a decrease in gastric discomfort, increasing bioavailability and saturation solubility and significant taste masking.

Pulmonary Delivery When compared to micro-suspension, the dosage and toxicity are reduced, superior to the oral solution in terms of tolerability, alveoli are not irritated, decreasing pharyngeal and bronchial deposition.

IV Administration Particle size uniformity and ultra-fineness can be achieved, larger dosages are better tolerated, isotonicity, sterility, and lack of pyrogens. They can be incorporated into various dosage forms as well, such as liquid nanosuspensions, lyophilized powder, pellets, tablets, capsules and films as well.

7 Application of Nanosuspension in Treatment of Tuberculosis

Nanosuspension can also be described as submicron colloidal dispersion of therapeutic substances inside a liquid. They are usually less than 1 m in size and are stabilized with the presence of surfactants and polymers. Many newly discovered medications are water-insoluble, resulting in low bioavailability, which can be easily answered using nanosuspension. For safe pulmonary delivery, an aqueous nanosuspension with appropriate droplet size in an aqueous nebulizer is to be utilized. A nebulizer's highly respirable droplets are fairly easy to transport inside the lungs, also the droplets can easily distribute across the lung surface, enhancing medication distribution. A nanosuspension can hold a large amount of medication. It helps in boosting the saturation solubility and dissolution rate of medications that are not very aqueous-soluble. It also increases biological performance by lowering the active pharmaceutical ingredient's toxicity and side effects. It increases both the chemical and physical stability of the active medicinal component [36, 37].

For local anti-inflammatory effect, inhalable fluticasone propionate nanosuspensions generated using an amalgamation of wet milling and high-pressure homogenization techniques may be utilized. In rat airways, fluticasone propionate nanosuspensions boosted mucociliary clearance, improved drug retention, lengthened pulmonary absorption time, and extended the duration of the local anti-inflammatory action [38]. Rundfeldt et al. developed an inhalable itraconazole nanosuspension that can be simply nebulized with common nebulizer techniques such as pressurized air nebulizers and mesh technology. In cystic fibrosis, patients with allergic bronchopulmonary aspergillosis, inhalable itraconazole nanosuspension exhibits higher and longer-lasting lung tissue concentrations. It is simple to make using a wet-milling method, and it is more effective than oral formulations for treating bronchopulmonary aspergillosis at low doses with few adverse effects. Because the nanosuspensions are non-irritating and tasteless, improved patient compliance is observed [39]. The nanosuspension having suitable particle size increases drug bioavailability and easy to target/administer in lungs for treatment of TB. The nanosuspension formulations are mostly preferred for water-insoluble drugs. The nanosuspension is recommended in TB because of their superior properties over the powder and solution formulations.

8 Conclusion and Future Perspectives

Targeting the lungs through inhalation is a route of choice to treat pulmonary tuberculosis. This is advantageous over other routes of administration as direct targeting of drugs to the site of action, less systemic toxicity and increased patient compliance with the high rate of cure. Nanosuspensions are an attractive and promising strategy to deliver the antitubercular drug by inhalations, nebulization, or aerosolized systems. Nanosuspension drug delivery for the treatment of TB is an emerging era. Drug delivery is improved by nanosuspension, which also improves the therapeutic efficacy of the medications. They are useful in the delivery of pulmonary drugs because of their positive charge. It improves the permeability of different medicines making them available for nasal administration. They are used as an adjuvant in the administration of vaccines. In a nutshell, nanosuspension is versatile and used for these purposes because of its physicochemical properties: biocompatibility, biodegradability, mucoadhesiveness, absorption boosting ability, and in situ gelling characteristics.

Acknowledgements The authors would like to thank Nirma University for providing financial assistance in the form of Nirma University fellowship-SRF to Atul Garkal (NU/Ph.D./IP/GAD /19-20/1496).

Conflict of Interest The authors report no conflict of interest.

References

1. Tripathi K. Essentials of medical pharmacology. 7th ed. New Delhi: Jaypee Medical Publishers; 2013.
2. Cohen A, Mathiasen VD, Schön T, Wejse C. The global prevalence of latent tuberculosis: a systematic review and meta-analysis. Eur Respir J. 2019;54:1900655. https://doi.org/10.1183/13993003.00655-2019.
3. Schrager LK, Vekemens J, Drager N, Lewinsohn DM, Olesen OF. The status of tuberculosis vaccine development. Lancet Infect Dis. 2020;20:e28–37. https://doi.org/10.1016/S1473-3099(19)30625-5.
4. Dheda K, Barry CE, Maartens G. Tuberculosis. Lancet. 2016;387:1211–26. https://doi.org/10.1016/S0140-6736(15)00151-8.
5. Churchyard G, Kim P, Shah NS, Rustomjee R, Gandhi N, Mathema B, Dowdy D, Kasmar A, Cardenas V. What we know about tuberculosis transmission: an overview. J Infect Dis. 2017;216:S629–35. https://doi.org/10.1093/infdis/jix362.
6. Natarajan A, Beena PM, Devnikar AV, Mali S. A systemic review on tuberculosis. Indian J Tuberc. 2020;67:295–311. https://doi.org/10.1016/j.ijtb.2020.02.005.
7. Sinigaglia A, Peta E, Riccetti S, Venkateswaran S, Manganelli R, Barzon L. Tuberculosis-associated MicroRNAs: from pathogenesis to disease biomarkers. Cell. 2020;9:2160. https://doi.org/10.3390/cells9102160.
8. Lyon SM, Rossman MD. Pulmonary tuberculosis. Microbiol Spectr. 2017;5 https://doi.org/10.1128/microbiolspec.TNMI7-0032-2016.
9. Acharya B, Acharya A, Gautam S, Ghimire SP, Mishra G, Parajuli N, Sapkota B. Advances in diagnosis of tuberculosis: an update into molecular diagnosis of mycobacterium tuberculosis. Mol Biol Rep. 2020;47:4065–75. https://doi.org/10.1007/s11033-020-05413-7.
10. Furin J, Cox H, Pai M. Tuberculosis. Lancet. 2019;393:1642–56. https://doi.org/10.1016/S0140-6736(19)30308-3.
11. MacLean E, Kohli M, Weber SF, Suresh A, Schumacher SG, Denkinger CM, Pai M. Advances in molecular diagnosis of tuberculosis. J Clin Microbiol. 2020;58 https://doi.org/10.1128/JCM.01582-19.
12. Ankrah AO, Glaudemans AWJM, Maes A, Van de Wiele C, Dierckx RAJO, Vorster M, Sathekge MM. Tuberculosis. Semin Nucl Med. 2018;48:108–30. https://doi.org/10.1053/j.semnuclmed.2017.10.005.
13. Seung KJ, Keshavjee S, Rich ML. Multidrug-resistant tuberculosis and extensively drug-resistant tuberculosis. Cold Spring Harb Perspect Med. 2015;5:a017863. https://doi.org/10.1101/cshperspect.a017863.
14. Sur S, Rathore A, Dave V, Reddy KR, Chouhan RS, Sadhu V. Recent developments in functionalized polymer nanoparticles for efficient drug delivery system. Nano Struct Nano Objects. 2019;20:100397. https://doi.org/10.1016/j.nanoso.2019.100397.
15. Chavhan SS, Petkar KC, Sawant KK. Nanosuspensions in drug delivery: recent advances, patent scenarios, and commercialization aspects. Crit Rev Ther Drug Carr Syst. 2011;28:447–88. https://doi.org/10.1615/CritRevTherDrugCarrierSyst.v28.i5.20.
16. Wang L, Du J, Zhou Y, Wang Y. Safety of nanosuspensions in drug delivery, nanomedicine nanotechnology. Biol Med. 2017;13:455–69. https://doi.org/10.1016/j.nano.2016.08.007.
17. Bhatia S. Nanoparticles types, classification, characterization, fabrication methods and drug delivery applications. In: Natural polymer drug delivery system. Cham: Springer; 2016. p. 33–93. https://doi.org/10.1007/978-3-319-41129-3_2.
18. Garkal A, Kulkarni D, Musale S, Mehta T, Giram P. Electrospinning nanofiber technology: a multifaceted paradigm in biomedical applications. New J Chem. 2021;45:21508. https://doi.org/10.1039/D1NJ04159B.
19. Chivate A, Garkal AD, Dhas NL, Mehta DTA. Hot Melt extrusion: an emerging technique for solubility enhancement of poorly water soluble drugs. PDA J Pharm Sci Technol. 2021;pda-jpst.2019.011403. https://doi.org/10.5731/pdajpst.2019.011403.

20. Dhas N, Mehta T, Sharma S, Garkal A, Yadav D, Hariharan K, Shamjetshabam B, Khot S, Kudarha R, Bangar P, Arbade G, Kalyankar P. Intranasal gene therapy for the treatment of neurological disorders. In: Direct nose-to-brain drug Delivery. Elsevier; 2021. p. 351–87. https://doi.org/10.1016/B978-0-12-822522-6.00017-5.
21. Dhas N, Yadav D, Singh A, Garkal A, Kudarha R, Bangar P, Savjani J, Pardeshi CV, Garg N, Mehta T. Direct transport theory: from the nose to the brain. In: Direct nose-to-brain drug delivery. Elsevier; 2021. p. 15–37. https://doi.org/10.1016/B978-0-12-822522-6.00001-1.
22. Dhas N, Kudarha R, Garkal A, Ghate V, Sharma S, Panzade P, Khot S, Chaudhari P, Singh A, Paryani M, Lewis S, Garg N, Singh N, Bangar P, Mehta T. Molybdenum-based hetero-nanocomposites for cancer therapy, diagnosis and biosensing application: current advancement and future breakthroughs. J Control Release. 2021;330:257–83. https://doi.org/10.1016/j.jconrel.2020.12.015.
23. Dahanayake MH, Jayasundera ACA. Nano-based drug delivery optimization for tuberculosis treatment: a review. J Microbiol Methods. 2021;181:106127. https://doi.org/10.1016/j.mimet.2020.106127.
24. Ahmadi Tehrani A, Omranpoor MM, Vatanara A, Seyedabadi M, Ramezani V. Formation of nanosuspensions in bottom-up approach: theories and optimization. DARU J Pharm Sci. 2019;27:451–73. https://doi.org/10.1007/s40199-018-00235-2.
25. Patel D, Zode SS, Bansal AK. Formulation aspects of intravenous nanosuspensions. Int J Pharm. 2020;586:119555. https://doi.org/10.1016/j.ijpharm.2020.119555.
26. Jacob S, Nair AB, Shah J. Emerging role of nanosuspensions in drug delivery systems. Biomater Res. 2020;24:3. https://doi.org/10.1186/s40824-020-0184-8.
27. Chivate A, Garkal A, Dhas N, Mehta T. Three dimensional printing by hot-melt extrusion; new era for development of personalized medicines and continuous manufacturing of pharmaceuticals. Int J Pharm Investig. 2020;10:233–6. https://doi.org/10.5530/ijpi.2020.3.43.
28. Chivate A, Garkal A, Hariharan K, Mehta T. Exploring novel carrier for improving bioavailability of Itraconazole: solid dispersion through hot-melt extrusion. J Drug Deliv Sci Technol. 2021;63:102541. https://doi.org/10.1016/j.jddst.2021.102541.
29. Pradip Sonwane KS, Bhusnure G, Gaikwad V, Ali SS, Atul G. A review – formulation & development of Orodispersible tablet. Indo Am J Pharm Res. 2015;5:3868–81.
30. Amit C, Viral P, Om Prakash S, Atul G. Application and functional characterization of Kollicoat Smartseal 30D as a solid dispersion carrier for improving solubility. Asian J Pharm. 2020;14:220–8.
31. Deshmukh S, Avachat A, Garkal A, Khurana N, Cardot JM. Optimization of a dissolution method in early development based on IVIVC using small animals: application to a BCS class II drug. Dissolution Technol. 2016;23:34–41. https://doi.org/10.14227/DT230416P34.
32. Zhang J, Xie Z, Zhang N, Zhong J. Nanosuspension drug delivery system: preparation, characterization, postproduction processing, dosage form, and application. Elsevier Inc.; 2017. https://doi.org/10.1016/b978-0-323-46143-6.00013-0.
33. Comuzzo P, Calligaris S. Potential applications of high pressure homogenization in winemaking: a review. Beverages. 2019;5:56. https://doi.org/10.3390/beverages5030056.
34. Young TJ, Mawson S, Johnston KP, Henriksen IB, Pace GW, Mishra AK. Rapid expansion from supercritical to aqueous solution to produce submicron suspensions of water-insoluble drugs. Biotechnol Prog. 2000;16:402–7. https://doi.org/10.1021/bp000032q.
35. Ahire E, Thakkar S, Darshanwad M, Misra M. Parenteral nanosuspensions: a brief review from solubility enhancement to more novel and specific applications. Acta Pharm Sin B. 2018;8:733–55. https://doi.org/10.1016/j.apsb.2018.07.011.
36. Wan KY, Weng J, Wong SN, Kwok PCL, Chow SF, Chow AHL. Converting nanosuspension into inhalable and redispersible nanoparticles by combined in-situ thermal gelation and spray drying. Eur J Pharm Biopharm. 2020;149:238–47. https://doi.org/10.1016/j.ejpb.2020.02.010.
37. Garkal A, Avachat A. Development and in-vitro in-vivo characterization of in-situ gelling sustained-release nevirapine suspension. J Drug Deliv Sci Technol. 2022;67:102938. https://doi.org/10.1016/j.jddst.2021.102938.

38. Fu T-T, Cong Z-Q, Zhao Y, Chen W-Y, Liu C-Y, Zheng Y, Yang F-F, Liao Y-H. Fluticasone propionate nanosuspensions for sustained nebulization delivery: an in vitro and in vivo evaluation. Int J Pharm. 2019;572:118839. https://doi.org/10.1016/j.ijpharm.2019.118839.
39. Rundfeldt C, Steckel H, Scherliess H, Wyska E, Wlaź P. Inhalable highly concentrated itraconazole nanosuspension for the treatment of bronchopulmonary aspergillosis. Eur J Pharm Biopharm. 2013;83:44–53. https://doi.org/10.1016/j.ejpb.2012.09.018.

Alginate Nanoparticles: A Potential Drug Carrier in Tuberculosis Treatment

Rakesh P. Patel, Geeta K. Patel, Nikunjana Patel, Sudarshan Singh, and Chuda Chittasupho

Abstract Among all infectious diseases that afflict humans, tuberculosis (TB) remains the deadliest. At present, epidemiologists estimate that one-third of the world population is infected with *Mycobacterium tuberculosis*, which is responsible for 8–10 million new cases of TB and 3 million deaths annually throughout the world. Over the past 50 years, with medical treatment and standard public health practices, tuberculosis diminished in developed countries and resulted in a loss of interest and funding for research in improving diagnostic and treatment options. In developing countries, efforts including BCG vaccination have failed to control tuberculosis, and the disease continues to spread as the world becomes more globalized. At the same time, multidrug-resistant tuberculosis has emerged, challenging even the most advance treatment centres. Various unique antibodies have been developed to overcome drug resistance, reduce the treatment regimen, and elevate the compliance to treatment. Therefore, we need an effective and robust system to subdue technological drawbacks and improve the effectiveness of therapeutic drugs which still remains a major challenge for pharmaceutical technology. Polymeric nanoparticulate carriers have shown convincing treatment and promising outcomes for chronic infectious diseases. Different types of nanocarriers have been evaluated as promising drug delivery systems for various administration routes. Controlled and sustained release of drugs is one of the advantages of nanoparticle-based anti-

R. P. Patel (✉)
Shree S. K. Patel College of Pharmaceutical Education and Research, Ganpat University, Kherva, Gujarat, India

Pharmaceutics & Pharmaceutical Technology Department, S. K. Patel College of Pharmaceutical Education and Research, Ganpat University Ganpat Vidyanagar, Mehsana, Gujarat, India
e-mail: rakesh_patel@ganpatuniversity.ac.in

G. K. Patel · N. Patel
Shree S. K. Patel College of Pharmaceutical Education and Research, Ganpat University, Kherva, Gujarat, India

S. Singh · C. Chittasupho
Department of Pharmaceutical Sciences, Faculty of Pharmacy, Chiang Mai University, Chiang Mai, Thailand

R. Shegokar, Y. Pathak (eds.), *Tubercular Drug Delivery Systems*,
https://doi.org/10.1007/978-3-031-14100-3_11

tuberculosis drugs over free drug. It also reduces the dosage frequency and resolves the difficulty of low poor compliance. This chapter reviews sodium alginate-based nanotechnology therapies which can be used for the treatment of TB, with a short summary on bibliometric analysis that could provide significant information to researcher about ongoing research.

Keywords Mycobacterium tuberculosis · Tuberculosis · Sodium alginate · Nanoparticles · Treatment

1 Introduction

Tuberculosis (TB) is an infectious disease caused due to *Mycobacterium tuberculosis* (Mtb). TB is the foremost infectious disease affecting human beings since early civilization. However, a notable decline in the number of TB cases has been observed for two decades excepting sub-Saharan Africa [1, 2]. Despite the reduced trend in TB, the disease still causes considerable health issues and death globally. Previous to the COVID-19 pandemic, tuberculosis was the prime root for death worldwide, above acquired immunodeficiency syndrome and human immunodeficiency virus [3]. In low-income countries and in developed countries, people have suppressed immune systems; TB still remains the main source of morbidity and mortality and the goal of TB-free world is far achieved. Bacillus of TB can infect the whole body, while it is a majorly respiratory pathogen. Still, huge investment is essential for thorough research in TB to eradicate the disease, as it is the tenth leading reason for global deaths. Easy access to its diagnosis, primary care and control interventions are major necessity in fighting against TB [4, 5].

As stated by the World Health Organization (WHO) report 2021, 10 million people developed TB in 2020. The large worldwide drop in the number of cases decreased to 5.8 million from 7.1 million in 2020 [3]. A small decline in trends of TB in 2020 was observed compared to 2019 as fewer people were provided with a preventive treatment of TB in 2020, and the positive trend was reversed compared to 2019 (Fig. 1). This disruption in TB services was probably caused by the prevalence of the COVID-19 pandemic. Different 16 countries considered for 93% of this reduction, including badly affected countries like India, Indonesia and Philippines. The major impact of this shortfall in 2020 for getting treatment is due to reduction in global spending on TB diagnostic, treatment and prevention services.

TB is transmitted by air and spreads by people infected with TB throw out bacteria into the air through coughing; susceptibility, infection, environment and exposure are the critical parameters for determining the transmission probability of *M. tuberculosis*. No environmental reservoir is known for pathogens of TB, 'humans are its only known reservoir' [6].

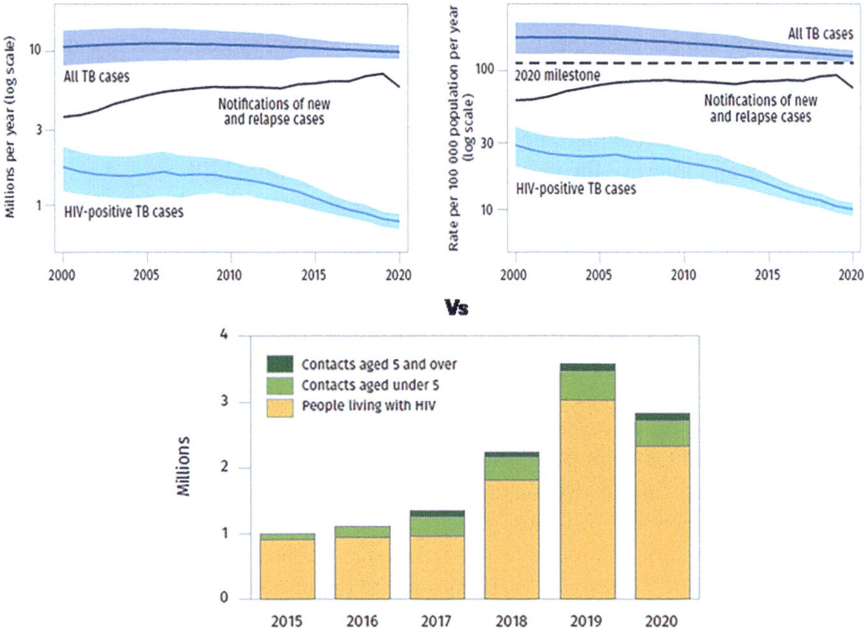

Fig. 1 Slow down trend in estimated number of incident cases verses reverse trends in number of people provided preventive treatment of TB from 2019–2020

Data suggests that majority of tuberculosis bacterial isolates show resistance against currently available antituberculosis drugs. Multidrug-resistant TB and extensively drug-resistant TB are the driving reasons for spreading TB. The development of novel effective drugs is necessary to get rid of TB. Drug resistance is also not solved by newly developed drugs like bedaquiline and delamanid [7–9]. Some of the new promising drugs are under phase II and III clinical trials. Moreover, more investigation and thorough research is required to identify and discover potential new drugs and their combination to effectively treat TB [10]. Also, new treatment strategies with newer approaches are strongly required to combat TB [11].

Recently, numerous nanocarriers have been developed by entrapping antitubercular drugs to fight against the endemic of TB worldwide [12]. Various nano-sized carriers like liposomes, niosomes, nanoparticles, micelles, nanoemulsions, dendrimers, nanospheres, nanocapsules, carbon nanotubes and nanoconjugates have been reported to target drugs at specific sites in order to achieve effective control over TB. These advanced novel drug deliveries offer great alternative to conventional therapy of TB as it is associated with certain drawbacks that lead to failure in therapy [13]. However, high concentrations of polymers, low drug loading, costly formulations, use of organic solvents are several associated demerits of novel systems [14–18].

Currently, one of the best ways to achieve higher drug levels in the lungs has been developed as a nanoparticulate system to deliver the right amount of drug directly to the lungs. Nanoparticles are taken up more efficiently by cells than other larger carriers which make them a promising drug delivery system to release drug in a controlled manner [19]. Furthermore, nanoparticles are adapted to control drug release in a persistent way; having a customized surface, improved solubility and multifunctionality allow them to increase bioavailability, reduce dosing frequency and toxicity and improve the therapeutic efficacy of the drug [20].

Nanoparticles are mainly fabricated using natural materials such as alginate, chitosan, gelatin, and albumin, polysaccharides and synthetic like poly(lactide) (PLA), poly(lactide-co-glycolide) (PLGA) copolymers, poly (ε-caprolactone) (PCL), poly(amino acids), cellulose acetate phthalate and Poly (βhydroxybutyrate) (PHB) [21–25]. Among the natural polymers, alginate is a considerably promising polymer of interest in pharmaceutical formulations as a matrix-forming polymer due to its intrinsic biological characteristics such as biocompatibility and biodegradability [26, 27]. Various studies revealed the potentiality of alginate as nanocarriers mainly in cancer treatment acquire popularity [28, 29]. Alginate has also been explored for drug targeting for various diseases [30–33]. Moreover, alginate contains carboxyl and hydroxyl functional moieties, which can be modified easily to get desirable properties to fabricate nanoparticles [34]. Among various nanocarriers investigated for TB as a controlled drug delivery system, alginate nanoparticles gained more attraction by researchers [35].

The potential methods for the fabrication of nanoparticles are emulsification/gelation, emulsification-solvent displacement, complexation, desolvation technique, dialysis technique, ionic gelation technique, nanoprecipitation technique, salting out technique, solvent evaporation technique, spray drying technique, polymerization, electrospray and supercritical fluid technique [36–44].

Anti-TB drugs entrapped in alginate nanoparticles have been studied for various routes of drug administration like oral, inhalation and intravenous. Alginate, along with its derivatives with other polymers such as chitosan, cyclodextrin, PLGA and PLA, has been explored for the fabrication of nanoparticles for TB [45, 46].

2 Challenges and Opportunities of Nanotechnology-Based Approaches in Treatment of TB

TB is an irresistible infection brought about by microscopic organisms of the Mtb complex, of which *M. tuberculosis* is the main and significant species responsible for human sickness. Similar types of infections are caused by related mycobacteria, *M. bovis*, *M. africanum*, and *M. microti*. In 1882, Robert Koch proved that the tubercle bacillus was the main reason for tuberculosis, and for this revelation, he received the Nobel Prize in 1905. When an individual inhales in bead cores containing tubercle bacilli, they reach the alveoli of the lungs. The majority of these tubercle bacilli are eliminated by macrophages present in alveoli. When the macrophages bite the dust, some *tubercle bacilli* are delivered intracellularly. In that scenario, if that bacillus is alive, they might reach further tissues and organs like local lymph hubs, the peak of the lung, kidneys, mind, and bone via lymphatic channels or through the circulatory system. This cycle of scattering prepares a framework for a fundamental reaction. Transmission of *M. tuberculosis* generally occurs via air, not by surface contact [47, 48].

The drawback of conventional therapy for tuberculosis includes some life-threatening adverse effects with complicated long-term treatment, which results in patient non-adherence to certain therapy with developing drug resistance; hence, novel therapy is introduced. Several drug delivery systems have been developed for the managemnt of TB that target at specific sites in a controlled manner; thus, it helps in patient adherence to the pharmacotherapy as the dose and dosing frequency reduces [49]. Using nanoparticles as targeted drug delivery can provide site-specific targeting, delivery of poorly water-soluble drugs, transportation across epithelial and endothelial barriers, retaining drug for longer time for efficacy, combining of diagnostic and therapeutic modalities into one agent, and many more [50].

Nanoparticulate technologies are new and rapidly developing science in which the particles of nanoscale are used to serve as therapeutic moiety or diagnostic tools to specifically targeted sites in a sustained or controlled manner. Site specificity in chronic human diseases and precise medicines in targeted drug delivery are some of the benefits of nanotechnology [51]. It is noticed that nanotechnology like nanostructures and nanophases acts to overcome the barrier of physical and biological sciences in various fields of sciences [52]. For prolonged period, nanostructures stay in the blood plasma, thus enabling the release of amalgamated drugs as per the specified dose. This can result into reduced adverse effects with less plasma fluctuations [53]. Drug designing has been a key aspect that marks the discovery of novel lead medications based on biological targets [54, 55]. Strategies to improve the drug release profiles of nanostructures [56], and to decrease the immunogenicity with several substances like natural polysaccharides [57, 58], cell membrane [59], peptides [60], polymers [61], tunable surfactants [62], etc., are being formulated.

Some of the excipients that are involved in the fabrication of nanoparticles are shown in Table 1.

2.1 *Alginate and its Derivatives Explored in the Fabrication of Nanoparticles*

Alginic acid, natural copolymer of mannuronic acid and guluronic acid, is obtained by the treatment of alkali with brown seaweed with some mineral acids. It can be commercially obtained in a pure form, and to improve its compatibility with a broad range of substances can be modified as neutral/charged forms as mentioned in Table 2 [64, 65]. The ability of alginate to gel in the presence of divalent cations is exploited in the development of alginate-based drug delivery systems. The original method of producing nanoparticles can be modified in two steps: a reduction in the

Table 1 Excipients used in the fabrication of nanoparticles [63]

Types of nanoparticles	Excipients used
Polymeric nanoparticles	Polylactic acid, Polyglycolic acid, Gelatin, Sodium alginate, Dextran
Solid lipid nanoparticles	Glyceryl monostearate, Cetyl palmitate, Palmitic acid Tricaprin, Tristearin
Emulsions	Poloxamer 188, Egg lecithin, Soybean lecithin, polysorbate 80, Tyloxapol
Stabilizers	Polyacrylic acid, Polyvinyl alcohol, Hydroxypropylmethyl cellulose, Polyvinyl pyrrolidine
Vesicle-based systems	Phosphotidylcholine, Dipalmitoyl phosphatidic acid, Distearoyl, Cholesterol

Table 2 Various chemical modifications on alginate with the specific advantages [68]

Chemical modification of alginate	Advantages
Oxidation	Improves biodegradability of alginate
Reductive amination of oxidized alginate	Controlled release rate, amphiphilic characteristics, lower surface tension, solubilization of solid azobenzene and heavy metal adsorption in practical application
Sulphation	High blood compatibility
Copolymerization	Good swelling capability, metal ion uptake, flocculating and low biodegradability in comparison to alginate
Cyclodextrin linked alginate	Permeability
Esterification	Increases hydrophobic nature
Ugi reaction	Modifies hydrophobicity
Amidation	Provides a host matrix with biocompatibility that holds enzyme molecules via gellification and electrochemical cross-linking

polymer to drug ratio and the replacement of chitosan for poly-L-lysine [66]. Alginate nanoparticles have been reported to permeate through the intestinal barrier and reach various organs. The encapsulation efficiency for various drugs for these types of alginate-based nanoparticles ranged between 70% and 90% [66, 67].

In a study, researchers developed alginate cellulose nanocrystal hybrid nanoparticles. At pH 7.4, initial drug release was slow at only 10–15% in 2 h which then gradually showed 100% release in 12 h [69].

The treatment of TB is done by routes such as oral, IV or IM. The main challenges for the treatment of tuberculosis are in vitro cellular uptake, solubility, bioavailability, drug-drug interactions, and needs higher dose for normal bioavailability. Using nanotechnology-based treatment for TB can overcome these challenges as there will be minimum drug-drug interaction due to the nano size, bioavailability would be improved as the surface area will increase by nanocarriers, they might be available in low prices in future perspectives, also the time of treatment can be shortened [70].

Table 3 Types of NPs

Types of NPs	Target	Therapeutic agent	Reference
PLGA, PLGA-PEG	Mtb cells	Rifampicin	[71]
AuNP	Mtb cells	SMVLD	[72]
SPIO	Mtb cells	Isoniazid	[73]
Silica NP, beta glucan	Mtb cells	Isoniazid	[74]
Mesoporous silica NP	Mtb cells	Isoniazid	[75]
PLG	Mtb cells	Isoniazid, rifampicin, pyrazinamide	[76–79]
Dendrimer	Mtb cells	Rifampicin	[80–82]
Chitosan	Mtb cells	Rifampicin	[83–85]
Aptamer	Ag85A	Apt22	[86]

There have been several approaches in nanotech for TB. Most NPs are having chemotherapy as mode of action. NPs like PLGA, PEG-PLGA, SPIO, beta glucagon, MSNP, PLG, dendrimers, chitosan, aptamers can be used for the treatment of tuberculosis by their specific therapeutic agents shown in Table 3.

3 Bioengineered Nanoparticles Fortified with Bioactive Compound in the Treatment of Tuberculosis

Most punctual known treatment of TB comprised of normal strategies to oversee TB and contained different customs obtained from old customary practices [87–90]. Age-old practices connected with TB treatment that has been given over from one age to another are still essential for TB treatment countries such as South Africa and Asia [91–95]. These conventional treatments, despite the fact that they are not prepared to totally kill the infection, however, are very powerful in treating the respiratory issues related to TB and lessening the harmfulness related to anti-tuberculosis treatment (ATT) as seen by the treatment given to the patients by the customary healers. Analysts all around the globe are attempting to observe new antimicrobials from rich and therapeutically huge plant auxiliary metabolites having a place with the Indian subcontinent, South Africa, and other eastern nations.

The traditionally used plant only removes, does not kill the microorganisms generally all alone; yet, they appear to assume an extremely significant part in overseeing side effects connected with TB, for example, delayed hack, chest torments, weakness, and fever that expands the degree of uneasiness in patients. This plant concentrate or optional metabolites apply expectorant, bronchodilator, mitigating, and antipyretic impacts. Since hundred years, plants, for example, *Tussilago farfara* and *Pulmonaria officinals*, have been investigated for these properties in the treatment of TB [96]. Treatment of TB using plants/phytochemicals comprises an extensive methodology, which could give TB treatment a further developed standpoint.

The phytochemicals are fundamentally the synthetic substances delivered in plants generally as a feature of assurance instrument used by the plants to assist them with flourishing hunters and microorganisms. They are result of the essential or auxiliary digestion of the plant [97]. These mixtures have a background marked by being utilized in biomedical treatments and their restorative properties and are principally connected with the presence of various mixtures like isothiocyanates, carotenoids, flavonoids, monoterpenes, indoles, and phenolic acids [98].

Till now, an adequate number of phytochemicals have been screened; however, because of the absence of nitty-gritty examination, this area of exploration has been constantly ignored and hence should be reinforced by new exploration. Restricted investigations have been led a long way in investigating the possible job of phytochemicals in antituberculosis treatment. Numerous analysts have presently been concentrating on the role of phytochemicals in TB treatment, as this treatment would assist with working on the impacts of DOTS treatment. A few examinations report the promising impact of the phytochemicals against *M. tuberculosis* bacteria [99, 100]. However, these examinations neglect to incorporate the host defensive components or the immunotherapeutic utilization of these mixtures. The space of phytochemical studies against *M. tuberculosis* is exceptionally wide and promising and is in this way in uncommon need of additional investigation. We realize that DOTS treatment while wiping out the microscopic organisms hoses the host insusceptible framework. A subordinate medication or compound that could forestall the hosing of the resistant cells will end up being a shelter for TB treatment. In the accompanying segment of Table 4, we have attempted to sum up the known phytochemicals, which have been utilized in TB treatment. Researchers have likewise reviewed the antimicrobial movement of these phytochemicals with extraordinary driving force on their utilization in the anticipation and treatment of tuberculosis.

Directly observed treatment short-course (DOTS), an internationally accepted therapy for TB, is effective but has several limitations [119, 120]. Even after a short duration of treatment, regular TB requires treatment for 6–9 months, whereas drug-resistant TB requires 12–24 months or more. This prolonged remedy length can also additionally result in the era of drug-resistant versions of the mycobacterial organisms. In addition, this treatment leads to treatment cessation by patients due to the serious toxicity of a mixture of antibiotics [121]. Moreover, DOTS-treated patients also showed disease reactivation and reinfection due to toxicity of isoniazid antigen-activated T cells during treatment [122, 123]. The antibiotic remedy which includes an immunity-boosting agent can also additionally reduce the length of remedy and facilitates in regaining shielding immunity that is useful in enhancing the remedy final results and in lowering the probabilities of producing a couple of extraordinarily drug-resistant (MDR and XDR) variations of TB [124, 125].

Curcumin (CUR) is additionally an inhibitor of Kv1.3 [126] and displays restorative advantages in a few fiery and irresistible diseases [127]. Curcumin analogs were tested and found with promising in vitro antimycobacterial activity towards drug-resistant traces of Mtb [128]. Recently, CUR becomes exhibited to increase mycobacterial killing in having macrophages with the aid of using inciting apoptosis [129]. Furthermore, CUR strongly represses hepatotoxicity, which is a simultaneous issue in numerous anti-microbial treatments that incorporate TB treatment. Taken together, this information has given solid proof that CUR is an astounding

Table 4 Phytoconstituents reported for tuberculosis treatment

Phytoconstituent	Class	Source	Mechanism	MIC	Ref
Allicin	Diallylthio sulphinate	*Allium sativum* (garlic)	Reduce replication of bacterial cell by reacting with SH group of enzymes	0.97–1.95 mcg/ml (Allicin reach Ext) 1–3 mg/ml (Garlic extract)	[101–105]
Andrographolide	Bicyclic diterpenoid	*Andrographis paniculata*	Inhibition of aminoglycoside 2-N-acetyltransferase (AAC)	Between 100 and 11.11 μg/ml against *M. microti* (ethanol extract of leaves)	[106–109]
Bergenin (Cuscutin)	Trihydroxybenzoic acid glycoside	Different source	Activation of the MAP kinase and ERK pathways, leading TNF-α, nitric oxide (NO), and Interleukin-12 (IL-12) production	–	[110, 111]
Curcumin	Diarylheptanoid	*Curcuma longa*	Upregulation in the expression of apoptosis and autophagy genes	128 μg/ml (synergistic activity with amikacin, ciprofloxacin, clarithromycin, and linezolid)	[112–114]
Pasakbumin A		*Eurycoma longifolia*	Activation of ERK1/2-intermediated signalling pathways, autophagy and high TNF-α production		[115, 116]
Thimoquinone	2-isopropyl 5-methyl-1, 4-benzoquinone	*Nigella sativa*	Inhibition of formation of biofilms		[117, 118]

calming immunomodulator and has restorative potential in an assortment of illnesses. Regardless of the heap of exercises detailed, CUR is yet to be endorsed as a restorative specialist because of bioavailability problems. Numerous clinical studies have obviously settled that customary curcumin has extremely poor bioavailability and is consequently unacceptable for delayed use [130, 131]. To conquer this restriction of curcumin, nanoparticles of curcumin with 200 nm in size were developed which showed fivefold upgraded bioavailability in mice over normal CUR. During treatment in mice, CUR nanoparticles reduced hepatotoxicity induced by antitubercular anti-microbials. Most curiously, co-treatment of nanoparticle-formed CUR alongside antitubercular anti-infection agents drastically diminished the danger for illness reactivation and reinfection, which is the significant setback of the current anti-toxin treatment embraced by DOT Short course. Moreover, nanoparticle-formed CUR fundamentally diminished the time required for anti-microbial treatment to get sterile insusceptibility, in this way decreasing the chance of creating drug-safe variations of the living beings. Subsequently, assistant treatment of nano-formed curcumin with upgraded bioavailability might be gainful to the treatment of tuberculosis and potentially other diseases [132].

In a further comparative study, Sing et al. had synthesized silver nanoparticles (AgNPs) and gold nanoparticles (AuNPs) by two different methods: chemical and biological [133]. These nanoparticles were evaluated against mycobacteria and AgNPs and found to have high myco-bactericidal potency at minimum inhibitory concentrations (MIC) of <3 μg/ml, while the AuNPs showed no such activity up to a concentration of 100 μg/ml. In addition, in vitro and ex vivo THP1 infection model tests showed higher potency of chemical AgNPs compared to biogenic AgNPs in inhibiting the growth of active and dormant mycobacteria. In human cell lines, an AgNP in concentration of 10× MIC (30 μg/ml) showed 40% cytotoxicity after 48 h. Silver nanoparticles exhibited higher efficacy against mycobacteria than other Gram-negative and Gram-positive pathogenic bacteria. The selectivity index was in the range of 11–23, indicating the potential of these nanoparticles to be used in the development of new therapeutics for tuberculosis.

Silver (AgNPs), gold (AuNPs), and gold-silver bimetallic nanoparticles (Au–AgNPs) were synthesized by Richa Singh and coworkers using three different medicinal plants (*Barleria prionitis*, *Plumbago zeylanica*, and *Syzygium cumini)*. The effectiveness of these nanoparticles was then checked against *Mtb* and *Mycobacterium bovis* BCG. These nanoparticles were evaluated for macrophage infection model assays in vitro as well as ex vivo by studying MIC and half MIC. Microscopic analyses were conducted to understand the intracellular uptake of nanoparticles in macrophages. In addition, nanoparticles showed good biocompatibility, specificity, and selectivity in human cell lines. Out of the tested nanoparticles, the highest antitubercular activity was shown by Au–AgNPs with minimum inhibitory concentration of 2.56 μg/ml, followed by Silver-NPs whereas AuNPs did not show any significant antitubercular activity up to 100 μg/ml. In macrophage infection model assays, Au–AgNPs have the capacity to inhibit active as well as dormant stage mycobacteria. These nanoparticles were capable of achieving concentration of 30 μg/ml (10 times MIC) inside macrophage cells and showed marked cytotoxicity up to 45% at 48 h. Out of prepared and tested nanoparticles, gold-silver bimetallic nanoparticles synthesized from *Syzygium cumini* were exhibited more activity and found more specific and selective towards mycobacteria, with their selectivity index in the range of 94–108 [134].

Gupta et al. has used *Ocimum gratissimum* linn (ethanolic extract [EE] and hydroalcoholic extracts [HAE]) for synthesizing gold nanoparticles (GNPs) and evaluated against the H37RV strain of *Mtb*. The prepared GNPs were found spherical in structure with a diameter of about 10–25 nm having λ max at 348 nm. The coating of GNPs with phytoconstituents (terpenoids) was confirmed with FTIR indicated that biomolecules may lead to efficient stabilization and capping of the gold nanoparticles. The ethanolic and hydroalcoholic extracts showed the presence of Ursolic acid at 2.89% and 1.97%. GNPs of these extracts showed superior antitubercular activity, with minimum inhibitory concentration of 2.5 μg/ml and 20 μg/ml, whereas EE and HAE exhibited MIC at concentrations 50 μg/ml and 75 μg/ml, respectively. Gold nanoparticles prepared from EE showed higher efficacy to inhibit mycobacteria. The synthesized GNPs have the potential for tuberculosis treatment as no chemical reagents were used in this method [135].

Ethanolic extract of *P. merkusii* was prepared by the maceration method. *P. merkusii* nanoparticles were prepared using tripolyphosphate (TPP) by employing ionotropic gelation method. Scanning electron microscope (SEM) was used for studying size and morphology of the *P. merkusii* nanoparticle. The MIC and MBC of *P. merkusii* nanoparticle on H37Rv strain of Mtb was performed by broth microdilution and micro diffusion methods. The SEM study of nanoparticle extract of *P. merkusii* revealed uniform sphericity with a rough surface and a solid dense cubical or rectangular structure. The size distribution of these nanoparticles ranged from10 to 800 nm with an average of 500 nm. *P. merkusii* nanoparticle showed significant antimycobacterial effects with an MIC value of 1000 mg/ml and MBCs value of 2000 mg/ml for the H37Rv strain of Mtb. *P. merkusii* extract nanoparticle has the potential to be developed further for tuberculosis treatment [136].

4 Biogenic Metallic Nanoparticles in Inhibition of *TB*

Several issues are associated with the efficacy of antibiotics used in the treatment of TB including correlation between blood plasma of antibiotics with intracellular concentration and their intracellular activity [137]. The anti-TB antibiotics have flexible internalization and intracellular accumulation path [137]; however, these antibiotics may not penetrate granulomas [138]. The intracellular survival of *Mtb* in the cell pertains to its adaptation-evading of the immune system and dissemination that leads to the selection of aggressive multidrug resistance [137]. Several studies indicated that the azithromycin could accumulate completely in phagolysosomes; however, its antimicrobial efficacy is limited. But, moxifloxacin could not accumulate but demonstrate effective antimicrobial efficacy on pathogens

[139]. Thus, cellular accumulation and related local environmental conditions (pH) are important parameters for administered antibiotic activity. This specific condition of anti-TB therapy via antibiotics could be overcome by the use of metallic nanoparticles. Several studies showed that green-synthesized silver, gold, and zinc oxide nanoparticles with dimensions less than 10 nm can directly penetrate through the bacterial cell membrane and macrophage bilayer membrane with excellent efficacy [140, 141].

Metallic nanoparticles have varied applications in various areas including chemical engineering [142], textile manufacturing [143], electronics, tissue engineering [144], clinical diagnostics [145], biological imaging, nanomedicine [146], organ implantations, biosensor [147], biomarkers, cell labelling, active packaging of food [148], etc. However, chemical-based synthesis of metallic nanoparticles is costly and involves the use of toxic chemicals that causes serious biological and environmental risks. Recently, several studies reported that the green synthesized metallic nanoparticles demonstrate potential efficacy and biocompatibility with excellent antibacterial activity against infectious microorganisms [147–149]. Although several approaches are being utilized to produce such entities, the use of a natural product such as phenolic compounds, pharmaceutical excipients, and the biological organism is a novel and biocompatible approach. Silver nanoparticles increase the permeability of membrane by motiving the aggregation of proteins in the periplasmic space and forming a required sized pore. However, zinc oxide nanoparticles can interact with the membrane of *Mtb*, leading to the formation of surface pores and the release of intracellular nucleotides [150]. However, the colloidal mixture of silver and zinc oxide nanoparticles can eliminate *Mtb* with reduced cytotoxic effect on human monocytic cells and breast cancer cell lines. Moreover, co-delivery of mixed metallic nanoparticles such as silver/gold or silver/zinc oxide through passive targeting releases metallic ions with reduced cytotoxicity and improved efficacy, compared to lone nanoparticles [150, 151]. In addition, studies reported that gallium nanoparticles can effectively import active pharmaceuticals to the macrophage, inhibit *Mtb* growth and reduce the inhibition of phagosome maturation [152]. Furthermore, magnetic encapsulate nanoparticles exhibit excellent drug release properties to the targeted site and might be suitable as the carrier of anti-TB drugs. The suggested mode of action for antibacterial activity of gold nanoparticles is attributed to the generation of reactive oxygen species that increase oxidative stress of microbial cells and result in the release of intracellular lactate dehydrogenase enzymes into extracellular medium forming vacuole leading to cell death [153]. Although the mechanisms behind the biocidal activity of copper nanoparticles are not yet fully understood, however some studies reported Cu ions nanoparticles might interact with phosphorus and sulphur-containing biomolecules such as DNA and protein to distort their structures and thus disrupt biochemical [154].

5 Current Clinical Management Outlook in Tuberculosis Treatment

Science and technology in the treatment and management of severe diseases have significantly developed with advancements in diagnostic instruments and target drug delivery pharmaceuticals. However, multidrug-resistance TB remains still a serious public health crisis and a health security threat. The WHO indicated that only one in three people with drug resistance TB accessed treatment in 2020. Moreover, current WHO strategies for the formulation and treatment regimens for multidrug resistance TB pay diminutive attention to the microbiological activity for anti-TB drugs [155]. In 2020, an estimated 10 million people fell ill with TB and 1.5 million people died from TB. In addition, 1.1 million children fell ill with TB globally in 2020. Furthermore, the 30 high TB burden countries accounted for 86% of new TB cases [156]. However, eight countries account for two-thirds of the total, with India leading the count, followed by China, Indonesia, the Philippines, Pakistan, Nigeria, Bangladesh, and South Africa [156]. Tuberculosis occurs in every part of the world, WHO reported that 43% of new cases in the Southeast Asian region, followed by African with 25% and Western Pacific with 18% [156]. Though several issues are associated with the treatment and management of TB, globally incidence of TB is dropped at about 2% per year with a cumulative reduction of 11% from 2015 to 2020. However, low- and middle-income countries that account for 98% of the reported TB cases fall short of basic medications might be due to a decline of 8.7% in spending in the last 2 years [157]. World Health Organization set TB strategy target of 'No TB patients' adopted in 2015 to reach the milestone of 0% with a target of 90% reduction in TB case and death by 2030 or 2035 [156].

Anti-TB drugs have been used for decades and strains that are resistant to one or more of these medicines have been well documented in every country. At present stage, the principal drugs for drug-resistant TB include rifampicin as first line for 6 months followed by fourth-generation fluoroquinolones and bedaquiline in regimens. Other drugs are 'companion drugs with high bactericidal activity' used to avert treatment failure due to acquired drug resistance against the basic drugs [158]. Globally in 2018, the success rate of multidrug/rifampicin resistance TB patients was 59% [156]. Recently several studies indicated delivery of anti-TB drugs via modified dosage or targeted delivery system can reduce the treatment region. Furthermore, the application of nano-medicine is increasing rapidly with the promise of targeted and efficient drug delivery in the treatment and management of TB. Nanomedicine addresses the lacunae within conventional therapy, as evidenced by several preclinical and clinical investigations indicating site-specific drug delivery with reduced side effects and better treatment outcomes. The development of such suitable and biocompatible targeted drug delivery vehicles include liposomes, mesoporous particles, micelles, dendrimers, and nanoparticles. The pulmonary administration of rifampicin and ascorbic acid co-loaded alginate-chitosan nanoparticles reported with prolonged antibacterial properties against *Staphylococcus aureus* due to surface functionalization with chitosan [159]. Biosynthesized metal nanoparticles including gold and silver fabricated by employing alginate followed by co-encapsulating doxorubicin and rifampicin demonstrated pH-dependent release with excellent in vivo imaging analysis [160]. Similarly, several studies indicated that alginate could be a potential carrier for anti-TB drug's ineffective treatment and management of tuberculosis.

6 Bibliometric Analysis of Alginate-Based Nanoparticles in the Treatment of Tuberculosis

Over the years, there has been increasing research interest in alginate-based nanomaterials considering tuberculosis management. Therefore, it will be enthralling to examine the overall trend and the new research direction in this field. A bibliometric analysis is an appropriate technique to utilize in this case. Additionally, bibliometric analysis has gained immense popularity in various research segments, and popularity attributed to the (I) advancement, availability, and accessibility of bibliometric software such as Gephi, Leximancer, VOSviewer, and Scientific databases such as Web of Science and Scopus, and (II) the cross-disciplinary pollination of the bibliometric methodology from information science to research in health science. The bibliometric analysis enables scholars to identify patterns of publication, authors prolific, top affiliation, year trends, citation tends, document type, productive source outlet over the period [161]. Moreover, in our study, we emphasized the significant contribution of authors, organizations, countries, and document sources with linkage among others.

Study Design

This study reported here follows Preferred Reporting Items for Systematic Reviews and Meta-Analyses (PRISMA) extension statement for network meta-analysis [162].

Data Collection

The data set presented in this study was obtained from the 'Scopus' database, with approximately 69 million records and access to Elsevier and the author's citation database [163]. In addition, to maintain consistency and to avoid data overlap, data sets were collected simultaneously from the Web of Science. Initially, data was obtained from the Web of Science and Scopus to compare the volume of the material available on each database. The Scopus database had 3776 published scientific papers relating to the topic, while the Web of Science yielded 419 published research papers. The data retrieval steps and the inclusion and exclusion criteria are depicted in Fig. 2. In this study, the data set was collected from 1994 to 2020 and accessed on 1 December 2021. Data were retrieved using "alginate nanoparticles", "sodium alginate nanoparticles", and "alginate nanoparticles in tuberculosis" as core keywords. The strategic search in retrieving original research published during 1994 to

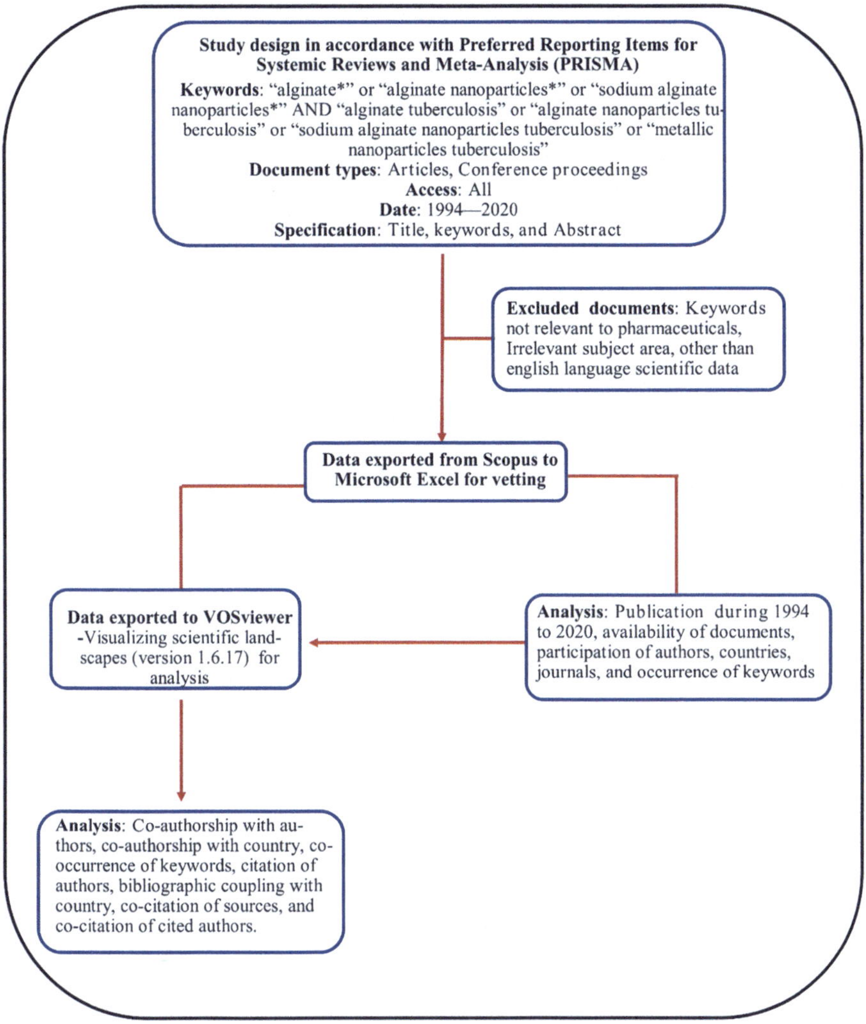

Fig. 2 Process flowchart for systemic bibliometric analysis of data retrieved from Scopus database on "alginate*" or "alginate nanoparticles*" or "sodium alginate nanoparticles*" AND "alginate tuberculosis" or "alginate nanoparticles tuberculosis" or "sodium alginate nanoparticles tuberculosis" or "metallic nanoparticles tuberculosis", dated 1 December 2021

2020 is as follows: ("alginate*" or "alginate nanoparticles*" or "sodium alginate nanoparticles*") AND ("alginate tuberculosis" or "alginate nanoparticles tuberculosis" or "sodium alginate nanoparticles tuberculosis" or "metallic nanoparticles tuberculosis") were entered with the specific document type. The quotation marks ("") were used for specifying the exact required phrase, whereas the asterisk (*) was a shortcut for retrieving both the singular and plural versions of required keywords. Document type was limited to original article or conference proceedings paper only and publication year ranged from 1994 to 2020, to internment original research from the beginning. The 1780 data obtained from the Scopus database following exclusion criteria were imported into VOSviewer-Visualizing scientific landscapes (version 1.6.17) software and Microsoft Excel (version 2010) for further analysis. Data exported to Excel were edited, sorted, and categorized based on inclusion and exclusion criteria, year of publications, countries, region, and fields. The data set was later imported to VOSviewer and used to create network maps of co-authorships of authors and countries, co-occurrence of indexed keywords, citation of authors, citations of countries, bibliographic coupling of authors and countries, and co-citations of cited source with authors.

The results for the data collected for the bibliometric analysis on alginate-based nanomaterials in the treatment and management of tuberculosis from the stipulated period of 1994 to 2020 are presented in Fig. 3. The following analyses were studied for the trend of publication, the interaction between countries/regions, institutions, authors, and journals. Key indicators were allocated to highlight the collaborative efforts and highlight key participants at the forefront of alginate nanoparticles in the treatment of tuberculosis research.

Initially, a total of 3776 documents were retrieved using the search keywords; however, several documents were excluded considering the exclusion criteria of the study. The included documents were recorded 1664 and 130 for original articles and conference proceedings, respectively. Figure 3 conveys the increase in publication trends of scientific data that indicate the severity of disease and resistance against potential antibiotics used during the management of TB. Clustering and network

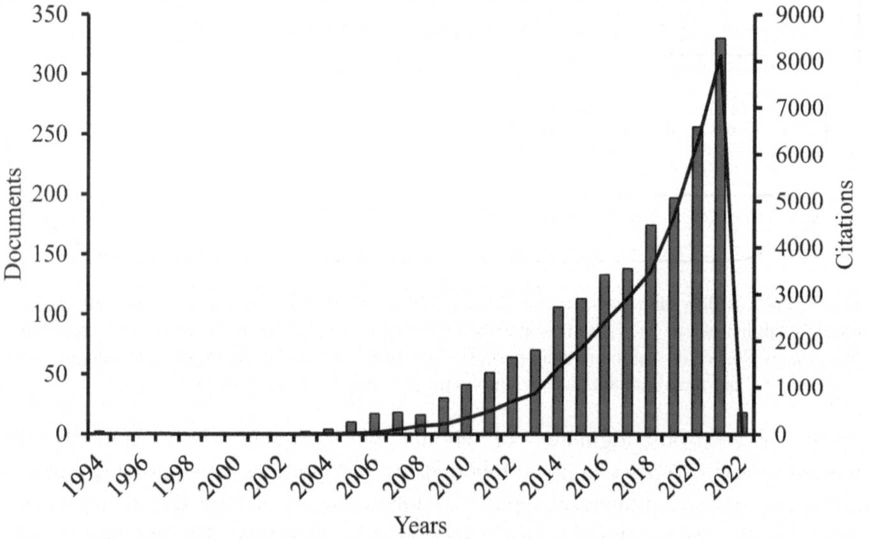

Fig. 3 The figure denotes the publication trend of scientific papers published during 1994 to 2021 for "alginate nanoparticles*" "tuberculosis*" during 1994–2021 imported from Scopus on 6 December 2021. (For interpretation of the results to colour in this figure legend, the reader is referred to either web version of this chapter or colour print)

visualization mapping is an important tool in bibliometric analysis that provides enrichment to the exported dataset with a primary goal to create thematic or social clusters with overlay in the visualization network.

Co-authorship analysis represents the interaction among scholars in a research field. Since co-authorship is a formal way of intellectual collaboration among scholars, therefore it is important to understand how scholars interact amongst themselves including associated authors with their institutional affiliation [164]. In fact, collaboration among scholars can lead to improvements in research, for example, contributions from different scholars can contribute to greater clarity with ironic perceptions. Figure 4a conveys the overlay visualization for co-authorship of authors considering the number of citation received from 2015 to 2019 with a maximum number of five documents of an author. Out of 6657 authors, 174 meet the thresholds, and among them, the top authors were Khuller G K, Sarmento B, Sharma S, Chen S, Zhang y, Liu J, Primo A, and Liu y with citation and total linkage strength of 928, 816, 810, 680, 635, 611, 608, and 601 and 16, 4, 12, 43, 66, 76, 6, and 61, respectively. However, Sukhodub IF received the least citation and linkage strength of 11 and 0, respectively. The citations-based overlay visualization between co-authors of different countries with a minimum number of five documents per

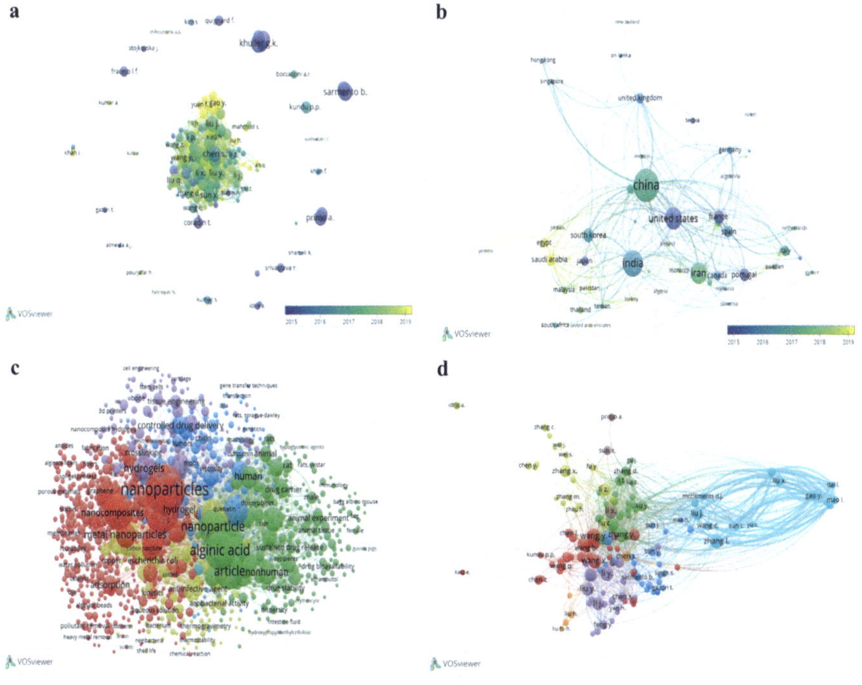

Fig. 4 Overlay visualization for co-authorship of author citation (**a**), overlay visualization for co-authorship of authors amid countries (**b**), co-occurrence of all keywords with their total link strength (**c**), co-citation of authors' published documents (**d**). (For interpretation of the results to colour in this figure legend, the reader referred to either web version of this chapter or colour print)

country is presented in Fig. 4b. A total of 57 meet the thresholds amid 88 countries; among them top countries were China, India, United States, and Iran with citation and total linkage strength of 10,566, 6452, 4704, and 4333 and 173, 83, 136, and 45, respectively.

The Scopus database gave a holistic view of the various categories that publication come under. However, it must be noted that there are overlaps with these key research areas, and it is only a primary indication of the prevalent categories. Figure 4c demonstrates the occurrence of all keywords with a minimum number of five occurrences. The distinctive clusters in Fig. 4c indicate the usage of similar, most commonly, and the least commonly used keywords in their publications. A total of 1441 keywords meet the thresholds among 13,264 keywords. The topmost keywords overall in this study were Nanoparticles, Alginate, Alginic acid, and Sodium alginate with an occurrence and total link strength of 814, 578, 481, and 433 and 13,202, 8933, 12,864, and 5724, respectively. However, the least transpired keywords were Antitubercular Drugs, Osteoblast, Skin Fibroblast, *Mycobacterium Bovis BCG*, and Lung Alveolus Macrophage with an occurrence of 5. The citation of authors with a maximum number of authors of 25 per paper and a minimum number of 5 documents were presented as overlay visualization in Fig. 4d. The top five authors document for the treatment of TB using alginate nanoparticulate carrier were Khuller GK, Sarmento B, Sharma S, Chen S, Zhang y with citations and linkage strength of 928, 816, 810, 680, and 635 and 35, 35, 29, 88, and 46, respectively. However, aside from total publications, the total linkage strength is another important indicator to the study the author Gao Y toped among 6657 authors with total linkage strength of 339.

Bibliographic coupling is a technique for screening and mapping operating on the assumption that two publications share common references with similar in their content. The analysis concentrate on the division of publications into thematic clusters based on shared references, and its best used within a specific timeframe [165]. Bibliographic coupling or participation between countries with a maximum number of ten documents and a minimum number of five documents per country indicated 57 countries meet the thresholds among 88 countries as presented in figure 8. The bibliographic coupling of documents among top counties was China, India, United States, Iran, and South Korea with total linkage strength of 31,841, 20,892, 18,019, 13,685, and 9399 respectively. Figure 5a also expresses that authors from Asia and its continents contribute 64% of publications for this topic, followed by the United States, Iraq, and South Korea. Moreover, these initial studies suggested that these regions/countries are leaders in the research relating to alginate nanoparticles in the management of TB. However, an important factor to also consider is the population of the leading country is comparatively high than other countries.

Co-citation analysis is a technique for scientific data mapping that assumes publications are cited together frequently with similar thematically [166]. This analysis is used to reveal the intellectual structure of a research field including its underlying theme. Moreover, the co-citation network indicates interconnection between two publications when they co-occur in the reference list of another publication. However, co-citation analysis more emphasizes on highly cited publications, and

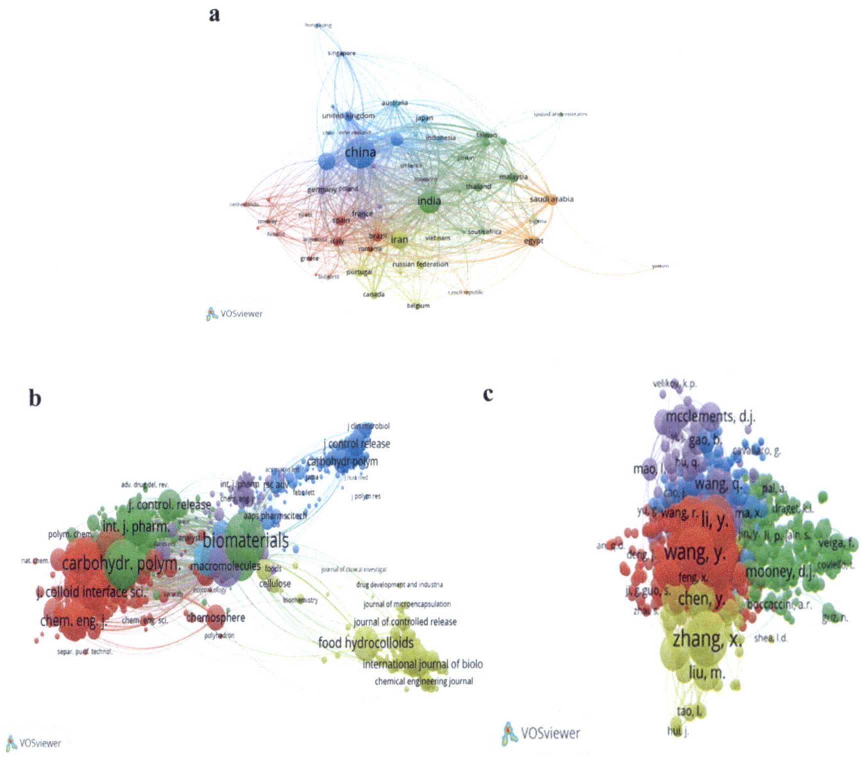

Fig. 5 Bibliographic coupling of countries with total link strength (**a**), co-citation of cited sources with total link strength (**b**), co-citation of cited authors with total link strength (**c**). (For interpretation of the results to colour in this figure legend, the reader referred to either web version of this chapter or colour print)

leave that are recent especially thematic issues. Figure 5b conveys the seminal publication and knowledge foundation of co-citation of cited sources with a minimum number of ten citations per source. A total of 15,580 journals participated in the publication; among them 1002 meet the inceptions. The journals were grouped into four distinct clusters. Biomaterials have the greatest mapping in cluster 1 with green colour, carbohydrate polymers showed mapped clusters in red colour, and food hydrocolloids showed mapped clusters in yellow colour. However, the *Journal of Controlled Release* showed clusters connection in blue and green. The results of the bibliographic analysis indicated the top five co-citations of source as Biomaterials, Carbohydrate Polymers, Langmuir, Biomacromolecules, and *International Journal of Biological Macromolecules* with citation and total linkage strength of 1475, 1041, 914, 735, and 715 and 48,940, 37,007, 31,629, 26,575, and 25,727, respectively. However, the *Journal of Nuclear Medicine* obtained the least citations and linkage strength of 10 and 84. The current impact factors (for the year 2021) of the journals compared to the number of total publications were also analysed. Impact factors are a commonly used tool to evaluate the relative significance of an article within a particular field, with the frequency of papers published weighed against the citations per document within a stipulated period. The top five cited journals in this study were ranked under the top ten in biomedicine, polymer science, and allied health science with an impact factor of 12.479, 9.381, 6.988, 6.95, and 3.882 for Biomaterials, Carbohydrate Polymers, Biomacromolecules, *International Journal of Biological Macromolecules*, and Langmuir, respectively.

The results of bibliographic mapping of the minimum number of 10 citations of authors are presented in Fig. 5c. A total number of 121,886 authors participated in the study, among them 4076 meet the thresholds. Wang Y, Zhang Y, Liu Y, and Li Y was the top-cited author with citation and linkage strength of 740, 697, 683, and 668 and 58,596, 52,355, 50,519, and 50,692, respectively, were among the top 1000 authors in the study. However, Zhang X documents gained linkage strength and citation of 63,304 and 661, respectively.

7 Conclusion

Tuberculosis has been intimidating for a decade, and the discovery of antibiotics with progressive development and transformation of drug delivery from conventional to nanoparticulate systems has provided the chance for recovery. However, the progression of multidrug resistance with strains continuously create challenges to the current treatment strategies. Development and discovery of novel antibiotics against TB remains a priority; however, the development of nanoparticles of existing drugs using biodegradable polymers and bio-reduction of metallic nanoparticles may represent a cost-effective and promising alternative. Sodium alginate fortified anti-TB agent as nanoparticles demonstrated significant efficacy with several advantages including improvement of drug bioavailability with reduction in dosing frequency. Moreover, considering the efficacy and biocompatibility, sodium alginate nanomaterials create an effective basis for the management of the diseases and making treatment regime more practical and affordable. Additionally, a short bibliometric analysis enables scholars to identify patterns of publication, prolific authors, top affiliation, year trends, citation tends, document type, productive source outlet over the period that can influence future research trends. Undoubtedly, success of such nanomaterials-based anti-TB agent depends on pre-clinical and clinical studies, which can bring more insight into their potential to commercialize it.

References

1. Dheda K, Barry CE 3rd, Maartens G. Tuberculosis. Lancet. 2016;387:1211–26.
2. Dheda K, Gumbo T, Maartens G, et al. The epidemiology, pathogenesis, transmission, diagnosis, and management of multidrug-resistant, extensively drug-resistant, and incurable tuberculosis. Lancet Respir Med. 2017;S2213-2600(17):30079–6.

3. Global tuberculosis report 2021, World Health Organization. https://www.who.int/publications/i/item/9789240037021.
4. Sulis G, Roggi A, Matteelli A, Raviglione MC. Tuberculosis: epidemiology and control. Mediterr J Hematol Infect Dis. 2014;6(1):e2014070.
5. Pai M, Behr MA, Dowdy D, et al. Tuberculosis. Nat Rev Dis Primers. 2016;2:16076. https://doi.org/10.1038/nrdp.2016.76.
6. Boudville DA, Joshi R, Rijkers GT. Migration and tuberculosis in Europe. J Clin Tuberc Other Mycobact Dis. 2020;18:100143.
7. Cui Y, Shen H, Wang F, Wen H, Zeng Z, Wang Y, Yu C. A long-term trend study of tuberculosis incidence in China, India and United States 1992–2017: a Joinpoint and age-period-cohort analysis. Int J Environ Res Public Health. 2020;17(9):3334.
8. Global Tuberculosis Report 2017: World Health Organisation. Available from: apps. who.int/iris/bitstream/hamdle/10665/259366/9789241565516-eng.pdf.
9. Comas I, Coscolla M, Luo T, et al. Out-of-Africa migration and Neolithic coexpansion of *Mycobacterium tuberculosis* with modern humans. Nat Genet. 2013;45:1176–82.
10. Pietersen E, Ignatius E, Streicher EM, et al. Long-term outcomes of patients with extensively drug-resistant tuberculosis in South Africa: a cohort study. Lancet. 2014;383:1230–9.
11. Dheda K, Gumbo T, Gandhi NR, et al. Global control of tuberculosis: from extensively drug-resistant to untreatable tuberculosis. Lancet Respir Med. 2014;2:321–38.
12. Dheda K, Migliori GB. The global rise of extensively drug-resistant tuberculosis: is the time to bring back sanatoria now overdue? Lancet. 2012;379:773–5.
13. Sotgiu G, Centis R, D'ambrosio L, Migliori GB. Tuberculosis treatment and drug regimens. Cold Spring Harb Perspect Med. 2015;5(5):a017822.
14. Chan ED, Iseman MD. Current medical treatment for tuberculosis. BMJ. 2002;325(7375):1282–6.
15. Patil K, Bagade S, Bonde S, Sharma S, Saraogi G. Recent therapeutic approaches for the management of tuberculosis: challenges and opportunities. Biomed Pharmacother. 2018;99:735–45.
16. Gilani SJ, Ameeduzzafar, Jafar M, Shakil K, Imam SS. Nano-carriers for the treatment of tuberculosis. Recent Pat Antiinfect Drug Discov. 2017;12(2):95–106.
17. Ain QU, Sharma S, Khuller GK, Garg SK. Alginate-based oral drug delivery system for tuberculosis: pharmacokinetic and therapeutic effects. J Antimicrob Chemother. 2003;51:931–8.
18. Labana S, Pandey R, Sharma S, Khuller GK. Chemotherapeutic activity against murine tuberculosis of once weekly administered drugs (INH and RIF) encapsulated in liposomes. Int J Antimicrob Agents. 2002;20:301–4.
19. Bala I, Hariharan S, Ravi Kumar MNV. PLGA nanoparticles in drug delivery: the state of the art. Crit Rev Ther Drug Carrier Syst. 2004;21:387–422.
20. Shah SP, Misra A. Development of liposomal amphotericin B dry powder inhaler formulation. Drug Deliv. 2004;11:247–53.
21. Hurley L, Andersen BR. Biodegradable implants from poly-(alphahydroxy acid) polymers for isoniazid delivery. Int J Tuberc Lung Dis. 1999;3:1015–24.
22. Wissing SA, Kayser O, Müller RH. Solid lipid nanoparticles for parenteral drug delivery. Adv Drug Deliv Rev. 2004;56(9):1257–72.
23. Dahanayake MH, Jayasundera AC. Nano-based drug delivery optimization for tuberculosis treatment: a review. J Microbiol Methods. 2021;181:106127.
24. Tabata J, Ikada Y. Protein pre-coating of polylactide microspheres containing a lipophilic immune potentiator for enhancement of macrophage phagocytosis and activation. Pharm Res. 1989;6:296–301.
25. Bodmeier R, Chen H. Indomethacin polymeric nanosuspensions prepared by micro- fluidization. J Control Release. 1990;12:223–33.
26. Zahoor A, Sharma S, Khuller GK. Inhalable alginate nanoparticles as antitubercular drug carriers against experimental tuberculosis. Int J Antimicrob Agents. 2005;26:298–303.

27. Lemarchand C, Gref R, Passirani C, Garcion E, Petri B, Muller R. Influence of polysaccharide coating on the interactions of nanoparticles with biological systems. Biomaterials. 2006;27:108–18.
28. Koosha F, Muller RH, Davis SS, Davies MC. The surface chemical structure of poly (hydroxybutyrate) microparticles produced by solvent evaporation process. J Control Release. 1989;9:149–57.
29. Choukaife H, Doolaanea AA, Alfatama M. Alginate nanoformulation: influence of process and selected variables. Pharmaceuticals. 2020;13:335.
30. Rinaudo M. Main properties and current applications of some polysaccharides as biomaterials. Polym Int. 2007;57:397–430.
31. Sorasitthiyanukarn FN, Muangnoi C, Bhuket PR, Rojsitthisak P. Chitosan/alginate nanoparticles as a promising approach for Oral delivery of curcumin Diglutaric acid for cancer treatment. Mater Sci Eng. 2018;93:178–90.
32. Markeb AA, El-Maali NA, Sayed DM, et al. Synthesis, structural characterization, and preclinical efficacy of a novel paclitaxel-loaded alginate nanoparticle for breast cancer treatment. Int J Breast Cancer. 2016;2016:7549372.
33. Li M, Sun Y, Ma C, Hua Y, Zhang L, Shen J. Design and investigation of penetrating mechanism of Octaarginine-modified alginate nanoparticles for improving intestinal insulin delivery. J Pharm Sci. 2021;110(1):268–79.
34. Baek S, Joo SH, Toborek M. Treatment of antibiotic-resistant bacteria by encapsulation of ZnO nanoparticles in an alginate biopolymer: insights into treatment mechanisms. J Hazard Mater. 2019;373:122–30.
35. Scolari IR, Páez PL, Musri MM, Petiti JP, Torres A, Granero GE. Rifampicin loaded in alginate/chitosan nanoparticles as a promising pulmonary carrier against staphylococcus aureus. Drug Deliv Transl Res. 2020;10:1403–17.
36. Venkatesan J, Anil S, Kim SK, Shim MS. Seaweed polysaccharide-based nanoparticles: preparation and applications for drug delivery. Polymers. 2016b;8(2):30.
37. Leong JY, Lam WH, Ho KW, et al. Advances in fabricating spherical alginate hydrogels with controlled particle designs by ionotropic gelation as encapsulation systems. Particuology. 2016;24:44–60.
38. Pestovsky YS, Martínez-Antonio A. The synthesis of alginate microparticles and nanoparticles. Drug Des Intellect Prop Int J. 2019;3:293–327.
39. Mendoza-Muñoz N, Alcalá-Alcalá S, Quintanar-Guerrero D. Preparation of polymer nanoparticles by the emulsification-solvent evaporation method: from Vanderho's pioneer approach to recent adaptations. In: Polymer nanoparticles for nanomedicines. Cham: Springer; 2016. p. 87–121.
40. Muhaimin BR. Effect of solvent type on preparation of ethyl cellulose microparticles by solvent evaporation method with double emulsion system using focused beam reflectance measurement. Polym Int. 2017;66:1448–55.
41. Rajaonarivony M, Vauthier C, Couarraze G, Puisieux F, Couvreur P. Development of a new drug carrier made from alginate. J Pharm Sci. 1993;82:912–7.
42. Saether HV, Holme HK, Maurstad G, Smidsrød O, Stokke BT. Polyelectrolyte complex formation using alginate and chitosan. Carbohydr Polym. 2008;74:813–21.
43. Jardim KV, Palomec-Garfias AF, Andrade BYG, et al. Novel magneto-responsive nanoplatforms based on $MnFe_2O_4$ nanoparticles layer-by-layer functionalized with chitosan and sodium alginate for magnetic controlled release of curcumin. Mater Sci Eng. 2018;92:184–95.
44. Krishnamoorthy K, Mahalingam M. Selection of a suitable method for the preparation of polymeric nanoparticles: multi-criteria decision making approach. Adv Pharm Bull. 2015;5(1):57–67.
45. Abdelghany S, Parumasivam T, Pang AB, et al. Alginate modified-PLGA nanoparticles entrapping amikacin and moxifloxacin as a novel host-directed therapy for multidrug-resistant tuberculosis. J Drug Deliv Sci Technol. 2019;52:642.

46. Gelperina S, Kisich K, Iseman MD, Heifets L. The potential advantages of nanoparticle drug delivery systems in chemotherapy of tuberculosis. Am J Respir Crit Care Med. 2005;172:1487–90.
47. Etiology and transmission of tuberculosis. Module 1. Global TB NJMS.
48. Transmission and Pathogenesis of Tuberculosis. Chapter 2. Centers for disease control and prevention (CDC) available on https://www.cdc.gov/tb/education/corecurr/pdf/chapter2.pdf
49. Patil K, Bagade S, Bonde S, Sharma S, Saraogi G. Recent therapeutic approaches for the management of tuberculosis: challenges and opportunities. J Biomed Pharmacother. 2018;99:735–45.
50. Nanotechnology for Drug Delivery Applications. Azonano. 2017; 11. https://www.azonano.com/article.aspx?ArticleID=4668
51. Patra J, Das G, Fraceto LF, Ramos Campos EV, et al. Nano based drug delivery systems: recent developments and future prospects. J Nanobiotechnol. 2018;16:71.
52. Liu Z, Tabakman S, Welsher K, Dai H. Carbon nanotubes in biology and medicine: in vitro and in vivo detection, imaging and drug delivery. Nano Res. 2009;2:85–120.
53. de Villiers MM, Aramwit P, Kwon GS, editors. Nanotechnology in drug delivery. New York: Science & Business Media; 2008.
54. Lounnas V, Ritschel T, Kelder J, McGuire R, Bywater RP, Foloppe N. Current progress in structure-based rational drug design marks a new mindset in drug discovery. Comput Struc Biotechnol J. 2013;5:e201302011.
55. Mavromoustakos T, Durdagi S, Koukoulitsa C, Simcic M, Papadopoulos M, Hodoscek M, Golic Grdadolnik S. Strategies in the rational drug design. Curr Med Chem. 2011;18:2517–30.
56. Torchilin VP. Multifunctional nanocarriers. Adv Drug Deliv Rev. 2012;64:302–15.
57. Almalik A, Benabdelkamel H, Masood A, et al. Hyaluronic acid coated chitosan nanoparticles reduced the immunogenicity of the formed protein corona. Sci Rep. 2017;7:10542.
58. Martens TF, Remaut K, Deschout H, et al. Coating nanocarriers with hyaluronic acid facilitates intravitreal drug delivery for retinal gene therapy. J Control Release. 2015;202:83–92.
59. Gao WW, Zhang LF. Coating nanoparticles with cell membranes for targeted drug delivery. J Drug Target. 2015;23:619–26.
60. Gao H, Yang Z, Zhang S, Cao S, Shen S, Pang Z, Jiang X. Ligand modified nanoparticles increases cell uptake, alters endocytosis and elevates glioma distribution and internalization. Sci Rep. 2013;3:2534.
61. Pelaz B, del Pino P, Maffre P, et al. Surface functionalization of nanoparticles with polyethylene glycol: effects on protein adsorption and cellular uptake. ACS Nano. 2015;9:6996–7008.
62. Muller J, Bauer KN, Prozeller D, et al. Coating nanoparticles with tunable surfactants facilitates control over the protein corona. Biomaterials. 2017;115:1–8.
63. Bansal AK. Excipients used in nano-technology assisted drug delivery systems. J Excipeint and Food Chem. 2014;5(4):173–6.
64. Smidsrod O, Draget KI. Chemistry and physical properties of alginates. Carbohydr Eur. 1996;14:6–13.
65. Skjak-Braek G, Espevik T. Application of alginate gels in biotechnology and biomedicine. Carbohydr Eur. 1996;14:19–25.
66. Ahmad Z, Pandey R, Sharma S, et al. Pharmacokinetic and pharmacodynamic behaviour of antitubercular drugs encapsulated in alginate nanoparticles at two doses. Int J Antimicrob Agents. 2006;27:420–7.
67. Ahmad Z, Pandey R, Sharma S, et al. Alginate nanoparticles as antitubercular drug carriers: formulation development, pharmacokinetics and therapeutic potential. Indian J Chest Dis Allied Sci. 2006;48:171–6.
68. Yang JS, Xie YJ, He W. Research progress on chemical modification of alginate: a review. Carbohydr Polym. 2011;84:33–9.
69. Thomas D, Latha MS, Thomas KK. Synthesis and in vitro evaluation of alginate-cellulose nanocrystal hybrid nanoparticles for the controlled oral delivery of rifampicin. J Drug Deliv Sci Technol. 2018;46:392–9.

70. Shehzada A, Rehmana G, Ul-Islamb M, Khattakb WA, Leea YS. Challenges in the development of drugs for the treatment of tuberculosis. Braz J Infect Dis. 2013;17(1):74–81.
71. Mani G, Jainuddin Y, Vellaichamy E, et al. Gold nanoparticle conjugated PLGA–PEG–SA–PEG–PLGA multiblock copolymer nanoparticles: synthesis, characterization, in vivo release of rifampicin. J Mater Chem B. 2014;2:418–47.
72. Jennifer CG, Jamee B, Carly JC, et al. Thiol-modified gold nanoparticles for the inhibition of Mycobacterium smegmatis. Chem Commun. 2014;50:15860–3.
73. Chinmayee S, Anowar H, Anand R, et al. Crosslinked thiolated starch coated Fe_3O_4 magnetic nanoparticles: effect of montmorillonite and crosslinking density on drug delivery properties. Starch Biosynth Nutr Biomed. 2014;66(7–8):760–71.
74. Hwang J, Son J, Seo Y, et al. Functional silica nanoparticles conjugated with beta-glucan to deliver anti-tuberculosis drug molecules. J Ind Eng Chem. 2018;58:376–85.
75. Clemens DL, Lee BY, Xue M, et al. Targeted intracellular delivery of antituberculosis drugs to mycobacterium tuberculosis-infected macrophages via functionalized mesoporous silica nanoparticles. Antimicrob Agents Chemother. 2012;56(5):2535–45.
76. Pandey R, Khullar GK. Oral nanoparticle-based antituberculosis drug delivery to the brain in an experimental model. J Antimicrob Chemother. 2006;57:1146–52.
77. Pandey R, Khullar GK. Nanoparticle-based oral drug delivery system for an injectable antibiotic–streptomycin. Chemotherapy. 2007;53:437–41.
78. Pandey R, Zahoor A, Sharma S, et al. Nanoparticle encapsulated antitubercular drugs as a potential oral drug delivery system against murine tuberculosis. Tuberculosis. 2003;83:373–8.
79. Sharma A, Sharma S, Khullar GK. Lectin-functionalized poly (lactide-co-glycolide) nanoparticles as oral/aerosolized antitubercular drug carriers for treatment of tuberculosis. J Antimicrob Chemother. 2004;54:761–6.
80. Kumar PV, Asthana A, Dutta T, et al. Intracellular macrophage uptake of rifampicin loaded mannosylated dendrimers. J Drug Target. 2006;14(8):546–56.
81. Bellini RG, Guimaraes AP, Pacheco MAC, et al. Association of the anti-tuberculosis drug rifampicin with a PAMAM dendrimer. J Mol Graph Model. 2015;60:34–42.
82. Saeid R, Luca C, Lam JKW. Pulmonary delivery of rifampicin microspheres using lower generation polyamidoamine dendrimers as a carrier. Powder Technol. 2016;291:366–74.
83. Raval T, Parmar R, Tyagi RK, et al. Rifampicin loaded chitosan nanoparticle improved dry powder presents a therapeutic approach for alveolar tuberculosis. Colloids Surf B: Biointerfaces. 2017;154:321–30.
84. Rajan M, Raj V. Encapsulation, characterisation and in-vitro release of anti-tuberculosis drug using chitosan – poly ethylene glycol nanoparticles. Int J Pharm Sci. 2012;4(4):255–9.
85. Oliveira PM, Matos BN, Pereira PAT, et al. Microparticles prepared with 50–190 kDa chitosan as promising non-toxic carriers for pulmonary delivery of isoniazid. Carbohydr Polym. 2017;174:421–31.
86. Ansari N, Ghasvini K, Ramezani M, et al. Selection of DNA aptamers against mycobacterium tuberculosis Ag85A, and its application in a graphene oxide-based fluorometric assay. Microchim Acta. 2018;185:01–21.
87. Ryan KJ, Ray CG, editors. Sherris Medical Microbiology. 4th ed. McGraw Hill; 2004.
88. World Health organization 2018, Global Tuberculosis Report. https://www.who.int/tb/publications/global_report/en/
89. Barberis I, Bragazzi NL, Galluzzo L, Martini M. The history of tuberculosis: from the first historical records to the isolation of Koch's bacillus. J Prev Med Hyg. 2017;58:E9–E12.
90. Sieniawska E, Maciejewska Turska M, Świątek L, Xiao J. Plantbased food products for antimycobacterial therapy. eFood. 2020;1(3):199–216.
91. Madikizela B, Kambizi L, McGaw LJ. An ethnobotanical survey of plants used traditionally to treat tuberculosis in the eastern region of O.R. Tambo district, South Africa. S Afr J Bot. 2017;109:231–6.

92. Lawal IO, Grierson DS, Afolayan AJ. Phytotherapeutic information on plants used for the treatment of tuberculosis in Eastern Cape Province. South Africa Evid Based Complement Alternat Med. 2014;2014:735423.
93. Semenya SS, Maroyi A. Ethnobotanical survey of plants used by Bapedi traditional healers to treat tuberculosis and its opportunistic infections in the Limpopo Province. South Africa S Afr J Bot. 2019;122:401–21.
94. Bunalema L, Obakiro S, Tabuti JR, Waako P. Knowledge on plants used traditionally in the treatment of tuberculosis in Uganda. J Ethnopharmacol. 2014;151:999–1004.
95. Gupta VK, Kaushik A, Chauhan DS, Ahirwar RK, Sharma S, Bisht D. Anti-mycobacterial activity of some medicinal plants used traditionally by tribes from Madhya Pradesh, India for treating tuberculosis related symptoms. J Ethnopharmacol. 2018;227:113–20.
96. Buszczyński S. Opis 125 ziół używanych w lecznictwie z podaniem ich uprawy i zastosowania. Berlin: "Przewodnik Zdrowia.". Berlin: Health Guide; 1905.
97. Molyneux RJ, Lee ST, Gardner DR, et al. Phytochemicals: the good, the bad and the ugly? Phytochemistry. 2007;68(22–24):2973–85.
98. Goldberg G. The report of a British nutrition foundation task force. Plants: diet and health. Oxford/London: Blackwell Publishing Ltd.; 2003. p. 347.
99. Kerr P. Plants and tuberculosis: phytochemicals potentially useful in the treatment of tuberculosis. Amsterdam: Academic; 2013. p. 45–64.
100. Veluthoor S, Anil P, Mandal V, Mukherjee K. Chapter 15: Phytochemicals: in pursuit of antitubercular drugs. In: Rahman A-U, editor. Studies in natural product chemistry, vol. 38. Elsevier Publications; 2012. p. 417.
101. Miron T, Rabinikov A, Mirelman D, et al. The mode of action of allicin: its ready permeability through phospholipid membranes may contribute to its biological activity. Biochim Biophys Acta. 2000;1463:20–30.
102. Rao RR, Rao SS, et al. Inhibition of Mycobacterium tuberculosis by garlic extract. Nature. 1946;157:441.
103. Ratnakar P, Murthy S. Purification and mechanism of antitubercular principle from garlic (Allium sativum) active against isoniazid susceptible and resistant Mycobacterium tuberculosis H37Rv. Ind J Clin Biochem. 1995;10(1):34–8.
104. Viswanathan V, Phadatare AG, Mukne A. Antimycobacterial and antibacterial activity of Allium sativum bulbs. Ind J Pharm Sci. 2014;76(3):256–61.
105. Dwivedi VP, Bhattacharya D, Singh M, et al. Allicin enhances antimicrobial activity of macrophages during Mycobacterium tuberculosis infection. J Ethnopharmacol. 2019;243:111634.
106. Prabu A, Hassan S, Prabuseenivasana AS, et al. Andrographolide: a potent antituberculosis compound that targets aminoglycoside 2-N- acetyltransferase in Mybacterium tuberculosis. J Mol Graph Mod. 2015;61:133–40.
107. Vetting MW, Hegde SS, Javid-Majd F, et al. Aminoglycoside 2′- N-acetyltransferase from Mycobacterium tuberculosis in complex with coenzyme A and aminoglycoside substrates. Nat Struct Biol. 2002;9(9):653–8.
108. Garg HK, Shrivastava A. Cytotoxic potential of Andrographolide against Bovine Tuberculosis. IOSR J Pharm Biol Sci. 2013;8(5):1–4.
109. Garg HK, Shrivastava A. Clinical use of andrographolide as a potential drug against vole tuberculosis. Int J Pure Appl Zool. 2013;1(3):223–6.
110. Dwivedi VP, Bhattacharya D, Yadav V, et al. The phytochemical bergenin enhances T helper responses and antimycobacterial immunity by activating the map kinase pathway in macrophages. Front Cell Infect Microbiol. 2017;7:149.
111. Kumar S, Sharma C, Kaushik SR, et al. The phytochemical Bergenin as an adjunct immunotherapy for tuberculosis in mice. J Bio Chem. 2019;294(21):8555–63.
112. Gupta PK, Kulkarni S, Rajan R. Inhibition of intracellular survival of multi drug resistant clinical isolates of Mycobacterium tuberculosis in macrophages by curcumin. Op Anti Agt J. 2013;4:1–5.

113. Bai X, Oberley-Deegan RE, Bai A. Curcumin enhances human macrophage control of Mycobacterium tuberculosis infection. Respirology. 2016;21:951–7.
114. Barua N, Buragohain A. Therapeutic potential of curcumin as an Antimycobacterial agent. Biomol Ther. 2021;11(9):1278.
115. Rehman SU, Choe K, Yoo HH. Review on a traditional herbal medicine, Eurycoma longifolia jack (tongkat ali): its traditional uses, chemistry, evidence-based pharmacology and toxicology. Molecules (Basel, Switzerland). 2016;21(3):331.
116. Lee HJ, Ko HJ, Kim SH, Jung YJ. Pasakbumin A. Controls the growth of Mycobacterium tuberculosis by enhancing the autophagy and production of antibacterial mediators in mouse macrophages. PLoS One. 2019;14(3):e0199799.
117. Chaieb K, Kouidhi B, Jrah H, et al. Antibacterial activity of Thymoquinone, an active principle of Nigella sativa and its potency to prevent bacterial biofilm formation. BMC Complement Altern Med. 2011;11:29.
118. Gupta R, Thakur B, Singh P, et al. Anti-tuberculosis activity of selected medicinal plants against multi-drug resistant Mycobacterium tuberculosis isolates. Indian J Med Res. 2010;131:809–13.
119. Pablos-Mendez A, Raviglione MC, Laszlo A, et al. Global surveillance for antituberculosis-drug resistance, 1994–1997. World Health Organization-International Union against tuberculosis and lung disease working group on anti-tuberculosis drug resistance surveillance. N Engl J Med. 1998;338(23):1641–9.
120. World Health Organization. Global Tuberculosis Report 2015. Geneva: World Health Organization; 2015. https://apps.who.int/iris/bitstream/handle/10665/191102/9789241565059_eng.pdf?sequence=1&isAllowed=y
121. Adhvaryu MR, Reddy N, Vakharia BC. Prevention of hepatotoxicity due to anti tuberculosis treatment: a novel integrative approach. World J Gastroenterol. 2008;14(30):4753–62.
122. Skakun NP, Shman'ko VV. Synergistic effect of rifampicin on hepatotoxicity of isoniazid. Antibiot Med Biotekhnol. 1985;30(3):185–9.
123. Fountain FF, Tolley E, Chrisman CR, Self TH. Isoniazid hepatotoxicity associated with treatment of latent tuberculosis infection: a 7-year evaluation from a public health tub tuberculosis clinic. Chest. 2005;128(1):116–23.
124. van Rie A, Warren R, Richardson M, et al. Exogenous reinfection as a cause of recurrent tuberculosis after curative treatment. N Engl J Med. 1999;341(16):1174–9.
125. Cox HS, Morrow M, Deutschmann PW. Long term efficacy of DOTS regimens for tuberculosis: systematic review. BMJ. 2008;336(7642):484–7.
126. Lian YT, Yang XF, Wang ZH, Yang Y, Yang Y, Shu YW, et al. Curcumin serves as a human kv1.3 blocker to inhibit effector memory T lymphocyte activities. Phytother Res. 2013;27(9):1321–7.
127. Aggarwal BB, Sundaram C, Malani N, Ichikawa H. Curcumin: the Indian solid gold. Adv Exp Med Biol. 2007;595:1–75.
128. Changtam C, Hongmanee P, Suksamrarn A. Isoxazole analogs of curcuminoids with highly potent multidrug-resistant antimycobacterial activity. Eur J Med Chem. 2010;45(10):4446–57.
129. Bai X, Oberley-Deegan RE, Bai A, et al. Curcumin enhances human macrophage control of Mycobacterium tuberculosis infection. Respirology. 2016;21(5):951–7.
130. Hatcher H, Planalp R, Cho J, Torti FM, Torti SV. Curcumin: from ancient medicine to current clinical trials. Cell Mol Life Sci. 2008;65(11):1631–52.
131. Strimpakos AS, Sharma RA. Curcumin: preventive and therapeutic properties in laboratory studies and clinical trials. Antioxid Redox Signal. 2008;10(3):511–45.
132. Tousif S, Singh DK, Mukherjee S, et al. Nanoparticle-formulated curcumin prevents Posttherapeutic disease reactivation and reinfection with mycobacterium tuberculosis following isoniazid therapy. Front Immunol. 2017;8:1–12.
133. Singh R, Nawale LU, Arkile M, Shedbalkar UU, Wadhwani SA, Sarkar D, Chopade BA. Chemical and biological metal nanoparticles as antimycobacterial agents: a comparative study. Int J Antimicrob Agents. 2015;46(2):183–8.

134. Singh R, Nawale L, Arkile M, et al. Phytogenic silver, gold, and bimetallic nanoparticles as novel antitubercular agents. Int J Nanomedicine. 2016;11:1889–97.
135. Gupta A, Pandey S, Variya B, Shah S, Yadav JS. Green synthesis of gold nanoparticles using different leaf extracts of Ocimum gratissimum Linn for anti-tubercular activity. Curr Nanomed. 2019;9(2):146–57.
136. Sudjarwo SA, Wardani G, Eraiko K. The potency of Pinus merkusii extract nanoparticles as anti Mycobacterium tuberculosis: an in vitro study. Int J Nutr Pharmacol Neurol Dis. 2019;9:48–52.
137. Welin A. Survival strategies of Mycobacterium tuberculosis inside the human macrophage. Linköping University Electronic Press; 2011. https://www.diva-portal.org/smash/get/diva2:395814/FULLTEXT02.pdf
138. Davis JM, Ramakrishnan L. The role of the granuloma in expansion and dissemination of early tuberculous infection. Cell. 2009;136(1):37–49.
139. Seral C, Carryn S, Tulkens PM, Van Bambeke F. Influence of P-glycoprotein and MRP efflux pump inhibitors on the intracellular activity of azithromycin and ciprofloxacin in macrophages infected by listeria monocytogenes or Staphylococcus aureus. J Antimicrob Chemother. 2003;51(5):1167–73.
140. Jayeoye TJ, Eze FN, Olatunde OO, Singh S, Zuo J, Olatunji OJ. Multifarious biological applications and toxic Hg^{2+} sensing potentiality of biogenic silver nanoparticles based on Securidaca inappendiculata Hassk stem extract. Int J Nanomedicine. 2021;16:7557.
141. Syukri DM, Nwabor OF, Singh S, Voravuthikunchai SP. Antibacterial functionalization of nylon monofilament surgical sutures through in situ deposition of biogenic silver nanoparticles. Surf Coat Technol. 2021;413:127090.
142. Eze FN, Nwabor OF. Valorization of Pichia spent medium via one-pot synthesis of biocompatible silver nanoparticles with potent antioxidant, antimicrobial, tyrosinase inhibitory and reusable catalytic activities. Mater Sci Eng C. 2020;115:111104.
143. Syukri DM, Nwabor OF, Singh S, Ontong JC, Wunnoo S, Paosen S, et al. Antibacterial-coated silk surgical sutures by ex situ deposition of silver nanoparticles synthesized with Eucalyptus camaldulensis eradicates infections. J Microbiol Methods. 2020;174:105955.
144. Marques L, Martinez G, Guidelli É, Tamashiro J, Segato R, Payão SLM, et al. Performance on bone regeneration of a silver nanoparticle delivery system based on natural rubber membrane NRL-AgNP. Coatings. 2020;10(4):323.
145. Huang H, Du X, He Z, Yan Z, Han W. Nanoparticles for stem cell tracking and the potential treatment of cardiovascular diseases. Front Cell Dev Biol. 2021;9:662406.
146. Singh S, Chunglok W, Nwabor OF, Ushir YV, Singh S, Panpipat W. Hydrophilic biopolymer matrix antibacterial peel-off facial mask functionalized with biogenic nanostructured material for cosmeceutical applications. J Polym Environ. 2022;30(3):938–953.
147. Jayeoye TJ, Eze FN, Singh S, Olatunde OO, Benjakul S, Rujiralai T. Synthesis of gold nanoparticles/polyaniline boronic acid/sodium alginate aqueous nanocomposite based on chemical oxidative polymerization for biological applications. Int J Biol Macromol. 2021;179:196–205.
148. Nwabor OF, Singh S, Paosen S, Vongkamjan K, Voravuthikunchai SP. Enhancement of food shelf life with polyvinyl alcohol-chitosan nanocomposite films from bioactive eucalyptus leaf extracts. Food Biosci. 2020;36:100609.
149. Ontong JC, Singh S, Nwabor OF, Chusri S, Voravuthikunchai SP. Potential of antimicrobial topical gel with synthesized biogenic silver nanoparticle using Rhodomyrtus tomentosa leaf extract and silk sericin. Biotechnol Lett. 2020;42(12):2653–64.
150. Ellis TD. Multi-metallic microparticles for the treatment of pulmonary tuberculosis. ACS Nano. 2018;12(6):5228–40.
151. Singh R, Nawale L, Arkile M, Wadhwani S, Shedbalkar U, Chopade S, et al. Phytogenic silver, gold, and bimetallic nanoparticles as novel antitubercular agents. Int J Nanomedicine. 2016;11:1889.

152. Choi S-R, Britigan BE, Moran DM, Narayanasamy P. Gallium nanoparticles facilitate phagosome maturation and inhibit growth of virulent Mycobacterium tuberculosis in macrophages. PLoS One. 2017;12(5):e0177987.
153. Mishra A, Mehdi SJ, Irshad M, Ali A, Sardar M, Moshahid M, et al. Effect of biologically synthesized silver nanoparticles on human cancer cells. Sci Adv Mater. 2012;4(12):1200–6.
154. Wu X, Ye L, Liu K, et al. Antibacterial properties of mesoporous copper-doped silica xerogels. Biomed Mater. 2009;4(4):045008.
155. Van Deun A, Decroo T, Piubello A, De Jong B, Lynen L, Rieder H. Principles for constructing a tuberculosis treatment regimen: the role and definition of core and companion drugs. Int J Tuberc Lung Dis. 2018;22(3):239–45.
156. Organization WH. Tuberculosis. Geneva: World Health Organization; 2021 [cited 2021 10.12.2021]. Available from: https://www.who.int/news-room/fact-sheets/detail/tuberculosis
157. Diel R, Sotgiu G, Andres S, Hillemann D, Maurer F. Cost of multidrug resistant tuberculosis in Germany—an update. Int J Infect Dis. 2021;103:102–9.
158. Naz F, Ahmad N, Wahid A, Ahmad I, Khan A, Abubakar M, et al. High rate of successful treatment outcomes among childhood rifampicin/multidrug-resistant tuberculosis in Pakistan: a multicentre retrospective observational analysis. BMC Infect Dis. 2021;21(1):1–11.
159. Scolari IR, Páez PL, Musri MM, Petiti JP, Torres A, Granero GE. Rifampicin loaded in alginate/chitosan nanoparticles as a promising pulmonary carrier against Staphylococcus aureus. Drug Deliv Transl Res. 2020;10(5):1403–17.
160. Il Kim M, Park CY, Seo JM, et al. In situ biosynthesis of a metal nanoparticle encapsulated in alginate gel for Imageable drug-delivery system. ACS Appl Mater Interfaces. 2021;13(31):36697–708.
161. Ezenwoke OA, Ezenwoke A, Eluyela F, Olusanmi O. A bibliometric study of accounting information systems research from 1975–2017. Asian J Sci Res. 2019;12(2):167–78.
162. Brian HG, Caldwell DM, Chaimani A, et al. The PRISMA extension statement for reporting of systematic reviews incorporating network meta-analyses of health care interventions: checklist and explanations. Ann Intern Med. 2015;162(11):777–84.
163. Moral-Muñoz JA, Herrera-Viedma E, Santisteban-Espejo A, Cobo MJ. Software tools for conducting bibliometric analysis in science: an up-to-date review. Profesional De La Informacion. 2020;29:4.
164. Cisneros L, Ibanescu M, Keen C, Lobato-Calleros O, Niebla-Zatarain J. Bibliometric study of family business succession between 1939 and 2017: mapping and analyzing authors' networks. Scientometrics. 2018;117(2):919–51.
165. Zupic I, Čater T. Bibliometric methods in management and organization. Organ Res Methods. 2015;18(3):429–72.
166. Combs JG, Ketchen J, David J, Crook TR, Roth PL. Assessing cumulative evidence within 'macro' research: why meta-analysis should be preferred over vote counting. J Manag Stud. 2011;48(1):178–97.

Niosomes in Tuberculosis

Geeta Patel, Bhupendra Prajapati, and Yashwant Pathak

Abstract Tuberculosis is a significant public health issue prevailing globally. Despite the progress in science and technology, TB is still an ongoing challenge. The discovery and development of novel molecules with potential antimicrobial activity against *Mycobacterium tuberculosis* and delivering them via appropriate drug delivery system is a promising tool in treating TB effectively. The applications of nanocarriers for the delivery of drugs to target organs and the modification of drug disposition have been thoroughly studied and investigated. Due to some demerits of liposomes, niosomes paid more attention as a vesicular system for entrapment of drugs, especially anti-TB drugs. These novel vesicles offer reductions in dosing frequency, increasing therapeutic effect in lungs by more accumulation, controlled and prolonged release of drugs. Adequate experimental evidence and data revealed the fact that nanomedicines for anti-TB drugs are significantly effective way to go compared to available conventional chemotherapy practices of TB. Currently, scientific literature and data for niosomes bring attention about its potentialities for development of anti-TB formulations. Furthermore, in-depth interest is required in development of newer and specific surfactants to produce specialized niosomes for drug targeting. Currently, improvement has made upon time to reduce burden of TB worldwide. Still, it requires deep understanding, education and awareness, thorough research, more financial investment to build up necessary infrastructure, improved standard of living to maintain the gains and attain advancement in the objective of TB-free world.

Keywords *Mycobacterium tuberculosis* · Nanocarriers · Niosomes · Conventional therapy · Drug resistance · Nanomedicines · TB-free world

G. Patel (✉) · B. Prajapati
Shree S K Patel College of Pharmaceutical Education and Research, Ganpat University, Mehsana, Gujarat, India
e-mail: geeta.patel@ganpatuniversity.ac.in

Y. Pathak
Taneja College of Pharmacy, University of South Florida, Tampa, FL, USA

Faculty of Pharmacy, Airlangga University, Surabaya, Indonesia

R. Shegokar, Y. Pathak (eds.), *Tubercular Drug Delivery Systems*,
https://doi.org/10.1007/978-3-031-14100-3_12

1 Introduction

Tuberculosis (TB) is becoming a worldwide problem due to *M. tuberculosis*, mainly affecting lungs [1]. The increased frequency of TB among the poor is exacerbated by war, poverty, homelessness, and a lack of medical treatment [2]. Because tuberculosis is easily transferred from person to person, a rise in TB in any section of the population puts everyone at risk. This is evident by World Health Organization Report on TB 2021 as TB is 13th prime disease responsible for deaths and second highest infectious disease after COVID-19. Without proper treatment, 45% of HIV-negative people with TB on average and nearly all HIV-positive people with TB will die [3]. This necessitates the implementation and maintenance of important public health measures such as screening, vaccination (if necessary), and treatment. A lack of public health measures will lead to increase the number of cases. The inadequate treatment promotes the spread of tuberculosis-resistant strains [4].

The incidences of multidrug-resistant tuberculosis are in an alarming state [5]. TB is a leading and difficult disease to be solved on urgent basis. The frequent and inadequate administration of antibiotic drugs results in the development of the disease that is not easily curable. Threatening bacteria and viruses resistant to various antibiotics are increasing day by day from last decades evident by pandemic effects of COVID-19. Exposure and disturbance of resistant strains have been observed for long [6]. Inappropriate treatment of antibiotics results in the increasing intensity of the disease. The moderate antibiotic susceptibility of the bacteria of tuberculosis results in treatment failure [7].

Still, 10 million people developed TB in 2020. The large worldwide drop in the number of cases decreased to 5.8 million from 7.1 million in 2020 compared to 2019. Moreover, TB still remains main source of morbidity and mortality, and goal of TB-free world is far from achieved. This indicates still huge investment is required for in-depth research in TB to eradicate it by driving newer technologies and rapid innovations to reach targets. In spite of huge funding per year for prevention and treatment of TB, the number of cases in 2020 was somewhat high compared to 2019 as less number of people provided TB treatment. This disruption was probably due to COVID-19. As per target in 2022, 40 million will be provided with TB preventive treatment (Fig. 1).

TB is transmitted by air and spreads when people infected with TB throw out bacteria into the air through coughing; susceptibility, infectiousness, environment and exposure are the critical parameters for determining transmission probability of *M. tuberculosis.* No environmental reservoir is known for pathogen of TB, 'humans are its only known reservoir' [8]. Small particles containing *M. tuberculosis* expelled in the air can remain suspended in the air for several hours, responsible for infection [9].

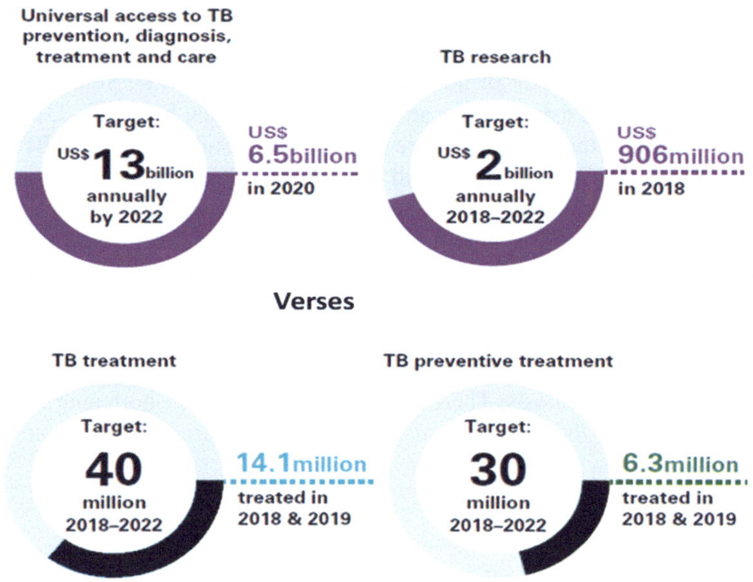

Fig. 1 Targets for increasing funding for research versus number of people provided with preventive treatment by 2022

2 Conventional Therapy Versus Future Need

Various conventional antitubercular drugs such as first-line oral anti-TB drugs (Isoniazid, Pyrazinamide, Ethambutol, Rifampicin, Streptomycin) injectable drugs (Kanamycin, Amikacin, Capreomycin, Vancomycin), newer Fluoroquinolones (Levofloxacin, Ciprofloxacin, Ofloxacin, Moxifloxacin, Gatifloxacin), oral second-line anti-TB drugs (Thiacetazone, P-amino salicylic acid, Ethionamide, Cycloserine) and other newer drugs with limited data such as Bedaquiline, Delamanid, Linezolid, Imipenem and certain other drugs with unclear action and function are available in the treatment of TB. Adverse effects and poor patient compliance are major drawbacks associated with conventional antitubercular therapy. Furthermore, development of drug-resistant strains against major antitubercular leads to failure in the treatment of TB [10, 11]. The development of effective drug combinations is always required in all the regimens to avoid the development of the drug-resistant TB.

Multidrug-resistant TB and extensively drug-resistant TB are the driving reasons of spreading TB. The development of novel effective drugs is necessary to get rid of TB. Drug resistance is also not solved by newly developed drugs like bedaquiline and delamanid [12–14]. Some of the new promising drugs are under phase II and III clinical trials. Moreover, more investigation and detail research is required to identify and discover potential new drugs and their combination to effectively treat TB [15]. An efficacious drug regimen should be investigated thoroughly. Also, new treatment strategies with newer approaches are strongly required to combat TB [16,

17]. Numerous antibodies having distinctive properties have been discovered to control drug resistance and reduce dosage regimens [18]. Many researchers are engaged in the development of novel anti-TB drugs worldwide. Still, efficacious drugs active against resistant strains are needed on an urgent basis [19–23].

The formation of the Global Alliance for TB Drug Development (GATB) altered the scenario on TB. The GATB will fill spaces in the R&D by collaborating with groups like university institutions, government and non-governmental organizations, the pharmaceutical sector and research firms on a contract basis. It will also encourage medication development for tuberculosis by providing a framework for bringing together the many aspects of the process [24].

Any new medicine for tuberculosis must be superior to the present treatment. The GATB has set numerous objectives for the development of novel medicines in the fight against tuberculosis. These include lowering doses, shortening treatment duration, improving the treatment of MDR-TB and offering better treatment of latent TB [25].

3 Current Challenges/Knowledge Gap

In spite of the various ventures to control and standardize TB prevention strategies, wide variation is observed globally in the selection and implementation of control strategies among different countries. These variations are mainly availability of resources, prevalence of TB, incidence of TB, rate of HIV co-infection, priorities of TB control programs, controversy in TB treatment, volume of migrants, TB diagnosis among immigrants, etc. [26].

The other main challenges in controlling TB in underdeveloped and poor nations are mainly poor healthcare systems in rural areas, private health sectors without regulation, unreasonable use of anti-TB drugs, prevalence of HIV infection, poverty; lesser involvement of government, corrupt government, negligence towards TB by rich people [27]. Use of unpasteurized milk and dairy products is one more likely source of TB for humans [28, 29].

Development and expansion of powerful National Tuberculosis Programmes (NTPs) to defeat the social and economic limitations of the nations are needed. The government of the individual nation should improve the diagnostic and therapeutic facilities of the nation. Education among healthcare personal, improving awareness for lifestyle and cough etiquettes in the community are the key factors to eradicate TB [30]. 'Indeed, there is a need in compiling viewpoints and beliefs about TB by various stakeholders working in the healthcare field globally especially related to status and challenges of TB'. Moreover, successful drug delivery systems and potent strategies for the development and administration of novel TB vaccines should be considered as prime current needs.

4 Novel Trends for the Treatment of TB Using Nanomedicines

Earlier to the discovery of nanomedicines, several traditional techniques and methods were utilized for the diagnosis of TB (Fig. 2). Several newer techniques and nanomedicines have been investigated and reported for the treatment of TB. Numerous delivery systems are used to target anti-TB drugs at a specific site. In spite of currently available effective drugs and regimens, urgent focus should emphasize in order to target anti TB drugs with correct therapeutic effects. For this, nanotechnology and nanocarriers are promising for the development of an effective drug delivery system for TB treatment [18].

Recently, various nanocarriers have been developed to entrap antitubercular drugs to deliver them to targeted sites effectively [31]. These nanocarriers are designed to release the drug in a controlled pattern to a specific target, increase the bioavailability of drugs, reduce dosing frequency, improve patient compliance and solve the issue associated with nonadherence of therapy. This leads to the success of therapy of TB [32, 33]. Various nano-sized carriers like liposomes, niosomes, nanoparticles, liquid crystals, micelles, nanoemulsions, dendrimers, nanospheres, nanocapsules, carbon nanotubes and nanoconjugates have been reported to target drugs at specific sites in order to achieve effective control over TB. However, high concentrations of polymers, low drug loading, costly formulations, use of organic solvents are several associated demerits of novel systems [34–36].

Certain newer implants such as microparticulate and carrier-based drug delivery systems to deliver anti-TB drugs, decrease dosing frequency and improve patient responses have been reported. Among these nanocarriers, niosomes can be considered as one of the best carriers to deliver anti-TB drugs at target site [37].

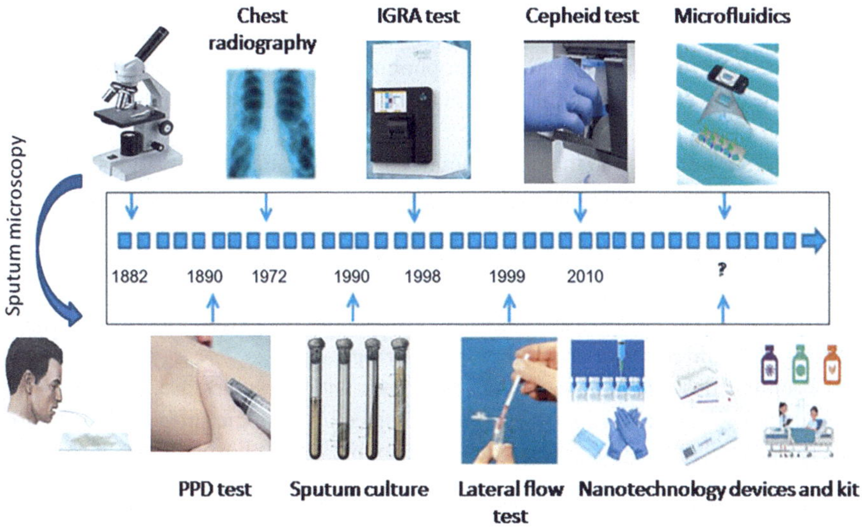

Fig. 2 Conventional methods and nanotechnology used for diagnosis of tuberculosis

Anti-TB drug (ATD) formulations containing entrapped isoniazid using polymers using polymethyl methacrylate, polyvinyl chloride, and carbomer have been reported [38]. In the early 1990s, the release kinetics of encapsulated isoniazid was studied using Eudragit RS 100 [39]. After that, significant advancement has been reported to incorporate ATDs in nanoformulation for two decades [40–44].

Enough experimental data revealed that nanomedicines containing ATD is significantly better way to treat and control TB compared to conventional therapy and practices. The analysis of recent data showed that there are still unfulfilled milestones in this area that obstruct clinical trials for further development of nanomedicines entrapping antitubercular. Still, several considerable challenges should be analysed and addressed for the routes of administration of ATD during nanomedicines development as oral treatment would be more comfortable and patient-friendly. An injectable formulation requires medical supervision and expertise and also has other challenges in pulmonary drug delivery. However, significant improvement in methods and techniques of manufacturing allows the formulation of ATD nanomedicines with ease, gaining interest in pharmaceutical industries in this domain [45].

5 Niosome as a Novel Vesicular System

Currently, in the era of nanomedicines and drug delivery, more attention has been given to vesicular systems, specially niosomes and liposomes [46–50]. Among reported nanovesicles, liposomes are spherical vesicles consisting of phospholipids bilayers, was first characterized as drug transporters in the early 1960s [51, 52]. Nonionic surfactant-based vesicles are investigated for applications for drug delivery or gene transfer. Nevertheless, characteristics and features of niosomes compared to liposomes are not documented well [53]. From the three decades, unique advantages of niosomes pay more observation as potential nanocarriers [54]. Nonionic surfactants having tails (hydrophobic part) and heads (hydrophilic part) show properties of self-assembling, varieties of shapes such as micelles and lamellar bilayer [55].

Several studies reported that niosomes act in vivo identical to liposomes [56, 57]. Niosomes are more stable and slightly leakier than liposomes. In contrast to liposomes, niosomes can create certain structures such as proniosome, discome and aspasome to entrapped drug [58]. Liposomes are prone to oxidative degradation, but niosomes are more stable. Therefore, niosomes are an easier and inexpensive way of carrier systems. Furthermore, the size of niosomes decreases significantly upon freezing in liquid nitrogen and following thawing.

Dissimilarities in attributes among niosomes and liposomes are mainly due to formulation components as niosomes are produced by nonionic surfactants and cholesterol; however, liposomes are produced by double-chain phospholipids. Niosomes contain a lesser amount of cholesterol compared to liposomes. Consequently, niosomes have greater drug entrapment efficiency to that of liposomes. Moreover, liposomes are more costly than niosomes and require special handling and storage conditions. The purity of phospholipids varies as they are obtained from natural sources.

5.1 Structural Components of Niosomes

Niosomes consists of drugs, cholesterol or its derivatives, nonionic surfactants, hydration medium and sometimes ionic amphiphiles. Both the water-soluble and insoluble drugs can be encapsulated in niosomal form [59, 60]. The amphiphilic character of molecules is an unexplained requirement for them to form vesicular assemblies, and the type of enabling hydrophilic head groups is varied. Alkyl ether lipids are the most common nonionic molecules that form vesicles.

5.1.1 Nonionic Surfactant

The prime components of niosomes are nonionic surfactants. Niosomes entrapment efficiency of drugs is mainly affected by head and tail of nonionic surfactants. The structure of surfactant and its HLB value plays a crucial role in the stability and size of niosomes, respectively. Mainly HLB value of nonionic surfactants ranging from 14 to 17 is not selected for the formulation of vesicles as it forms a bigger size of niosomes [61, 62].

Alkyl ethers are most stable and better vesicle forming nonionic surfactants and are able to entrap macromolecules like peptides and proteins [63, 64]. Polyoxyethylene 4 Lauryl Ether (Brij 30-HLB value-9.7) forms large unilamellar vesicles [65]. Cetyl derivatives of polyoxyethylene (Polyoxyethylene Cetyl Ether (Brij 52, 54, 58)) can be used for vesicle formation. Polyoxyethylene ether derivatives like Polyoxyethylene Stearyl Ethers (Brij 72 and Brij 76) form vesicles especially Brij 72 having HLB 4.9 produce multilamellar vesicles [66].

Sorbitan fatty acid esters mainly span [20, 40, 60, 65, 79, 84] are used to produce vesicles. Vesicles forming from higher molecular weight spans are most stable to osmotic grades [67]. Polyoxyethylene sorbitan fatty acid esters like Tween [20, 40, 60, 65, 79, 84] are used for niosomes formation [68]. Block copolymers like Pluronic L64 and Pluronic p105 are hydrophilic nonionic surfactants used for niosomes formation [69]. Fatty alcohols and fatty acids are also used to form 'Ufasomes' vesicles. Newer classes of surfactants like Bola surfactants and Gemini are utilized to formulate novel and specialized noisomes. Tyloxapol has been used to formulate niosomes to load antitubercular drugs [70].

5.1.2 Cholesterol

Cholesterol is used to provide mechanical rigidity, orientational order and stabilize membrane of vesicles. This results in formation of less leaky niosomes having increased entrapment efficiency [71]. Cholesterol is mixed with bilayer membranes to control the structure and flexibility of membrane as cholesterol alone cannot form a bilayer structure.

5.1.3 Charge-Inducing Molecule

Several charge-inducing molecules are incorporated in the formulation of niosomes to improve the stability of niosomes by electrostatic repulsion and prevent aggregation and coalescence. Negative charge inducers and positive charge inducers develop surface charges on vesicles and enhance their stability.

5.1.4 Hydration Medium

Phosphate buffer is mainly used as a hydration medium for the preparation of niosomes. Buffers of different pH values 5 and 7 are used according to solubility of entrapped drugs like pH 5 phosphate buffer is used for niosomes containing ascorbic acid [72], and pH 7 is used for niosomes containing aceclofenac [69].

5.2 *Factors Influencing Niosomes' Formulation*

The production of niosomes mainly relies on proper knowledge and a detailed understanding of the various properties of formulation ingredients. Various factors affecting the formation of niosomes are discussed as follows:

5.2.1 Type of Surfactants

The selection of surfactants is greatly affected by formation of niosomes. Nonionic surfactants are widely used for production of niosomes due to its versatility. Wide varieties of nonionic surfactants, their derivatives with examples and selection of different nonionic surfactants along with their HLB values for the formation of unilamellar and multilamellar vesicles are discussed in structural components of niosomes earlier in this chapter.

5.2.2 Thermodynamic Feature

Thermodynamic stability of niosomes can be achieved by selecting appropriate concentration of surfactants and formulation additives. Different kinds of energies like mechanical energy, chemical potential excess energy and surface energy are required for the development of niosomes. Compared to submicron size niosomes, particles with size ranging from 1 to 10 mm are found more stable [73].

5.2.3 Hydrophilic–Lipophilic Balance

HLB value is a critical parameter affecting the size and entrapment efficiency of niosomes. Surfactants having HLB values ranging from 4 to 8 are more suitable for the formation of vesicles. Surfactants having HLB values greater than 6 are rarely used as they are not able to form vesicles properly; hence, cholesterol is an essential component for the formation of niosomes [74].

5.2.4 Geometric Features of Amphiphilic Molecule

The morphology of the niosomes is generally anticipated by geometric features of the surfactant used for the preparation. According to the critical packing parameter (CPP) value, the shape and size of the equilibrium aggregate will form spherical micelles (CPP $\leq$ 0.33) to cylindrical micelles (1/3 $\leq$ CPP $\leq$ 0.5), bilayers (0.5 $\leq$ CPP $\leq$ 1), or inverse micelles (CPP > 1) [75].

5.2.5 Gel–Liquid Transition Temperature (Tc)

Phase transition temperature (*T*c) greatly affects entrapment efficiency, membrane permeability, bilayer rigidity and stability of niosomes. *Tc* and alkyl chain length of nonionic surfactants are mutual parameters that affects niosomes formation. It has been reported that niosomes prepared from surfactants with less *Tc* value produced more flexible vesicles [76]. More *Tc* values improve encapsulation efficiency of niosomes. The hydration medium's temperature should be more than the *T*c of the system as it will directly affect the formation of niosomes and the development of changes in a bilayer structure.

5.2.6 Additive Agents

Additives are added in niosomes formation to alter the biodistribution, improve physical stability and permeability of niosomes. The addition of cholesterol affects the structural properties of vesicles. The amount of cholesterol required depends on the HLB value of the surfactant. The other common additives are ionic compounds

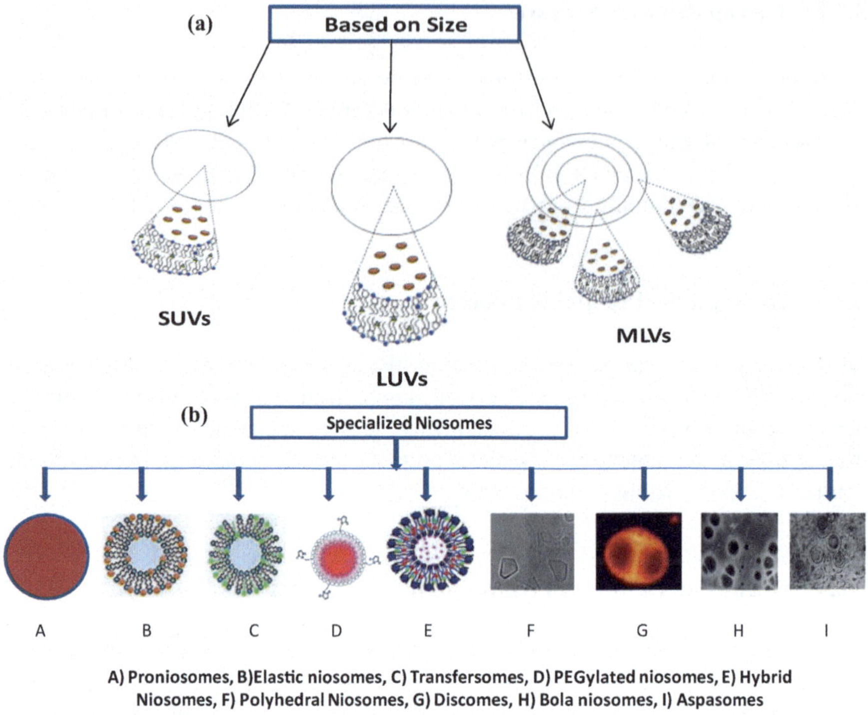

Fig. 3 Different types of niosomes: (**a**) based on size and (**b**) specialized niosomes: (A) proniosomes, (B) elastic niosomes, (C) transfersomes, (D) PEGylated niosomes, (E) hybrid niosomes, (F) polyhedral niosomes, (G) discomes, (H) bola niosomes, (I) aspasomes

such as dicetyl phosphate, stearyl amine, or stearyl pyridinium chloride are utilized in the formulation of niosomes to increase its stability [77].

5.3 Classification of Niosomes

Different types of niosomes are reported as follows and shown in Fig. 3:

5.3.1 Based Upon the Size and Lamella

- Small unilamellar vesicles (SUV – 0.025 – 0.05 μm)
- Large unilamellar vesicles (LUV – >0.10 μm)
- Multilamellar vesicles (MLV – >0.05 μm)

5.3.2 Specialized Niosomes

Proniosomes: These are prepared using water-soluble carriers by applying a thin coat of nonionic surfactant. The dry powder form of niosomes has certain merits compared to simple niosomes like better stability, less drug leakage and aggregates formation [78].

Elastic Niosomes: These are flexible kinds of niosomes; the formulation ingredients of these niosomes are surfactants, cholesterol, water and ethanol. Elastic niosomes are mainly used in transdermal/topical formulations [79, 80].

Discomes: These are large disc-like structures mainly formulated by hexadecyl diglycerol ether, cholesterol and dicetyl phosphate [81].

Bola Niosomes: Bola surfactants are used to formulate bola niosomes. This specific kind of surfactant was observed in the archaebacteria's membrane in 1980s. This surfactant having strong assembly-forming ability was reported by Zakharova in 2010 [82].

Transfersomes: These are newer deformable types of vesicular carrier systems majorly made up of phospholipids with self-assembly into a bilayer of lipid in an aqueous media and close to producing vesicle. Wide varieties of drugs can incorporate and entrap transfersomes as they contain both hydrophobic and hydrophilic functional groups [83].

Aspasomes: Ascorbyl palmitate forms vesicles with cholesterol and dicetyl phosphate called negatively charged lipid. Thin film hydration and sonication methods are used for preparation of aspasomes. These types of niosomes have been studied for drugs via transdermal delivery and offer greater permeation [84].

PEGylated Niosomes: PEGylated niosomes is a favourable vesicular system for anticancer compound having higher encapsulation capacity, good colloidal stability, more cellular uptake, greater bioavailability and chemical stability [85, 86].

Hybrid Niosomes: These niosomes are favourable for targeted drug delivery, co-delivery and multifunctional biomedicine applications. Furthermore, it provides increase therapy outcomes offering more drug entrapment efficiency [87, 88].

Polyhedral Niosomes: They are made by hexadecyl diglycerol ether replaced with one of the nonionic surfactants and polyoxyethylene24 cholesteryl ether in absence of cholesterol. Hydrophilic drugs can easily entrap unconventional structures of polyhedral niosomes [89].

5.4 Niosomes Preparation Methods

Various well-developed and studied methods and techniques are used for the formulation of niosomes. Many of them are briefly discussed below:

In the thin film hydration (hand shaking) method, dissolved surfactants and cholesterol formed thin film in a round-bottom flask upon vaporization of organic solvents. Afterward scrapping of film and allowing swelling it in aqueous medium above *Tc* of the surfactant for appropriate time interval by providing mild agitation leads to the formation of multilamellar vesicles followed by unilamellar vesicles production by suitable treatment [90–92].

In the Ether injection method, surfactant and drug are solubilized in diethyl ether. The solution is introduced to an aqueous phase at a low rate, then heated exceeding the boiling point of the solvent. LUVs are formed by this method, and their size is reduced further according to the requirement. During evaporation of the organic phase, single-layered vesicles formed [93, 94].

Sonication is one of the older methods for the fabrication of niosomes. The desired niosomes are obtained by sonicating the mixture at specific frequency, temperature and time. This method can be used to reduce the diameter of niosomes [95, 96].

The submerged jet principle is used in the microfluidization method. In this technique streams of drug and surfactant interact at ultra-high velocity at a 100 ml/min rate in the chamber. The niosomal vesicles are formed bypassing the solution from a cooling loop to remove heat generated by microfluidization [97].

In multiple membrane extrusion methods, thin film of the lipidic mixture is hydrated with water or phosphate buffer saline containing water-soluble drug and obtained mixture is extruded by membrane filter to obtain niosomes of required size [98–100].

In reverse phase evaporation method, excipients for niosomes are dissolved in ether and chloroform and incorporated to drug containing aqueous phase. These immiscible phases are then mixed followed by removal of organic solvents at lower pressure leads to formation of niosomes [101, 102].

Niosomes can be prepared using the bubble method, which is a single-step and solvent-free process. Buffer solution containing surfactants and cholesterol maintained at 70 °C is homogenized for 15 s by high shear homogenizer; afterward purging of nitrogen gas from the solution results in the formation of large unilamellar vesicles [103, 104].

By emulsion method, O/W emulsion is formulated using surfactant, cholesterol and drug in organic solvent. Later on, evaporation of the organic solvent forms niosomes dispersed in the aqueous media [105].

In transmembrane pH gradient, drug uptake method surfactant and cholesterol in solution form are evaporated under decreased pressure to form a film. A hydration medium containing citric acid solution pH 4 is selected to hydrate film by vortex mixing. The pH of the solution is maintained at 7.0–7.2 and heated at 60 °C to generate MLVs [106, 107].

The lipid injection method is an organic solvent-free technique used to produce niosomal suspension [108, 109].

Mixed micellar solution with enzymes can be used to prepare niosomes. Ester link breakage of polyethylene stearyl derivatives leads to generation of cholesterol and polyoxyethylene which can further produce niosomes [110].

Niosomes are also prepared from proniosomes. Coating of surfactant on the sorbitol or water soluble carrier forms proniosomes in dry form. Upon incorporation of water phase at higher temperature more than mean phase transition temperature forms niosomes from proniosomes [111, 112].

Supercritical reverse-phase evaporation is solvent-free technique for niosomes production. Niosomes can be easily scaled up to produce a large number of niosomes between 100 and 500 nm [113, 114].

5.5 Characterization of Niosomes

Evaluation of niosomes is required to control the quality of the formulations. Various parameters such as size and shape [115–117], morphology, bilayer formation [118], zeta potential [119, 120], membrane rigidity [121], entrapment efficiency [122], stability, in vitro drug release [123, 124] and other important parameters with details of instruments and features used to evaluate niosomes are described in Table 1.

5.6 Salient Features of Niosomes

Niosomes offers certain advantages such as they can entrap wide varieties of drugs and also increases steadiness of drugs. Niosomes are osmotically active and stable. They are biodegradable, biocompatible and non-immunogenic. Unacceptable solvents are not used for production of niosomes. No special handling and storage conditions are required as niosome has chemical stability due to its structural composition. Various properties, for instance shape, fluidity and size of niosomes, are simply controlled by altering its structural components and manufacturing methods. The high stability exhibited by long shelf life of niosomes permits the delivery of drugs at specific targets in a controlled way. Niosomes can be easily administered by different routes using different dosage forms. It improves the bioavailability of poorly soluble drugs administered orally and increases the permeability of drugs by topical application. They are a better alternative compared to oily formulations in terms of patient compliance [125–130].

On the negative side, niosomes show some stability issue during storage such as aggregation, fusion, drug leakage and hydrolysis of encapsulated drugs which limits the shelf life of the dispersion. Furthermore, sterilization of niosomes formulation is a challenging endeavour. Major sterilization processes are inadequate for sterilization of niosomes. Hence, more investigations may require in this area as future direction.

5.7 Purification of Niosomes

Generally, purification of niosomes is required to separate unentrapped drugs in vesicles. Some researchers used the dialysis method for the purification of vesicle dispersion and further removal of unentrapped materials [131, 132]. Gel-filtration

Table 1 Characterization parameters of niosomes

Characterization parameters	Description	Instrumentation	Specificity/additional features
Vesicle size	To check physical properties and stability of formulation	Photon correlation spectroscopy (dynamic light scattering) Laser diffraction Polydispersity index Electronic microscopy	Rapid and non-destructive method (measure size range 3–3000 nm) DLS not give shape of niosomes Micron size niosomes Distribution of niosome size Measure size of niosomes
Bilayer formation	To determine number of lamellae Bilayer's packing structure	Small angle X-ray scattering Nuclear magnetic resonance spectroscopy Atomic force microscopy Fluorescent polarization	Measure niosomal bilayer's thickness Microviscosity of niosomal membranes
Morphology		Transmission electronic microscopy Negative-staining transmission Electronic microscopy Freeze-fracture transmission electronic microscopy Scanning electron microscopy Atomic force microscopy	Liquid sample Solid sample Bilayer thickness of niosomes
Zeta potential (vesicle charge)	Physical stability of niosomes	Laser Doppler anemometry	Indication of the degree of electrostatic repulsion
Stability	For physical, chemical and biological stability	Measure by particle size and zeta potential UV irradiation and fluorescent light	Determined at 4 °C, 25 °C, 40 °C at 75% Use to study photodegradation of niosomes
Entrapment efficiency (EE)	Actual drug enclosement in vesicles	Spectrophotometry Gel electrophoresis UV densitometry Fluorescence marker (calcein) Separation of unentrapped drug	Number of marker molecules entrapped By dialysis, centrifugation, or gel filtration
In vitro drug release	Determination of percentage drug released	Dialysis membrane Franz diffusion cells	Release behaviour of niosomes
Spreadability	For topical niosome	Spreadability test apparatus, texture analyser	For uniform application on targeted site to ensure the suitable dose

(continued)

Table 1 (continued)

Characterization parameters	Description	Instrumentation	Specificity/additional features
Rigidity	Improves bilayer stability	AFM	
Separation/ purification		Ultracentrifugation Gel filtration	Separate unentrapped drug Sephadex-G-50 column
Skin/corneal permeation		Confocal laser scanning microscopy (CLSM)	

chromatography is a popular and versatile technique that allows the effective separation of free molecules in high yield. To purify niosomes, centrifugation and ultracentrifugation separation were also reported [133, 134].

6 Niosomes in TB

Till time, the number of anti-tubercular drugs has been entrapped in niosomal form successfully to deliver at targeted sites for the effective management of TB. Encapsulation of drugs showed more cellular uptake compared to plain drugs. Moreover, niosomal anti-TB drugs are able to deliver drug to where *M. tuberculosis* hide. Various researchers have investigated niosomes containing anti-TB drugs and achieved good cellular uptake of drugs, encapsulation efficiency and satisfactory results of various characterization parameters of niosomes. Many researchers tried to achieve the reduction of dose-related toxicity, drug targeting over *M. tuberculosis*, increased antimicrobial activities, prolonged drug release, etc. Moreover, encapsulated niosomal drugs showed improved bioavailability and stability, better pharmacokinetic profiles, and good localization in the lungs. Some of the outcomes and results are summarized in Table 2.

7 Future Perspective of TB Treatment and Niosomes in TB

As a bacterial disease, TB has taken the lives of about a billion patients for two decades. As per recent observation, there is a declining phase of TB globally, excepting nations having a greater number of HIV patients. Some of the critical parameters like proper diagnosis, rational use of drugs, short treatment time, and effective vaccine can play a crucial role to eradicate TB in near future. It is quite difficult to predict future of TB as various factors related to TB are beyond the control of researchers, pharmaceutical industries and clinicians. Currently, academy and industry collaborative projects and varying international bodies come ahead in TB research and development to get rid of TB. Still, rapid, accurate and cost-effective

Table 2 Niosomal anti-TB drugs investigated for different purposes

Anti-TB drugs	Nonionic surfactant	Preparation method	Particle size (nm)	Encapsulat-ion (%)	In vitro release	Purpose/applications/outcome/results	References
Isoniazid	Span 60, 20	Reverse phase evaporation	–	–	90%	61.8% cellular uptake of niosomes	[135]
	Span 60	Ethanol injection	2.28 ± 0.008 μm	80.23%	90%	Prolong release up to 48 h	[136]
	Tyloxapol	Sonication	150 nm	98.89 ± 0.2	>90%	Reduction in dose-related drug toxicity	[137]
Isoniazidisatin-INH hybrid	–	–	500–600 nm	74.2%	In vivo study	Fourfold increase in anti-mycobacterial activity by niosomes against plain drug	[138]
Rifampicin	Span-85	Handshaking and ether injection	8-15 nm	3.44 ± 0.0008 mol g	90% at 36 h	Effective localization of drug achieved by controlling size of niosomes	[139]
	Span 60 Triton X 100	Modified lipid layer hydration	11.8 μm mean diameter	34.2%	94% at 10 h	Improved bioavailability of drug was observed in niosomal form	[140]
	Span and tween	Sonication	7–8 μm	68.62 ± 0.24	In vivo study	AUC-51353.6 ± 2.3 μg/h/ml, MRT-12.30 ± 2.6 h, Cl-0.09736 ± 1.9(ml/h), $t_{1/2}$–8.52 ± 0.9 h indicated formulation having good accumulation in lungs	[141]
Rifampicin + Ofloxacin	Span 20,60,80	Ether injection	100–300 nm	81.76%	70% at 15 days	Entrapped drugs in niosomes showed improved bactericidal activity	[142]
Rifampicin + Gatifloxacin	Span and tween	Lipid hydration technique	100–300 nm	73%-RIF 70%-GAT	98.98 97.74	Improved bactericidal activity against the tubercle bacilli	[143]
Pyrazinamide	Span 60,85	Vortex dispersion method	255–701 nm	11.2 ± 2.9	83.3 ± 5.8% retained	Prolonged drug release up to 96 h Macrophage targeting	[144]

Anti-TB drugs	Nonionic surfactant	Preparation method	Particle size (nm)	Encapsulat-ion (%)	In vitro release	Purpose/applications/outcome/results	References
Isoniazid + Rifampicin + Pyrazinamide	Triton X 100, span 80	Sonication	500 nm	99.49 ± 0.2	4% at 5 h.	Improved drug entrapment and stability	[145]
Ethambutol Hcl	Span 60, 85	Thin-film hydration	324 nm	25.81_1.73	<20% at 20 h	Higher efficacy and safety of niosomal ethambutol hydrochloride	[146]
	Brij 72	Reverse phase evaporation	200–300 nm	86.73	92.09%	To achieve sustained activity and local release	[147]
Ethionamide	Span 40, 60,80	Thin film hydration	124.4 nm	88.9%	94.89% at 24 h	Effective formulation for the treatment of MDR TB	[148]
Ethionamide + D-cycloserin	Tween 80, 20, span 20,60, 80	Ethanol injection	137.4 nm	>70%	96% of ETH at 72 h 97% D-CS at 48 h	Prepared for combating drug-resistant TB	[149]
Levofloxacin	Span 60	Thin film hydration followed by sonication	303.5 nm	94%	86.89%	Prolonged release and longer duration of action	[150]
Ciprofloxacin HCl	Span 60 and tween 60	Drug loading on preformed niosomes	7.15 ± 0.14 µm	77.9 ± 2.8	–	Increased antimicrobial effects of drugs in niosomes form	[151]
Gatifloxacin	Span 40, 60,80	Thin film hydration	3.3–4.1 µm	55.77 ± 1.03%	96.17% at 24 h	Better stability and improved pharmacokinetic profiles	[152]
Streptomycin sulphate	Span 60 Tween 60	Thin layer hydration	97.8 nm	86.7%,	66.4 ± 1.3% at 72 h	Development of nanostructures for treatment of current bacterial infections	[153]

diagnosis, development of newer and effective drugs and vaccines are needed on urgent basis for the treatment and prevention of TB.

However, numbers of downsides are associated with the current treatment of TB like ineffective anti-TB drug regimens and development of multidrug resistance. The longer period of the treatment proves poor patient compliance. Increased resources and funding for TB research for the identification of new targets, development of newer technologies and drug delivery systems can effectively manage the decline phase of TB in a long run.

Though predicting the future of nanotechnology for the treatment of infectious diseases like TB is difficult due to issues associated with treatment can be addressed by increasing more focus on the development of multifunctional nanocarriers to deliver one or more anti-TB drugs intracellularly to solve drug resistance.

Researchers have to focus on nanocarriers mainly vesicular systems for specific drug targeting to achieve better therapeutic response. Also, potential nanocarriers such as niosomes should be explored more for TB therapy. In comparison to other nanocarriers investigated for TB therapy, niosomes showed superior chemical stability and offer numerous advantages.

Moreover, niosomes should be investigated thoroughly before they could take place in applications in drug delivery for TB. There is a lot of scope to encapsulate drug for infectious diseases, cancer, inflammation, AIDS, HIV and other categories of drug in niosomal form as a promising drug carrier to achieve drug targeting. Hence, niosomes can be successfully used as all-round tools in drug delivery system in near future to treat TB. However, 'the niosomal technology is still at a premature stage and abundant work is still required to guide their future applications in different clinical fields'. It is also important to focus more for developing newer nonionic surfactants able to form novel and multifunctional niosomes suitable for preclinical studies and later on to clinical studies for the successful and effective treatment of TB.

8 Conclusion

Applications of nanotechnology and the usefulness of various nanocarriers can solve the drawbacks of currently used anti-TB drugs. Nanocarriers can improve the therapeutic efficacy of drugs. Smart, well-studied, cost-effective, easily accessible, well-tolerated, safe and efficacious approaches should be developed to curb TB. Still, combination therapy based on nanotechnology for TB is in its infancy. Development of rational and effective combination therapy to treat multi-resistant TB and latent TB with greater cure rate, lesser risk of associated adverse effects and scalability is required on an urgent basis. Furthermore, easy access of various tools to fight TB, development of appropriate and efficacious anti-TB formulation with its right applications would be important parameters as a future preventive plan of action to eradicate TB. By considering important properties of niosomes, it would be the potential carrier to deliver antitubercular drugs at specific target sites. In spite of this, younger system niosomes still offer unique advantages to deliver drugs. To sum up, still more research requires understanding the true opportunities of these vesicles to be effective in delivering anti-TB drugs. Moreover, emphasis should be put on continued education, awareness programs and allocating high funding for research for prevention and control of TB.

References

1. Hayward S, Harding RM, McShane H, Tanner R. Factors influencing the higher incidence of tuberculosis among migrants and ethnic minorities in the UK [version 2; referees: 2 approved]. F1000Research. 2018;7:461. https://doi.org/10.12688/f1000research.14476.2.
2. Story A, Murad S, Roberts W. Tuberculosis in London: the importance of homelessness, problem drug use and prison. Thorax. 2007;62(8):667–71.
3. Global tuberculosis report 2021, World Health Organization. https://www.who.int/publications/i/item/9789240037021.
4. Shahiwala A, Misra A. Studies in topical application of niosomally entrapped nimesulide. J Pharm Pharmaceut Sci. 2002;5(93):220–5.
5. Seung KJ, Keshavjee S, Rich ML. Multidrug-resistant tuberculosis and extensively drug-resistant tuberculosis. Cold Spring Harb Perspect Med. 2015;5(9):a017863. https://doi.org/10.1101/cshperspect.a017863.
6. Fair RJ, Tor Y. Antibiotics and bacterial resistance in the 21st century. Perspect Medicin Chem. 2014;6:25–64. https://doi.org/10.4137/PMC.S14459.
7. Kim SJ. Drug-susceptibility testing in tuberculosis: methods and reliability of results. Eur Respir J. 2005;25:564–9. https://doi.org/10.1183/09031936.05.00111304.
8. Boudville DA, Joshi R, Rijkers GT. Migration and tuberculosis in Europe. J Clin Tuberc Other Mycobact Dis. 2020;18:100143.
9. Atkinson J, Chartier Y, Pessoa-Silva CL, et al. Natural ventilation for infection control in health-care settings. Geneva: World Health Organization; 2009. Annex C, Respiratory droplets. Available from https://www.ncbi.nlm.nih.gov/books/NBK143281.
10. Sotgiu G, Centis R, D'ambrosio L, Migliori GB. Tuberculosis treatment and drug regimens. Cold Spring Harb Perspect Med. 2015;5(5):a017822. https://doi.org/10.1101/cshperspect.a017822.
11. Kaur M, Garg T, Narang RK. A review of emerging trends in the treatment of tuberculosis. Artif Cells Nanomed Biotechnol. 2016;44(2):478–84. https://doi.org/10.3109/21691401.2014.962745.
12. Cui Y, Shen H, Wang F, et al. A long-term trend study of tuberculosis incidence in China, India and United States 1992-2017: a join point and age-period-cohort analysis. Int J Environ Res Public Health. 2020;17(9):3334. https://doi.org/10.3390/ijerph17093334.
13. Global Tuberculosis Report 2017: World Health Organization. Available from https://apps.who.int/iris/handle/10665/259366.
14. Comas I, Coscolla M, Luo T, et al. Out-of-Africa migration and Neolithic coexpansion of *Mycobacterium tuberculosis* with modern humans. Nat Genet. 2013;45:1176–82. https://doi.org/10.1038/ng.2744.
15. Pietersen E, Ignatius E, Streicher EM, et al. Long-term outcomes of patients with extensively drug-resistant tuberculosis in South Africa: a cohort study. Lancet. 2014;383:1230–9.
16. Dheda K, Gumbo T, Gandhi NR, et al. Global control of tuberculosis: from extensively drug-resistant to untreatable tuberculosis. Lancet Respir Med. 2014;2:321–38.
17. Sotgiu G, Centis R, D'ambrosio L, Migliori GB. Tuberculosis treatment and drug regimens. Spring Harb Perspect Med. 2015;5:a017822. https://doi.org/10.1101/cshperspect.a017822.

18. Nasiruddin M, Neyaz MK, Das S. Nanotechnology-based approach in tuberculosis treatment. Tuberc Res Treat. 2017:Article ID 4920209, 12 pages. https://doi.org/10.1155/2017/4920209.
19. Pavan FR, Poelhsitz GV, da Cunha LV, et al. In vitro and in vivo activities of ruthenium(II) phosphine/diimine/picolinate complexes (SCAR) against *Mycobacterium tuberculosis*. PLoS One. 2013;8(5):e64242. https://doi.org/10.1371/journal.pone.0064242.
20. Pavan FR, Poelhsitz GV, do Nascimento FB, et al. Ruthenium (II) phosphine/picolinate complexes as antimycobacterial agents. Eur J Med Chem. 2010;45(2):598–601. https://doi.org/10.1016/j.ejmech.2009.10.049.
21. Tomioka H, Namba K. Development of antituberculous drugs: current status and future prospects. Kekkaku. 2006;81(12):753–74.
22. do Nascimento FB, Poelhsitz GV, Pavan FR, et al. Synthesis, characterization, X-ray structure and *in vitro* antimycobacterial and antitumoral activities of Ru(II) phosphine/diimine complexes containing the "SpymMe(2)" ligand, SpymMe(2) =4,6-dimethyl-2-mercaptopyrimidine. J Inorg Biochem. 2008;102:1783.
23. Patricia Bento da S. Synthesis, characterization and investigation of the biological activities of copper (II) complexes containing bioactive molecules and nitrogenous ligands. 2012. p. 215 f. Thesis (Doctorate) – Universidade Estadual Paulista, Institute of Chemistry of Araraquara. Available at http://hdl.handle.net/11449/97936.
24. Kettler H. Valuing industry contributions to public-private partnerships for health product development, initiative on public-private partnerships for health. Global Forum for Health Research; 2003.
25. https://globalhealthprogress.org/collaboration/global-alliance-for-tb-drug-development-tb-alliance/.
26. Frothingham R, Stout JE, Hamilton CD. Current Issues in global tuberculosis control. Int J Infect Dis. 2005;9(6):297–311.
27. Sandhu GK. Tuberculosis: current situation, challenges and overview of its control programs in India. J Global Infect Dis. 2011;3(2):143–50.
28. Prasad H, Singhal A, Mishra A, et al. Bovine tuberculosis in India: potential basis for zoonosis. Tuberculosis. 2005;85:421–8.
29. Srivastava K, Chauhan DS, Gupta P, et al. Isolation of *Mycobacterium bovis* and *M. tuberculosis* from cattle of some farms in North India-possible relevance in human health. Indian J Med Res. 2008;128:26–31.
30. Griffiths C, Barne M, Saxena P, et al. Challenges of tuberculosis management in high and low prevalence countries in a mobile world. Prim Care Respir J. 2014;23:106–11. https://doi.org/10.4104/pcrj.2014.00019.
31. Dheda K, Migliori GB. The global rise of extensively drug-resistant tuberculosis: is the time to bring back sanatoria now overdue? Lancet. 2012;379:773–5.
32. Da Silva PB, de Freitas ES, Bernegossi J, Gonçalez ML, Sato MR, Leite CQ, Pavan FR, Chorilli M. Nanotechnology-based drug delivery systems for treatment of tuberculosis – a review. J Biomed Nanotechnol. 2016;12(2):241–60. https://doi.org/10.1166/jbn.2016.2149.
33. Kumar N, Kumar P, Kumar P, Kumar M, Kumar R. Nanotechnology: a focus on treatment of tuberculosis. Int J Drug Deliv. 2011;3:25.
34. Chan ED, Iseman MD. Current medical treatment for tuberculosis. BMJ. 2002;325(7375):1282–6. https://doi.org/10.1136/bmj.325.7375.1282.
35. Patil K, Bagade S, Bonde S, Sharma S, Saraogi G. Recent therapeutic approaches for the management of tuberculosis: challenges and opportunities. Biomed Pharmacother. 2018;99:735–45.
36. Gilani SJ, Ameeduzzafar JM, Shakil K, Imam SS. Nano-carriers for the treatment of tuberculosis. Recent Pat Antiinfect Drug Discov. 2017;12(2):95–106. https://doi.org/10.2174/1574891X12666170427120230.
37. Karim KM, Sattwa A, Mandal NB. Niosome: a future of targeted drug delivery systems. J Adv Pharm Technol Res. 2010;1(4):374–80.

38. Bulut-Oner F, Capan Y, Kas S, Oner L, Hincal AA. Sustained release isoniazid tablets. I. Formulation and in vitro evaluation. Farmaco. 1989;44(7–8):739–52.
39. Muhuri G, Pal TK. Computation of release kinetics of isoniazid microcapsules. Boll Chim Farm. 1991;130(5):169–71.
40. Barrow ELW, Winchester GA, Staas JK, Quenelle DC, Barrow WW. Use of microsphere technology for sustained and targeted delivery of rifampicin to *Mycobacterium tuberculosis*-infected mice. Antimicrob Agents Chemother. 1998;42(10):2682–9.
41. Dutt M, Khuller GK. Therapeutic efficacy of poly (dl-lactide-co-glycolide) encapsulated anti-tubercular drugs against *Mycobacterium tuberculosis* infection induced in mice. Antimicrob Agents Chemother. 2001;45(1):363–6.
42. Ul-Ain Q, Sharma S, Khuller GK. Chemotherapeutic potential of orally administered poly (lactide-co-glycolide) microparticles containing isoniazid, rifampicin and pyrazinamide against experimental tuberculosis. Antimicrob Agents Chemother. 2003;47(9):3005–7.
43. Bonifàcio BV, da Silva PB, dos Santos Ramos MA, et al. Nanotechnology based drug delivery systems and herbal medicines: a review. Int J Nanomedicine. 2014;9:1–15.
44. Gangadharam PR, Kailasam S, Srinivasan S, Wise DL. Experimental chemotherapy of tuberculosis using single dose treatment with isoniazid in biodegradable polymers. J Antimicrob Chemother. 1994;33(2):265–71.
45. Mubarak MM, Ahmad Z. Chapter 9 – nanotechnology-based approaches for tuberculosis treatment. Elsevier; 2020. p. 143–62. https://doi.org/10.1016/B978-0-12-819811-7.00009-6.
46. Mohamad EA, Fahmy HM. Niosomes and liposomes as promising carriers for dermal delivery of *Annona squamosa* extract. Braz J Pharm Sci. 2020;56 https://doi.org/10.1590/s2175-97902019000318096.
47. Choi MJ, Maibach HI. Liposomes and niosomes as topical drug delivery systems. Skin Pharmacol Appl Ski Physiol. 2005;18:209–19. https://doi.org/10.1159/000086666.
48. Mura S, Pirot F, Manconi M, Falson F, Fadda AM. Liposomes and niosomes as potential carriers for dermal delivery of minoxidil. J Drug Target. 2007;15(2):101–8. https://doi.org/10.1080/10611860600991993.
49. Nasr M, Mansour S, Mortada ND, Elshamy AA. Vesicular aceclofenac systems: a comparative study between liposomes and niosomes. J Microencapsul. 2008;25(7):499–512. https://doi.org/10.1080/02652040802055411.
50. Schreief H, Bouwstrab J. Liposomes and niosomes as topical drug carriers: dermal and transdermal drug delivery. J Control Release. 1994;30:l–15.
51. Akbarzadeh A, Rezaei-Sadabady R, Davaran S, et al. Liposome: classification, preparation, and applications. Nanoscale Res Lett. 2013;8(1):102. https://doi.org/10.1186/1556-276X-8-102.
52. Bozzuto G, Molinari A. Liposomes as nanomedical devices. Int J Nanomedicine. 2015;10:975–99. https://doi.org/10.2147/IJN.S68861.
53. Bartelds R, Nematollahi MH, Pols T, et al. Niosomes, an alternative for liposomal delivery. PLoS One. 2018;13(4):e0194179. https://doi.org/10.1371/journal.pone.0194179.
54. Muzzalupo R, Tavano L. Niosomal drug delivery for transdermal targeting: recent advances. Res Rep Transdermal Drug Deliv. 2015;4:23–33.
55. Uchegbu IF, Florence AT. Nonionic surfactant vesicles (niosomes): physical and pharmaceutical chemistry. Adv Colloid Interf Sci. 1995;58:1–55.
56. Florence AT, Baillie AJ. Nonionic surfactant vesicles- alternatives to liposomes. In: Prescott LF, Nimmo WS, editors. Novel drug delivery and its therapeutic application. New York: Wiley; 1989.
57. Thakur V, Arora S, Prashar B, Vishal P. Niosomes and liposomes-vesicular approach towards transdermal drug delivery. Int J Pharm Chem Sci. 2012;1:981–93.
58. Khoee S, Yaghoobian M. Chapter 6 – niosomes: a novel approach in modern drug delivery systems. In: Nanostructures for drug delivery. Elsevier Inc; 2017. p. 207–37.

59. Ritwiset A, Krongsuk S, Johns JR. Molecular structure and dynamical properties of niosome bilayers with and without cholesterol incorporation: a molecular dynamics simulation study. Appl Surf Sci. 2016;380:23–31.
60. Ge X, Wei M, He S, Yuan WE. Advances of non-ionic surfactant vesicles (niosomes) and their application in drug delivery. Pharmaceutics. 2019;11(2):55. https://doi.org/10.3390/pharmaceutics11020055.
61. Bachhav AA. Proniosome: a novel non-ionic Provesicules as potential drug. Carrier. 2016;10(3):1–10.
62. Kumar GP, Rajeshwarrao P. Nonionic surfactant vesicular systems for effective drug delivery – an overview. Acta Pharm Sin B. 2011;1(4):208–19.
63. Lavergne A, Zhu Y, Pizzino A, Molinier V, Aubry JM. Synthesis and foaming properties of new anionic surfactants based on a renewable building block: sodium dodecyl isosorbide sulfates. J Colloid Interface Sci. 2011;360(2):645–53.
64. Pardakhty A, Varshosaz J, Rouholamini A. In vitro study of polyoxyethylene alkyl ether niosomes for delivery of insulin. Int J Pharm. 2007;328(2):130–41.
65. Manosroi A, Wongtrakul P, Manosroi J, et al. Characterization of vesicles prepared with various non-ionic surfactants mixed with cholesterol. Colloids Surf B: Biointerfaces. 2003;30(1, 2):129–38.
66. Raymond CR, Paul JS, Sian CO. Handbook of pharmaceutical excipients. Washington, DC: American Pharmaceutical Association; 2006.
67. Oranuch Thanaketpaisarn. Niosome delivery systems in pharmaceutical applications. Isan J Pharm Sci. 2012;8(2):1226.
68. Akbari V, Abedi D, Pardakhty A, Sadeghi-Aliabadi H. Release studies on ciprofloxacin loaded non-ionic surfactant vesicles. Avicenna J Med Biotechnol. 2015;7(2):69–75.
69. Bodratti AM, Alexandridis P. Formulation of poloxamers for drug delivery. J Funct Biomater. 2018;9(1):11. https://doi.org/10.3390/jfb9010011.
70. Mehta SK, Jindal N. Formulation of Tyloxapol niosomes for encapsulation, stabilization and dissolution of anti-tubercular drugs. Colloids Surf B: Biointerfaces. 2013;101:34–441.
71. Gharbavi M, Amani J, Kheiri-Manjili H, Danafar H, Sharafi A. Niosome: a promising nanocarrier for natural drug delivery through blood-brain barrier. Adv Pharmacol Sci. 2018;2018:6847971. https://doi.org/10.1155/2018/6847971.
72. Varshosaz J, Taymouri S, Pardakhty A, Asadi-Shekaari M, Babaee A. Niosomes of ascorbic acid and α-tocopherol in the cerebral ischemia-reperfusion model in male rats. Biomed Res Int. 2014;2014:816103. https://doi.org/10.1155/2014/816103.
73. Moazeni E, Gilani K, Sotoudegan F, Pardakhty A, Najafabadi AR, Ghalandari R, Fazeli MR, Jamalifar H. Formulation and in vitro evaluation of ciprofloxacin containing niosomes for pulmonary delivery. J Microencapsul. 2010;27:618–27.
74. Yoshida H, Lehr CM, Kok W, et al. Niosomes for oral delivery of peptide drugs. J Control Release. 1992;21:145–53.
75. Israelachvili JN. Intermolecular and surface forces. 3rd ed. London: Academic; 2011. p. 536–42.
76. Alves FR, Loh W. Vesicles prepared with the complex salts dioctadecyl dimethyl ammonium polyacrylates. J Colloid Interface Sci. 2012;368:292–300.
77. Abdelbary G, El-gendy N. Niosome-encapsulated gentamicin for ophthalmic controlled delivery. AAPS Pharm Sci Technol. 2008;9(3):740–7.
78. Yuksel N, Bayindir ZS, Aksakal E, Ozcelikay AT. In situ niosome forming maltodextrin proniosomes of candesartan cilexetil: In vitro and in vivo evaluations. Int J Biol Macromol. 2016;82:453–63.
79. Jantrawuta MP, Manosroia J. Anti-inflammatory activity of gel containing novel elastic niosomes entrapped with diclofenac diethylammonium. Int J Pharm. 2008;360:156–63.
80. Chankhampan MC, Manosroi W, Manosroi J. Transdermal absorption enhancement of papain loaded in elastic niosomes incorporated in gel for scar treatment. Eur J Pharm Sci. 2013;48:474–83.

81. Abdelkader H, Wu Z, Al-Kassas R, Alany RG. Niosomes and discomes for ocular delivery of naltrexone hydrochloride: morphological, rheological, spreading properties and photo-protective effects. Int J Pharm. 2012;433:142–8.
82. Paolino D, Cosco D, Muzzalupo R, Trapasso E, Picci N, Fresta M. Innovative bola surfactant niosomes as topical delivery systems of 5-fluorouracil for the treatment of skin cancer. Int J Pharm. 2008;353:233–42.
83. Opatha SAT, Titapiwatanakun V, Chutoprapat R. Transfersomes: a promising nanoencapsulation technique for transdermal drug delivery. Pharmaceutics. 2020;12(9):855. https://doi.org/10.3390/pharmaceutics12090855.
84. Gopinath D, Ravi D, Rao BR, Apte SS, Renuka D, Rambhau D. Ascorbyl palmitate vesicles (Aspasomes): formation, characterization and applications. Int J Pharm. 2004;271:95–113.
85. He R-X, Ye X, Li R, et al. PEGylated niosomes-mediated drug delivery systems for Paeonol: preparation, pharmacokinetics studies and synergistic anti-tumor effects with 5-FU. J Liposome Res. 2017;27(2):161–70. https://doi.org/10.1080/08982104.2016.1191021.
86. Lin T, Fang Q, Peng D, et al. PEGylated non-ionic surfactant vesicles as drug delivery systems for Gambogenic acid. Drug Delivery. 2013;20:7277–84. https://doi.org/10.3109/10717544.2013.836618.
87. Maurer V, Altin S, Ag Seleci D, et al. In-vitro application of magnetic hybrid niosomes: targeted siRNA-delivery for enhanced breast cancer therapy. Pharmaceutics. 2021;13:394. https://doi.org/10.3390/pharmaceutics13030394.
88. Gharbavi M, Johari B, Mousazadeh N. Hybrid of niosomes and bio-synthesized selenium nanoparticles as a novel approach in drug delivery for cancer treatment. Mol Biol Rep. 2020;47:6517–29. https://doi.org/10.1007/s11033-020-05704-z.
89. Arunothayanun P, Uchegbu IF, Craig DQ, Turton JA, Florence AT. In vitro/in vivo characterisation of polyhedral niosomes. Int J Pharm. 1999;83(1):57–61. https://doi.org/10.1016/s0378-5173(99)00044-7.
90. Tavano L, Muzzalupo R, Picci N, Cindio BD. Co-encapsulation of antioxidants into niosomal carriers: gastrointestinal release studies for nutraceutical applications. Colloids Surf B: Biointerfaces. 2014;114:82–8.
91. Yeo LK, Chaw CS, Elkordy AA. The effects of hydration parameters and co-surfactants on methylene blue-loaded niosomes prepared by the thin film hydration method. Pharmaceuticals. 2019;12:46. https://doi.org/10.3390/ph12020046.
92. Ravalika V, Sailaja AK. Formulation and evaluation of etoricoxib niosomes by thin film hydration technique and ether injection method. Nano Biomed Eng. 2017;9(3):242–8. https://doi.org/10.5101/nbe.v9i3.p242-248.
93. Sailaja AK, Shreya M. Preparation and characterization of naproxen loaded niosomes by ether injection method. Nano Biomed Eng. 2018;10(2):174–80. https://doi.org/10.5101/nbe.v10i2.p174-180.
94. Rangasamy M, Ayyasamy B, Raju S, Gummadevelly S, Shaik S. Formulation and in vitro evaluation of niosome encapsulated Acyclovir. J Pharm Res. 2008;1(2):163–6.
95. Bansal S, Aggarwal G, Chandel P, Harikumar SL. Design and development of cefdinir niosomes for oral delivery. J Pharm Bioallied Sci. 2013;5(4):318–25.
96. Verma AK, Bindal M. A review on niosomes: an ultimate controlled and novel drug delivery carrier. Int J Nanoparticles. 2012;5(1):73–87.
97. Khandare JN, Madhavi G, Tamhankar BM. Niosomes novel drug delivery system. East Pharm. 1994;37:61–4.
98. Ag Seleci D, Maurer V, Stahl F, Scheper T, Garnweitner G. Rapid microfluidic preparation of niosomes for targeted drug delivery. Int J Mol Sci. 2019;20(19):4696. https://doi.org/10.3390/ijms20194696.
99. Obeid MA, Elburi A, Young LC, et al. Formulation of non-ionic surfactant vesicles (NISV) prepared by microfluidics for therapeutic delivery of siRNA into cancer cells. Mol Pharm. 2017;14:2450–8. https://doi.org/10.1021/acs.molpharmaceut.7b00352.

100. Obeid MA, Khadra I, Mullen AB, Tate RJ, Ferro VA. The effects of hydration media on the characteristics of non-ionic surfactant vesicles (NISV) prepared by microfluidics. Int J Pharm. 2017;516:52–60. https://doi.org/10.1016/j.ijpharm.2016.11.015.
101. Guinedi AS, Mortada ND, Mansour S, Hathout RM. Preparation and evaluation of reverse-phase evaporation and multilamellar niosomes as ophthalmic carriers of acetazolamide. Int J Pharm. 2005;306(1, 2):71–82. https://doi.org/10.1016/j.ijpharm.2005.09.023.
102. Rajni S, Singh DJ, Prasad DN, Sahil K, Anchal P. Formulation and evaluation of clindamycin phosphate niosomes by using reverse phase evaporation method. J Drug Deliv Ther. 2019;9(3-s):515–23.
103. Talsma H, van Steenbergen MJ, Borchert JC, Crommelin DJ. A novel technique for the one-step preparation of liposomes and nonionic surfactant vesicles without the use of organic solvents. Liposome formation in a continuous gas stream: the 'bubble' method. J Pharm Sci. 1994;83:276–80.
104. Durak S, Rad ME, Yetisgin AA, et al. Niosomal drug delivery systems for ocular disease-recent advances and future prospects. Nanomaterials. 2020;10(1191):1–29. https://doi.org/10.3390/nano10061191.
105. Marianecci C, Di Marzio L, Rinaldi F, et al. Niosomes from 80s to present: the state of the art. Adv Colloid Interf Sci. 2014;205:187–206. https://doi.org/10.1016/j.cis.2013.11.018.
106. Mayer LD, Bally MB, Hope MJ, Cullis PR. Uptake of antineoplastic agents into large unilamellar vesicles in response to a membrane potential. Biochim Biophys Acta. 1985;816(2):294–302. https://doi.org/10.1016/0005-2736(85)90497-3.
107. Rogerson A, Cummings J, Willmott N, Florence AT. The distribution of doxorubicin in mice following administration in niosomes. J Pharm Pharmacol. 1988;40(5):337–42. https://doi.org/10.1111/j.2042-7158.1988.tb05263.x.
108. Kremer JM, Esker MW, Pathmamanoharan C, Wiersema PH. Vesicles of variable diameter prepared by a modified injection method. Biochemistry. 1977;16(17):3932–5. https://doi.org/10.1021/bi00636a033.
109. Durga Bhavani G, Veera LP. Recent advances of non-ionic surfactant-based nano-vesicles (niosomes and proniosomes): a brief review of these in enhancing transdermal delivery of drug. Future J Pharm Sci. 2020;6:100. https://doi.org/10.1186/s43094-020-00117-y.
110. Uchegbu IF, Vyas SP. Non-ionic surfactant based vesicles (niosomes) in drug delivery. Int J Pharm. 1998;172:33–70.
111. Radha GV, Rani TS, Sarvani B. A review on proniosomal drug delivery system for targeted drug action. J Basic Clin Pharm. 2013;4(2):42–8. https://doi.org/10.4103/0976-0105.113609.
112. Khudair N, Agouni A, Elrayess MA, Najlah M, Younes HM, Elhis A. Letrozole-loaded non-ionic surfactant vesicles prepared via a slurry-based proniosome technology: formulation development and characterization. J Drug Deliv Sci Technol. 2020;58:1017212.
113. Manosroi W, Manosroi J, Manosroi A, Ruksiriwanich W, Abe M, Sakai H. Biological activities of the rice bran extract and physical characteristics of its entrapment in niosomes by supercritical carbon dioxide fluid. J Supercrit Fluids. 2010;54:137–44.
114. Manosroi J, Manosroi A, Chutoprapat R, Abe M. Characteristics of niosomes prepared by supercritical carbon dioxide (scCO 2) fluid. Int J Pharm. 2008;352:248–55.
115. Blazek-Welsh AI, Rhodes DG. Maltodextrin-based proniosomes. AAPS Pharm Sci. 2001;3(1):E1. https://doi.org/10.1208/ps030101.
116. Biswal S, Murthy PN, Sahu J, Sahoo P, Amir F. Vesicles of non-ionic surfactants (niosomes) and drug delivery potential. Int J Pharm Sci Nanotechnol. 2008;1:1–8.
117. Marianecci C, Di Marzio L, Rinaldi F, Celia C, Paolino D, Alhaique F, Esposito S, Carafa M. Niosomes from 80s to present: the state of the art. Adv Colloid Interf Sci. 2014;205:187–206.
118. Chandraprakash KS, Udupa N, Umadevi P, Pillai GK. Formulation and evaluation of Methotrexate niosomes. Indian J Pharm Sci. 1992;54(5):197.

119. Azarbayjani AF, Tan EH, Chan YW, Chan SY. Transdermal delivery of haloperidol by proniosomal formulations with non-ionic surfactants. Biol Pharm Bull. 2009;32(8):1453–8. https://doi.org/10.1248/bpb.32.1453.
120. Agrati C, Marianecci C, Sennato S, Carafa M, Bordoni V, Cimini E, Tempestilli M, Pucillo LP, Turchi F, Martini F, Borioni G, Bordi F. Multicompartment vectors as novel drug delivery systems: selective activation of Tγδ lymphocytes after zoledronic acid delivery. Nanomedicine. 2011;7(2):153–61. https://doi.org/10.1016/j.nano.2010.10.003.
121. Manosroi A, Wongtrakul P, Manosroi J, Sakai H, Sugawara F, Yuasa M, et al. Characterization of vesicles prepared with various non-ionic surfactants mixed with cholesterol. Colloids Surf B. 2003;30:129–38.
122. Balakrishnan P, Shanmugam S, Lee WS. Formulation and in vitro assessment of minoxidil niosomes for enhanced skin delivery. Int J Pharm. 2009;244:73–80.
123. Stafford S, Baillie AJ, Florence AT. Drug effects on the size of chemically defined non-ionic surfactant vesicles. J Pharm Pharmacol. 1988;40:26.
124. D'Souza SA, Ray J, Pandey S, Udupa N. Absorption of ciprofloxacin and norfloxacin when administered as niosome-encapsulated inclusion complexes. J Pharm Pharmacol. 1997;49(2):145–9. https://doi.org/10.1111/j.2042-7158.1997.tb06769.x.
125. Makeshwar K, Wasankar S. Niosomes: a novel drug delivery system. Asian J Pharm Res. 2013;3:16–20.
126. Sankhyan A, Pawar P. Recent trends in niosome as vesicular drug delivery system. J Appl Pharm Sci. 2012;2:20–32.
127. Gurjar P. Niosome: a promising pharmaceutical drug delivery. Int J Pharm Anal. 2014;2:425–31.
128. Mahale NB, Thakkar PD, Mali RG, Walunj DR, Chaudhari SR. Niosomes: novel sustained release nonionic, stable vesicular systems-an overview. Adv Colloid Interf Sci. 2012;183:46–54.
129. Sankar V, Ruckmani K, Durga S, Jailani S. Proniosomes as drug carriers. Pak J Pharm Sci. 2010;23:103–7.
130. Verma S, Singh SK, Syan N, Mathur P, Valecha V. Nanoparticle vesicular systems: a versatile tool for drug delivery. J Chem Pharm Res. 2010;2(2):496–509.
131. Manconi M, Sinico C, Fadda AM, Vila AO, Figueruelo J, Molina F. Colloids Surf A Physicochem Eng Asp. 2005;270–271:124.
132. Muzzalupo R, Nicoletta FP, Trombino S, Cassano R, Iemma F, Picci N. Colloids Surf B: Biointerfaces. 2007;58:197.
133. Carafa M, Di Marzio L, Marianecci C, Cinque B, Lucania G, Kajiwara K, et al. Eur J Pharm Sci. 2006;28:385–93.
134. Assadpour E, Jafari SM. Chapter one – an overview of lipid-based nanostructures for encapsulation of food ingredients. In: Lipid-based nanostructures for food encapsulation purposes, vol. 2. Nanoencapsulation in the Food Industry; 2019. p. 1–34.
135. Singh G, Dwivedi H, Saraf SK, Saraf SA. Niosomal delivery of isoniazid – development and characterization. Trop J Pharm Res. 2011;10(2):203–10. https://doi.org/10.4314/tjpr.v10i2.66564.
136. Karki R, Mamatha GC, Subramanya G, Udupa N. Preparation, characterization and tissue disposition of niosomes containing isoniazid. Rasayan J Chem. 2008;1(2):224–7.
137. Mehta SK, Jindal N. Formulation of Tyloxapol niosomes for encapsulation, stabilization and dissolution of anti-tubercular drugs. Colloids Surf B: Biointerfaces. 2013;101:434–41. https://doi.org/10.1016/j.colsurfb.2012.07.006.
138. Eldehna WM, El Hassab MA, Abdelshafi NA, Al-Zahraa Sayed F, Fares M, Al-Rashood ST, Elsayed ZM, Abdel-Aziz MM, Elkaeed EB, Elsabahy M, Eissa NG. Development of potent nanosized isatin-isonicotinohydrazide hybrid for management of *Mycobacterium tuberculosis*. Int J Pharm. 2021;612:121369. https://doi.org/10.1016/j.ijpharm.2021.121369.
139. Jain CP, Vyas SP. Preparation and characterization of niosomes containing rifampicin for lung targeting. J Microencapsul. 1995;12(4):401–7. https://doi.org/10.3109/02652049509087252.

140. Kamath MP, Shenoy BD, Tiwari SB, Karki R, Udupa N, Kotian M. Prolonged release of biodegradable vesicular carriers for rifampicin – formulation and kinetics of release. Indian J Exp Biol. 2000;38:113–8.
141. Mullaicharam AR, Murthy RSR. Lung accumulation of niosome-entrapped rifampicin following intravenous and intratracheal administration in the rat. J Drug Deliv Sci Technol. 2004;14:99–104. https://doi.org/10.1016/S1773-2247(04)50020-5.
142. Chowdhury P, Uma Shankar MS. Formulation and evaluation of Rifampicin and Ofloxacin niosomes for Drug-resistant TB on Logarithmic-phase cultures of *Mycobacterium tuberculosis*. Int J Rev Life Sci. 2016;3:628–33.
143. Rani NP, Suriyaprakash TN, Senthamarai R. Formulation and evaluation of rifampicin and gatifloxacinniosomes on logarithmic-phase cultures of *Mycobacterium tuberculosis*. Int J Pharm Bio Sci. 2010;1(4):379–87.
144. El-Ridy MS, Abdelbary A, Nasr EA, Khalil RM, Mostafa DM, El-Batal AI, Abd El-Alim SH. Niosomal encapsulation of the antitubercular drug, pyrazinamide. Drug Dev Ind Pharm. 2011;37(9):1110–8. https://doi.org/10.3109/03639045.2011.560605.
145. Mehta SK, Jindal N, Kaur G. Quantitative investigation, stability and in vitro release studies of anti-TB drugs in Triton niosomes. Colloids Surf B: Biointerfaces. 2011;87(1):173–9.
146. El-Ridy MS, Yehia SA, Kassem MA, Mostafa DM, Nasr EA, Asfour MH. Niosomal encapsulation of ethambutol hydrochloride for increasing its efficacy and safety. Drug Deliv. 2015;22(1):21–36. https://doi.org/10.3109/10717544.2013.868556.
147. Magdum AS. Niosomes: a promising vesicular drug delivery system for tuberculosis. Indo Am J Pharm Sci. 2017;4(08):2710–23.
148. Sadhu P, Saisivam S, Debnath S. Design and characterization of niosomes of ethionamide for multi drug resistance tuberculosis. World J Pharm Res. 2019;8(6):921–33. https://doi.org/10.20959/wjpr20196-1481.
149. Kulkarni P, Rawtani D, Barot T. Formulation and optimization of long acting dual niosomes using Box-Behnken experimental design method for combinative delivery of Ethionamide and D-cycloserine in Tuberculosis treatment. Colloids Surf A Physicochem Eng Asp. 2019;565(1):131–42.
150. Ranjini DS, Parthiban S, Senthil Kumar GP, Tamizh Mani T. Development of levofloxacin loaded nano-niosomes as nanoscale drug delivery for effective treatment of tuberculosis. World J Pharm Pharm Sci. 2017;6(4):1819–28. https://doi.org/10.20959/wjpps20174-8998.
151. Moazeni E, Gilani K, Sotoudegan F, et al. Formulation and in vitro evaluation of ciprofloxacin containing niosomes for pulmonary delivery. J Microencapsul. 2010;27(7):618–27. https://doi.org/10.3109/02652048.2010.506579.
152. Nethaji R, Gopal N, Surya Prakash TNK, Jayakar B, Subramanian N. Formulation and evaluation of gatifloxacin niosomes using sorbitan monoesters. Acta Pharm Sci. 2011;53:619–34.
153. Mansouri M, Khayam N, Jamshidifar E, et al. Streptomycin sulfate-loaded niosomes enables increased antimicrobial and anti-biofilm activities. Front Bioeng Biotechnol. 2021;9:745099. https://doi.org/10.3389/fbioe.2021.745099.

Surface-Modified Drug Delivery Systems for Tuberculosis Intervention

Simisola Ayodele, Pradeep Kumar, Armorel van Eyk, and Yahya E. Choonara

Abstract Tuberculosis (TB) presents as the second most lethal infectious disease after HIV/AIDS and has presented difficulty in treatment over the years, due to prolonged duration of therapy and side effects of the drugs resulting in patient non-compliance and the development of multidrug resistance (MDR) strains. Anti-TB drugs incorporated in nanosystems may reduce side effects by delivering the drug selectively into infection reservoirs such as macrophages, which may assist in clearing the TB bacilli faster and reducing the duration of therapy. The rapid development of nanosciences has improved the targeted delivery of therapeutics, offering great benefits in the treatment of chronic diseases. Nanosystems demonstrate great prospect by specific and selective targeting, supported by their ability to be surface functionalized with targeting ligands. This chapter will therefore demonstrate the state of the art in the development of surface-modified nanotechnology incorporated with anti-TB therapeutics.

Keywords Tuberculosis · Polymeric nanosystems · Surface modifications · MDR TB · Therapeutics

S. Ayodele · P. Kumar · Y. E. Choonara (✉)
Wits Advanced Drug Delivery Platform Research Unit, Department of Pharmacy and Pharmacology, School of Therapeutic Sciences, Faculty of Health Sciences, University of the Witwatersrand, Johannesburg, South Africa
e-mail: yahya.choonara@wits.ac.za

A. van Eyk
Department of Pharmacy and Pharmacology, University of the Witwatersrand, Johannesburg, South Africa

R. Shegokar, Y. Pathak (eds.), *Tubercular Drug Delivery Systems*,
https://doi.org/10.1007/978-3-031-14100-3_13

1 Introduction

Even with enhanced research and decades of dedication towards finding the solution for the best treatment of tuberculosis (TB) whilst ensuring efficacy and safety, TB is still considered cataclysmic in nature, ranking first in mortality by a single infectious agent in the world [34]. In 2019, it was reported that 1.4 million people died from the disease. The mortality rate has been found to drop by 9% between 2015 and 2019 by 2% annually, but is still less than half the milestone 20% target set between the years 2015 and 2020 in correlation with the United Nations Sustainable Development Goals (SDGs) and World Health Organization (WHO) goals to end TB by 2030 and 2035, respectively [75, 77]. With this target in place and our reports not on track, it is evident that more interventions and efforts are needed to achieve these goals. These interventions will need to improve the 6 monthly duration of treatment and patient compliance which are causes of treatment failure and MDR strains [15]. MDR strains require second-line drugs which take a longer period of 9–12 months to treat, increasing the possibility of patient incompliance and mortality.

With TB maintaining its notable mortality rate over the past decades, it has become more evident that the conventional anti-TB therapeutic regimen with no enhancing delivery system alone may not be sufficient to end TB in the set periods due to the limitations it poses. Targeted drug delivery systems are looked up to as a method to counter these limitations and enhance the therapeutic efficacy of these drugs by enhancing drug permeability, decreasing resistance, providing controlled drug release, increasing potency whilst decreasing dosage amount, frequency and the potential for adverse effects [56].

The conventional anti-TB drugs poorly permeate the hydrophobic lipoarabinomannan-rich envelope of *Mycobacterium tuberculosis* (*M.tb*), which also includes a hydroxyl fatty acid, mycolic acid which is bound to arabinogalactan, granting it a very hydrophobic character [56]. Out of the four standard anti-TB drugs, isoniazid (INH), pyrazinamide (PYZ) and ethambutol (EMB) are all hydrophilic in nature making them impermeable to these cell walls. Rifampicin (RIF) is the only lipophilic drug and most structurally complex of all standard anti-TB drugs. Studies suggest that lipophilic derivatives were more active against replicating mycobacteria than its hydrophilic companions [50]; it was found that the application of lipophilic compounds such as bedaquiline, rifapentine, clofazimine, PA-824, sutezolid and the new nitroimidazole called TBA-354 on rats resulted in sterilization from *M.tb* in 6–8 weeks in comparison to RIF+INH+PYZ which took 6 months [50, 74]. Additionally, only lipophilic compounds killed both dormant H12/H19 cells and A5/H5 cells [50]. Based on such a background, a drug delivery system that enhances permeability and internalization of hydrophilic drugs may be required to improve therapeutic outcomes.

Targeted drug delivery systems will allow the encompassing of pulmonary therapeutic advancement and also extrapulmonary tuberculosis therapeutic advancement which contributes to 16% of TB cases and is the most difficult to treat [77]. Central Nervous System (CNS) manifestations of TB is the most prominent form of extrapulmonary TB, with worse therapeutic outcomes than pulmonary TB due to a lack of an effective system to ensure drug delivery to require site of action. From the first line, only INH and PYZ show sufficient CSF bioavailability but RIF and EMB have been shown to have poorer CSF bioavailability, thus requiring a higher dose for therapeutic relevance consequently causing an increased severity in the possible adverse effects like cerebral haemorrhage and visual impairment, respectively. Fluoroquinolones, a second-line agent, also have good bioavailability in the CSF. New generational fluoroquinolones have also proven to be greatly efficacious in the treatment of MDR-TB, like gatifloxacin due to its low incidence of resistance [81]. On the other hand, streptomycin demonstrates minimal efficacy whereby at high doses it may increase the risk of nephrotoxicity and ototoxicity [81].

Nanotechnology is an extensively studied method of delivery in recent years. The nanospheres small-sized property of nanoparticles allow the particle to be able to travel uninterruptedly across the human body [47]. Nanomedicine has gained more medical relevance over time due to its ability to bind to a drug or even encapsulate it whilst simultaneously attaching unto its surface, functionalizing proteins to promote targeted controlled delivery of the drug. Surface modification of the delivery system brings about a more specific targeted delivery which is also accompanied by theranostics, the integration of therapy and diagnosis. Dendrimers are typical examples of nanoparticles well known for their entrapment and surface-attaching capabilities. Their very small size of 2–10 nm and multiple functional group sites on its surface allow binding of drugs and ligands for targeted-controlled drug delivery [27, 62, 63]. Carbon nanotubes are another synthetic nano-biotechnology of size 1–100 nm with great prospects in the nano-field due to the ease of surface modification with drug and functional ligands based on the desire for enhanced site-specific delivery [45].

Nanosensors modification on nanoparticles further allows diagnosis apart from treatment. In the case of poly (ethylene glycol)-modified poly (amidoamine) dendrimers encapsulating gold nanoparticles, a system that employs both imaging functions of cancer diagnosis and treatment [22], gold nanoparticles are used as biomarkers and tumour labels in detection assays [47]. Combination studies of nanosystems and natural products have gained prominence over the last decades for the treatment of various conditions. This green nanotechnology allows the incorporation of nanoparticles with natural products, decreasing the development of toxicity encountered from biosynthetic processes which result in side effects [31]. These methods have been greatly appreciated, leading to the FDA approval of liposomes and micelles, the fathers of nanoparticle-based therapy [58].

The CD206/mannose receptors are predominantly expressed on the surface of macrophages and dendritic cells acting as a pattern recognition receptor (PRR), which is recognized by mannose-binding lectin receptors that bind to glycoproteins found on the surface pathogens like *M.tb*. The M2 macrophages which are responsible for the resolution of inflammation, protection from excessive inflammation and tissue repair, express more CD206 receptors in comparison to M1 macrophages [23]. The CD206 receptor, unfortunately, stands as the site of exploitation by *M.tb* to facilitate infection in macrophages [25]. In addition to that, soluble forms of

CD206, (sCD206) are found in the periphery and can recognize mannosylated carbohydrates and cause an effect on the innate and immune response [67]. The CD206 receptor can therefore be used as a therapeutic target for the treatment of *M.tb*.

This concept has therefore created a path using mannosylated-nanostructured lipid carrier (NLC) for active targeting of the CD206 on macrophages [70, 71, 49] and mannose microstructured modified liposomes [11]. A definite increase in the uptake of the particles was observed in rats alveolar macrophages after pulmonary administration. Furthermore, plasma concentrations of encapsulated drugs are found to be decreased in comparison to free drugs indicating increased specificity and selectivity of the delivery system. In the study of liposomes at a size range of 90–125 nm, a positive correlation was observed between the concentrations of mannose on the surface and a more notable cellular uptake [73]. In the presence of excess solubilized mannan, the uptake of the nanoparticles also decreased due to receptor overload, further confirming the specificity of mannose to CD206 receptor [73].

HIV/AIDS is an important cause of the continuation of TB, which has contributed to the mortality and morbidity rates of TB. A slight contributing factor to the increased mortality and morbidity includes intestinal malabsorption of anti-TB drugs, mainly RIF and ETB [65]. Since RIF is taken over a period of 6 months, it is important for it to have good bioavailability, due to its lipophilic nature, but unfortunately demonstrates low aqueous solubility and low intestinal absorption [39]. This is because, in gastric acidic conditions, RIF is hydrolysed into even lesser soluble forms such as 3-formyl rifampicin and 1-amino-4-methyl piperazine [66]. In fixed-dose combinations, the intragastric reaction of RIF with INH also forms an insoluble hydrazone [64].

Enhancing drug bioavailability and targeted treatment will increase drug effectiveness and decrease treatment duration, improve patient compliance and minimize the dose required to obtain a therapeutic effect. These are the goals for the pharmaceutical industry making targeted encapsulated drug delivery systems great prospects in achieving this goal [66].

2 Nanotechnologies in Tuberculosis

Nanosystems generally have various structural and functional properties; therefore, modification is possible, in order to achieve the required outcome by changing the compositions of the system, which includes the drug, polymeric materials, functional groups and even excipients [19]. Nanosystems' success is easily measured by their physicochemical abilities and their permeation abilities across anatomical barriers that unloaded drugs hardly pass through [62, 63]. Drug encapsulation within nano-sized carriers reduces drug elimination from the intracellular medium facilitated by resistant-mediated efflux pumps that is prevalent in various pathologies including cancer, HIV, and hepatitis. Nanotechnology thus has the prospects to overcome or decrease drug resistance.

3 Polymeric Nanosystems

Polymeric nanoparticles (PNPs) present great prospects in the treatment of diseases due to their intricate structure that allows surface modification. They have been substantially investigated to improve drug solubility, stability and targeting, which ultimately improves drug bioavailability at the required site of action [66]. PNPs are mainly subclassified into two broad groups, namely, nanocapsules and nanospheres. Nanocapsules comprise drugs incorporated with hydrophilic/hydrophobic solvents bounded by a polymeric membrane, whilst nanospheres is a solid network, a model that allows even distribution of drug across variable pores in its structure. PNP's high loading capacity of both hydrophilic and hydrophobic drugs and its viability across various routes of administration makes it a prominent delivery system for drug entrapment [4]. Opsonization and phagocytosis are two mechanisms for eliminating PNP from the body. This clearance method could also present advantageously in the treatment of TB since the body clears TB in the same way. Furthermore, in order to prolong blood circulation periods, promising outcomes have been identified by the modification of nanosystems with highly hydrophilic chains such as poly(ethylene glycol) (PEG) [66].

3.1 Modified Polymeric Nanoparticles

Lectin surface-modified poly(lactic-co-glycolic acid) (PLGA) nanoparticles enhanced targeted delivery of anti-tubercular drugs across pulmonary and gastrointestinal mucosa [57]. The process employed the use of carbodiimide, which yields the formation of amide bonds between the amines of lectin; wheat germ agglutinin (WGA), and carboxylic acid groups of PLGA. This surface grafting method yielded a coupling efficiency of 60–70%, whereby 3–3.5 mg lectin bound per mg of PLGA by acting as bioadhesive drug carrier, causing the drug residence time in plasma of RIF, INH and PYZ to be improved. The WGA coating allowed RIF to be retained in plasma for 6–7 days, instead of 4–6 days in uncoated particles. PYZ and INH were retained for 13–14 days instead of 8–9 days. In the liver, spleen and lungs, the drugs were all detectable for 15 days, minimising the chance of drug resistance. In SGF and SIF, only 3–5% drug was released, stipulating its targeting ability and stability in the GIT, but the glycosylated structures in the GIT allowed immobility on the system using the surface-grafted lectin recognizable by GI mucosa and lung mucosa. The dosing frequency was also revealed to decrease the dose required to reach undetectable mycobacteria colony-forming units from 45 doses in the conventional free drugs compared to three doses of oral/nebulized lectin-coated nanoparticles administered fortnightly [57].

A more interesting concept of polymeric nanoparticles involves the use of the polymer poly(ethylene oxide) monomethyl ether-block-poly(ε - caprolactone) with a Förster resonance energy transfer system sensor which allows perceptible assessment of the delivery system drug release. There was an observed enhanced release of drug from polymer in vivo for cells that retained the system than in vitro dialysis tubes. Furthermore, in vitro assays showed that apart from the antimicrobial effect of RIF, blank NPs also exhibited antimicrobial effects against *Mycobacterium fortuitum* [69].

Polymeric nanoparticles together with liposomes (LPs) may form a hybrid nanocarrier, called solid lipid nanoparticles (SLNPs), a solid lipid core, surrounded by surfactants to improve the drug stability and loading capacity of LPs [1]. To practicalize these concepts, a WGA-conjugated SLNP has been studied for the enhancement of RIF delivery. The nanoparticles consisted of glyceryl monostearate (GMS) and stearic acid (SA), of which the SA introduced free carboxylic acid to the surface of the nanoparticles, whilst forming together with GMS, the internal part of the lipid. The nanosystem was synthesized by a process of simple emulsification, prior to solvent evaporation followed by a carbodiimide reaction for WGA conjugation as in Fig. 1 below [51].

The entrapment efficacy of RIF into the nanosystem was 68.71%; consequently, it was noted that the entrapment had no significant effect on the particle size and zeta potential of the system. The WGA allowed a high binding efficacy unto intestinal mucus glycoproteins of 93.06% compared to polymeric nanoparticles, since lipids have the ability to bind to mucin, which promoted a more controlled release, from 20 h in pure RIF to a biphasic release over a period of 120 h with an initial burst release of 37.1% in the first 24 h and 76.9% release after 120 h. The system presented toxicity by showing a positive haemagglutinin test, which could be explained by the binding of the lectin conjugated system (WGA) to the carbohydrate moiety on red blood cells (RBC). In the presence of excess N-acetyl glucosamine (NAG), this toxicity was prevented, altering binding of the system to RBCs [51].

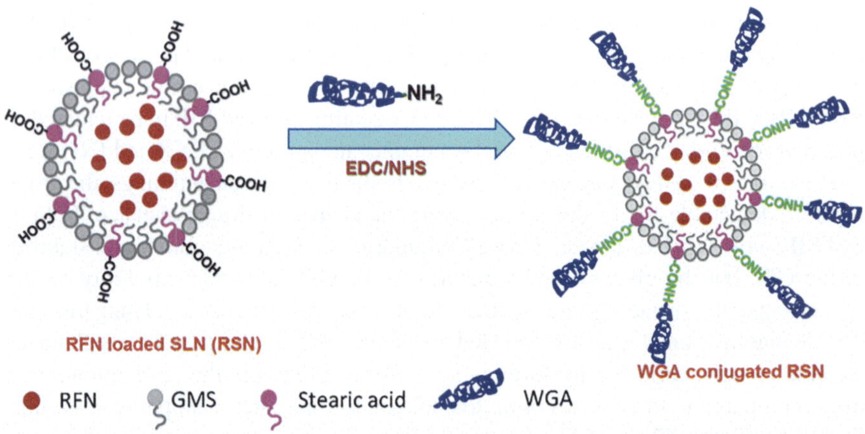

Fig. 1 Schematic diagram showing the arrangement of rifampicin (RFN) and lipids (stearic acid and glyceryl monostearate) in the nanoparticles followed by conjugation of wheat germ agglutinin (WGA) on the surface. Keywords: *EDC* ethyl carbodiimide, *NHS* N-hydroxysuccinimide. (Figure obtained with permission from Ref. [51])

In order to survive in GI acidic medium, the RIF degradation, in the presence of INH was decreased, SLNPs made up of Compritol® ATO 888/polysorbate 80. The nanosystem synthesized by microemulsification method was found to minimize the drug's interaction. Results obtained indicated that RIF-loaded SLNPs promoted 60% degradation protection induced by free INH at acidic pH while a 74.7% degradation protection was achieved when both the agents were incorporated separately into SLNPs [61].

RIF-loaded SLNPs coated with chitosan, with the lipid phase composed of cetyl palmitate synthesized by ultrasonication method, demonstrated improved mucoadhesive properties in vitro, increasing retention time of system in the alveolar and improving RIF permeability across alveolar epithelial cells due to the chitosan coating. This platform may be a prospective system for the enhancement of aerosol drug delivery [71]. Apart from RIF, INH has also been incorporated in similar strategies, whereby chitosan was more soluble in gastric pH of 1.2 compared to intestinal pH of 7.4, the acidic pH favoured the swelling of the polymer and solubility of the drug; due to the basic nature of INH, it solubilizes at acidic pH [2]. An initial burst release in the first 3 h of 50% of the drug due to the adsorbed drug on the surface of the nanoparticles was observed.

Employing biomaterial, PLGA and PEG, a PLGA-PEG-PLGA triblock copolymer synthesized by a water-oil-water double emulsification technique for the delivery of INH, loaded by sonication with a drug loading efficiency and drug content of (12.8–18.67%) and (6.4–8.9%) respectively, improved bioavailability of INH by 28 fold. The nitrogen atom of INH binds to the carbonyl oxygen of the polymer, causing INH to be loaded on the surface of the core shell. This effect resulted in an initial 2 h burst release of INH from the surface. Due to the emulsification technique of nanoparticle synthesis, an additional small burst release occurred due to the interstitial space in the core shell, and subsequently, a third stage controlled release from individual nanoparticles as in Fig. 2 below. The sustained release is said to be present for up to a period of 124 h [17].

Another approach looked towards the synthesis of RIF-, INH- and PYZ-loaded PLGA nebulized nanoparticles for pulmonary delivery, with a mass median aerodynamic diameter (MMDA) of 1.88 μm, indicating the respirability of the nanoparticles, showing plasma detection 6 h post-administration. Due to the prolonged half-life of the system, loaded anti-TB drugs had a higher AUC than both IV and oral free drugs bioavailability. When compared to nebulized free drugs, the bioavailability increased by 51.0-, 20.0- and 28.0-fold for RIF, INH and PYZ, respectively, indicating the significance of the PLGA nanoparticles. Furthermore, the drug residence time in the body was 11 days for drug-incorporated nanosystems as compared to 24 h for free unloaded drugs. The retarded elimination of the drug allowed a dosing regimen of five doses of nanoparticles administered every 10 days showing equal therapeutic efficacy as 46 oral daily doses [46]. A similar approach with the only difference of nanoparticles being alginate nebulized nanoparticles demonstrated faster plasma detection at 3 h and a drug residence time up to 14 days, whereby three doses of nanoparticles administered every 15 days for 45 days was as effective in clearing *M.tb* bacilli of lung and spleen as 45 oral daily doses of the free drugs [80].

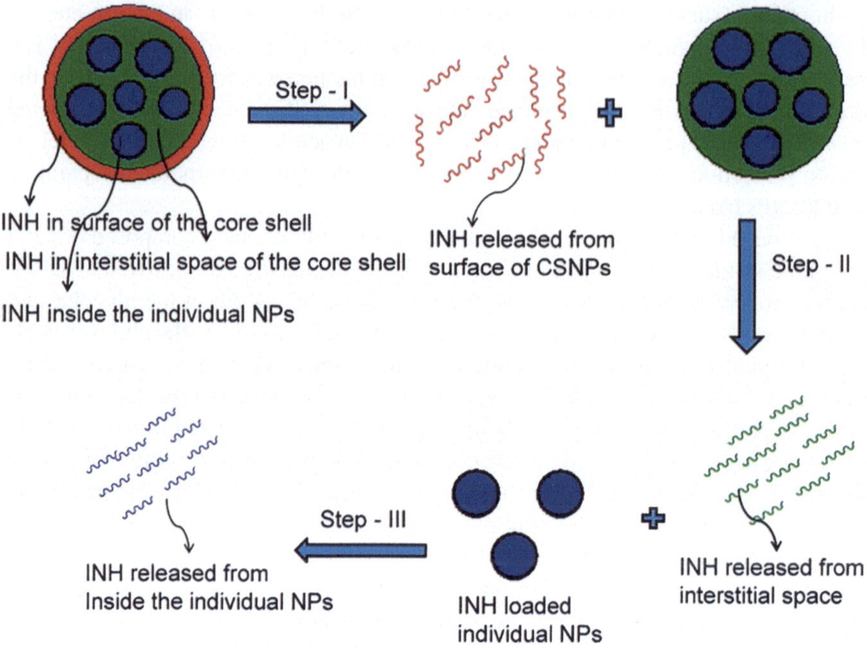

Fig. 2 Mechanism of drug release of PLGA-PEG-PLGA triblock copolymer synthesized by water-oil-water double emulsification. (Figure obtained with permission from Ref. [17])

A RIF/PLGA nanoparticle incorporated in mannitol microspheres enhanced the uptake by NR8383 macrophage cells in the lungs, with the mannitol microspheres affecting its uptake from 13.5% in RIF/PLGA nanoparticle to 47.1% in a microsphere, RIF/PLGA nanoparticle [44].

3.2 *Modified Polymeric Micelles*

Polymeric micelles are self-assembled nanocarriers, made of a hydrophobic core, representing the drug reservoir, and an outer hydrophilic shell, usually PEG, which acts on the surface as an opsonization and elimination preventative method [66]. The two components form an amphiphilic stimuli-responsive system that incorporates hydrophobic drugs, loading it, for improved targeted delivery [66]. Drugs loaded in the inner core are consequently protected from the environment, chemical and biological degradation. Polymeric micelles have a slow disassembly process depending on the molecular weight, HLB of the polymer and properties of the encapsulating drug, making it a more stable delivery system compared to conventional micelles.

In relation to TB, a study investigated the incorporation of RIF, a large hydrophobic drug within various poly(ethylene oxide)–poly(propylene oxide) (PEO–PPO) block copolymers, which includes linear poloxamers and branched poloxamines. The solubilization was found to be minimal (~2 folds of RIF alone), due to the size of the core in comparison to the RIF molecule. With the flexibility of synthesis, modification of the polymer by linking mono and bifunctional PEG precursors of different molecular weight with polycaprolactone which improves the HLB and enlarges the core cause improved solubilization to ~5–7 folds of RIF alone [66]. A thermo-responsive poly(ε-caprolactone-coglycolide)–poly(ethylene glycol)-poly(ε-caprolactone-co-glycolide) block synthesized by ring-opening polymerization copolymer loaded with RIF presented with an unimproved solubility of 2 mg/ml but presented the slow release of drug over 32 days which could present a pathway in the development of a depot injection [24].

The use of enantiomeric pure micelles was outclassed by an evenly mixed molar ratio of monomethoxy PEG poly (ethylene glycol)–poly(L-lactide) (MPEG-PLLA) and monomethoxy PEG poly(ethylene glycol)-poly(D-lactide) (MPEG-PDLA) block copolymers, improving RIF loading capacity and release in comparison to the single polymer micelles. In vitro studies showed a fast initial release of 50% after 4–8 h, then became sustained afterwards to release 100% of the drug in 48 h. It is important to also note that although there was a burst release, there was no free drug on the shell of the micelles. The release rate was inversely proportional to the contents of the polylactic acid (PLA) segment in the stereo complex block copolymers [9].

In another design, chitosan was employed to modify spherical PLA micelles, which upon entrapment of RIF resulted in a size increase from a range of (154–181) nm to (163–210) nm. In vitro release demonstrated a 35% burst release within the first 5 h, followed by a slow sustained release until 5 days [78].

Furthermore, in other studies involving other anti-TB drugs, encapsulating INH, PYZ and RIF, a poly(ethylene glycol)-poly(aspartic acid) (PEG-PASP) copolymer with INH embedded in it, resulted in a 5.6 times decrease in *Mycobacterium tuberculosis* MIC compared to the drug alone [60]. The same micelles had a level of drug conjugated between 65.02–85.70% and were with a size of 78.2 nm, 84.2 nm and 98.9 nm, for INH, PYZ and RIF, respectively. Due to a decreased diameter and hydrophilic surface, these systems have minimized renal filtration and reticular system elimination [66].

3.3 Modified Dendrimers

Dendrimers are exemplary types of lipid-based polymeric drug delivery systems that are monodispersed and can easily be surface modified to desired targeting goals. Dendrimers are very small macromolecular synthetic nanosystems ranging from 2 and 10 nm, which allow the attachment of various and multiple functional groups on their external surface that can be used for treatment and diagnosis [40, 62, 63]. This presence of multiple terminal groups causes dendrimer to be less viscous than other linear polymers, highly soluble, reactive and miscible [29, 62, 63].

Dendrimers are, therefore, candidates in reducing toxicity, enhancing solubility and bioavailability of anti-TB drugs. They are special because as their diameter increases linearly, the number of surface groups increase exponentially [40].

The dendrimers macromolecular structure can be divided into four main components, which includes (a) central core moiety, responsible for the molecular arrangement, size and shape, (b) interior layers which consist of patterns of repeating branches attached to the core, which increases based on the generation causing surface groups to increase exponentially. These branches form (c) void spaces which act as room for molecular cargos. The outer surface is made of (d) terminal functional groups used in targeted delivery [40].

A gen 1 dendrimer, would have eight terminal amino groups, whilst a gen 2 dendrimer would exponentially increase to 16 terminal groups and gen 5 dendrimer therefore would have 128 terminal groups [40]. These terminal groups, therefore, act as sites for the attachment of functionalizing groups. Dendrimers are adjustable pH-based systems, which retain the drug at a blood pH of 7.4 and release the drug at acidic pH, in correlation with a TB-infected macrophage which is a pH of ~4.5. For this reason, dendrimers present themselves as very prospective candidates for the incorporation and functionalization of anti-TB drugs presenting a form of targeted delivery. The various types of dendrimers differ based on their size and the number of terminal functional groups. The main materials used comprise PAMAM dendrimers, poly-ether hydroxyl-amine (PEHAM) dendrimers, PPI dendrimers, carbosilane dendrimers, and phosphorus-based dendrimers [35].

In relation to TB, studies incorporating anti-TB drugs as in Fig. 3 has been explored. Mannose functionalized ethylene diamine poly(propylene imine) (EDA-PPI) fifth gen dendrimers with 64 amino groups on the surface for targeted delivery of RIF to alveolar macrophages has been performed [30]. The core of the particle was composed of EDA and its surface was grafted with ~30-D-mannose groups. Since RIF hydrolysed in pH of 4–5 (gastric pH) into less soluble compounds, the use of dendrimers with mannose will allow leptin receptors on phagocytic cells the

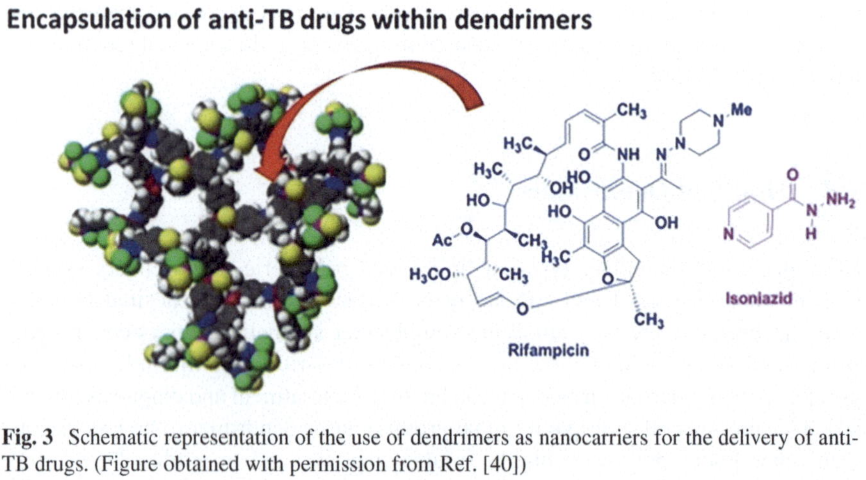

Fig. 3 Schematic representation of the use of dendrimers as nanocarriers for the delivery of anti-TB drugs. (Figure obtained with permission from Ref. [40])

recognition of the system and its uptake, consequently enhancing the uptake of the drugs found in the void space [35]. Although the solubility of the nanosystem is decreased from 50 mg/ml to 5 mg/ml in unmannosylated RIF-loaded dendrimers compared with mannosylated RIF. The mannose-functionalized dendrimer still possesses an aqueous solubility twice that of RIF alone. Noncovalent interactions were observed between RIF and the mannosylated dendrimer, whilst hydrogen and hydrophobic interactions occur between RIF and the core. Haemolytic toxicity limitations were observed in the G5 EDA-PPI dendrimers; though it was observed that the unmannosylated dendrimers were haemolytic, their toxicity was decreased by functionalizing with mannose from 15.6% to 2.8% due to a minimized interaction of the charged peripheral amino groups with cells. This serves as a prelude to the requirement of modification of their surface by functional groups to inhibit its cationic and subsequent haemolytic effect since these effects will preclude the effect from the use of unfunctionalized dendrimers [35].

In relation to anti-TB drugs, a study on vero cells used in cell studies indicated negligible cytotoxic effect as 100 μg/ml, with RIF-loaded mannosylated dendrimer (G5 EDA-PPI) having ~85% viability better than RIF alone at ~50%. In mannosylated dendrimers, RIF stayed entrapped for ~5 days in mannosylated dendrimers in comparison to less than 10 h in PPI dendrimers. Furthermore, mannosylated dendrimers undergo drug release at a pH of 5, the same required condition in phagolysosomes like infected macrophages. The target delivery of the system using mannose improves the intracellular concentration of the drug, improves cellular viability whilst improving the therapeutic efficacy of lipophilic drugs life RIF [35].

Taking into consideration the same method, the authors describe the formulation of G4 and G5 PPI dendrimers, which is PEGylated instead of mannose. In addition to the cytotoxic effects of primary amino group terminal bearing PPI and PAMAM dendrimers, they have also been found to be cleared rapidly if administered intravenously [32]. In contrast, PEG is water-soluble, thermostable, chemically inert, unhydrolysed, and not deteriorative, making it non-toxic and biocompatible [72]. Therefore, the incorporation of PEG into the surface of polymeric micelles such as dendrimers suppresses the interaction of the delivery system with plasma proteins and cells, prolonging their blood elimination half-life [72].

The incorporation of RIF, known for its short half-life, solubility difficulty into PEGylated ethylene diamine (EDA)-PPI dendrimers, resulted in a slower release of the drug, with G4 non-PEGylated releasing 97.3 % of RIF in 36 h, whilst G4 PEGylated released half of that in 36 h and 93.1% in 96 h. In G5 non-PEGylated released 94.6% of RIF in 48 h and G5 PEGylated released half of that in 48 h and 93.4% after 120 h. This is because as the generation increases, the hydrophobic interactions of the core increase; therefore, these interactions with RIF as well as the doubling up of the terminal groups from 32 to 64 contributed to its delayed release profile. With regard to its haemolytic toxicity, dendrimers generally have (~14.6–20.3%) RBC haemolysis, and PEG significantly decreased it to less than 3% due to minimized interaction between quaternary ammonium groups in the terminal end of dendrimers and RBCs [72].

Furthermore, on G5 PEGylated dendrimer, encapsulation efficacy of RIF more recently was found to be ~99% by P. Dineshkumar and co-workers, with a comparable drug release rate of 81% after 120 h and 98% after 72 h in non-PEGylated, with inhibition of RBC haemolysis by less than 2.5% [13]. Wistar albino rats studies performed in vivo comparing the pharmacokinetic properties of RIF encapsulated G5 PEGylated dendrimer and RIF alone showed that over a period of 120 h, the presence of RIF alone was not observed after 6 h. In correspondence with in vitro studies, RIF encapsulated G5 PEGylated dendrimer was retained in plasma for 120 h. Even after its storage for 3 months at −40 °C, the release of the drug system was unaltered, ensuring improved AUC and drug residence time [40, 62, 63].

INH is another drug that has been incorporated into dendrimers. An INH G1.5 PAMAM dendrimers were synthesized using the dialysis method. Drug release study showed controlled 93.25% release of INH over a period of 24 h. A zero-order kinetic was demonstrated [26, 62, 63]. Dendrimers show great prospect in the incorporation of both high molecular weight, hydrophilic and hydrophobic chemical entities that have low solubility and low half-life.

4 Nanotechnological Advances for MDR and XDR TB

TB resistance has emerged over the years and proven to be a challenging aspect of TB to manage, with patient incompliance in the first-line therapy, recurrently causing therapeutic failure. Furthermore, the resistant TB bacilli may also be contracted through person-to-person transmission. With nanotechnology showing strife over enhancing patient compliance in the first-line therapy, which reduces the emergence of resistant TB, a strong urge to improve treatment of present individuals with MDR and XDR-TB assures nanotechnology an opening to be a prospective breakthrough for this aim. According to the World Health Organisation, fluoroquinolones are considered the main therapeutic pathway to counter MDR, particularly moxifloxacin (MOX), which is recommended for the 'shorter MDR-TB regimen' by WHO, even though its shortcomings include lack of patient compliance due to its associated hepatotoxicity.

Nanoencapsulation may be a good attempt to counter the shortcomings of MOX, by reducing its side effect, with studies indicating that modification of NPs with hydrophilic polymers such as PEG and water-soluble chitosan deters the accumulation of MOX in the liver due to their 'brush-like' configuration [42]. Polyisobutylcyanoacrylate and gelatin synthesized nanoparticles incorporating RIF+MOX also targeted improved cellular immune response in murine alveolar macrophages by increasing intracellular pro-inflammatory cytokines [55].

Bedaquiline showed complete encapsulation as well as a surface-binding capacity on different diblock copolymers, which includes poly(lactic-co-glycolic acid) (PLGA), poly(lactic-co-hydroxymethyl glycolic acid) (PLHMGA) and poly(lactic-co-benzyloxymethyl glycolic acid) (PLBMGA) which are all surface modified by the hydrophilic mPEG [53]. This causes a burst release of 30–42% within the first hour, and a sustained release thereafter up to 100% within ~3–7 days [53]. Modification of NPs to make an LPNP also shows prospective advantages in second-line drugs. A study incorporating INH and CPX in lipid-polymer hybrid NPs (LPNPs), composed of soy lecithin (LC) and PEGylated DSPE (1,2-distearoyl-sn-glycero-3-phosphoethanolamine) resulted in improved in vitro cellular uptake of CPX into J774A.1 cell [3]. Soy LC was used to construct the shell of lipid particles for electrostatic interaction with oppositely charged polymers whilst PEGylated DSPE was responsible for modifying the system to escape the recognition by the reticuloendothelial system (RES) [68]. In vivo, the drug accumulation from nanocarrier in lungs was significantly more than unloaded free drugs.

5 Modified Nanosystem for Active Targeting of *M.tb* Reservoirs

As much as surface modification of a delivery system may be passive and be present to stabilize drugs, protect drugs, and prevent rapid elimination; it may also be actively directed to the specific site of action in a ligand-receptor interaction mechanism as in Fig. 4 that may exist between the nanocarrier and cell/tissue in the body. These ligands are generally coated, complexed or modified unto the surface of the nanosystem.

Since macrophages remain a significant factor in the pathogenesis of *M.tb*, acting as a reservoir of the *M.tb* bacilli, as well as in latent infections, active targeting would look to specify drug delivery towards alveolar macrophages in pulmonary TB, and lymph nodes in extrapulmonary TB; consequently, increasing site bioavailability and minimizing side effects of the drug without having to increase the dose.

The pattern recognition receptors on macrophages give a site of exploitation for a ligand for targeted delivery. These receptors include toll-like receptors (TLRs), C-type lectin receptors (CLRs), NOD-like receptors (NLRs), and DC-specific intercellular adhesion molecule-3-grabbing non-integrin (DC-SIGN). The most researched ligand is mannose, which is recognizable by CLRs found on macrophages surfaces. Mannose has presented prospective results, whereby cationic mannose lipid NPs significantly improve the uptake of the nanosystem in alveolar macrophages in vitro and in vivo. It is further understood that the cationic nature of the system allows NP-plasma protein interaction and aggregation in the alveolar macrophages, causing its improved lung deposition [21].

Solid lipid NPs functionalized with mannose and loaded with RIF and INH individually have shown similar outcomes. In a fluorescent RIF-loaded mannosylated SLNPs, made of glycerol tripalmitate (lipid phase) and surfactant (Tween® 80) in double deionized water (aqueous phase) with an entrapment efficacy of 90%, the internalization of RIF into THP-1 differentiated macrophages were improved, with the system remaining biocompatible at concentrations less than 57 μg/ml of which RIF bactericidal concentration is 3.2 μg/ml [79]. Eighty percent or more of the drug remained entrapped in the system after contact with pulmonary-simulated

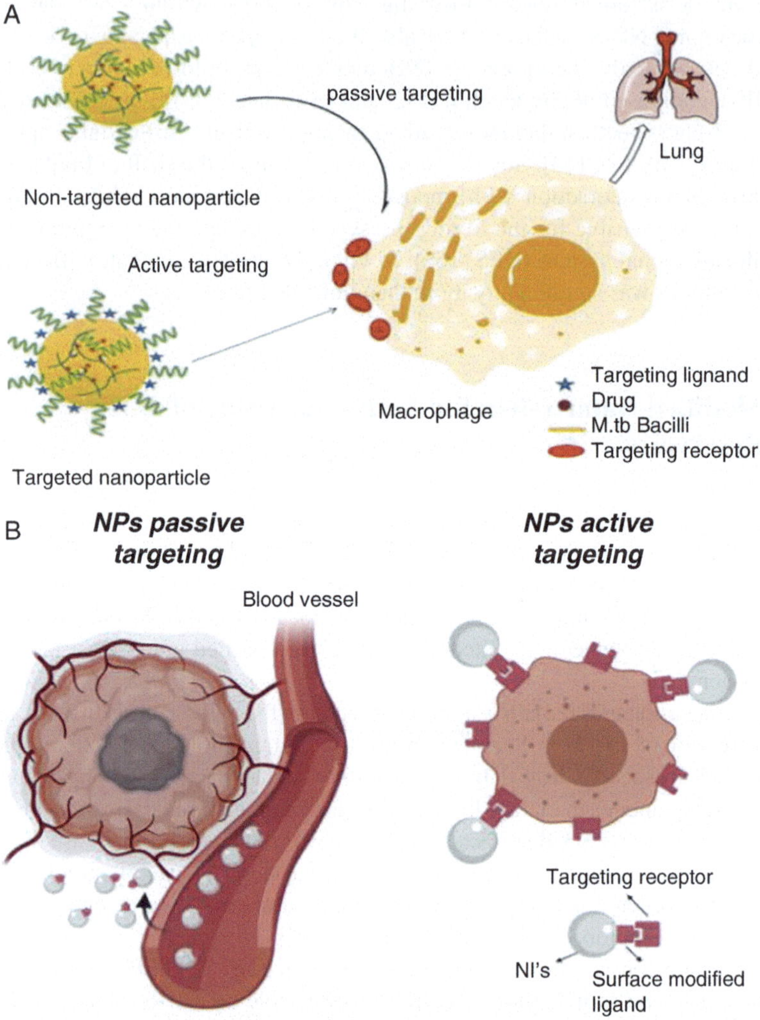

Fig. 4 (**a**) Schematic representation of macrophage targeting by surface-modified nanoparticles. (Figure adapted with permission from [48]). (**b**) Schematic diagram comparing active and passive uptake of nanoparticles. (Figure adapted with open access permission from Ref. [54]).

conditions for up to 8 h, and are able to release the drug in infected cells in a pH-dependent manner, with a faster drug release at acidic pH, which supports its suitability for pulmonary delivery, with diminished degradation and improvement in bioavailability. The NPs cell entry took place quickly within 15 min of administration [70]. With regard to INH, the SLNP were synthesized by a modified solvent emulsification-evaporation method employing a w/o/w double emulsion technique. The 500 nm system was composed of Witepsol® E85 forming the lipid phases and stearylamine (SA) which contributed to the system by surfacing amine groups for the attachment of mannose to form a lipid matrix which showed enhanced internalization of the system even based on mere adsorption of mannose to the surface of the particles. The fluorescent M-SLNP system enhances in vitro uptake of INH into dTHP-1 than the unfunctionalized SLNP [12].

In [49], a novel 50–300 nm mannosylated and PEGylated graphene oxide nanocarrier was synthesized for selective targeting and delivery of RIF to mucosal CD14+ macrophages (derived from *M.tb*-infected rhesus macaques) via mannose receptor-mediated endocytosis. Results indicated that selectivity for macrophages was established, with a higher concentration of RIF as compared to T and B cells due to low mannose-binding, c type lectin receptors on their surface. Subsequently, drug release in the macrophages was attributed to acidic lysosomic condition, causing a burst release to enhance intracellular killing of *M.tb*.

A 272 nm lactoferrin (Lf) conjugated RIF-loaded SLNPs with an encapsulation efficacy of 66% was synthesized. It is composed of tristearin, soya lecithin and stearylamine. Most drugs remained in the nanoparticles upon administration in the blood which decreases RIF toxicity. IV administration of free drugs resulted in 0.4% RIF concentration in the lungs after 4 h and undetectable levels after 24 h. In unfunctionalized SLNP, 15.6% and 2.1% RIF concentration was recovered at 4 h and 24 h, respectively, whilst in Lf-conjugated SLNPs, a staggering 47.7% and 35.3% RIF concentration was recovered after 4 h and 24 h respectively, a threefold increase in drug load in lung tissue in functionalized nanoparticle as compared to that of unfunctionalized nanoparticles, indicating the significance of the Lf specificity to the lung even when administered systemically and not locally/nebulized, due to the abundance of Lf recognizing receptors in the lungs. Upon fluorescence imaging, the alveolar macrophages were distinctly filled with the nanosystem [59].

A similar approach has been applied, whereby methyl α-d-mannopyranoside was engineered on an SLNP to deliver rifampicin, synthesized through melt emulsification technique combining cholesteryl myristate with palmitic acid (PA set) or tripalmitin (TP set) and functionalized with methyl α-d-mannopyranoside. Maretti *et al.* synthesized SLNP, demonstrated good cytotoxicity and cell internalization ability into J774 murine macrophage cell line, whilst making possible drug preservation in SLNPs along respiratory tract prior to macrophage internalization. Although favourable results were achieved, the particles cohesiveness hampered respirability, especially when SLNPs were functionalized, causing the need for better-balanced amphiphiles with -d-mannose residues prior to further investigation that can protect powder against moisture in the environment enhancing particle de-aggregation and lowering powder cohesiveness [38].

Aside from monosaccharides, mannose, other ligands on other polymeric platforms have been experimented with. The study of a flower-like poly(epsilon-caprolactone)-poly(ethylene glycol)-poly(epsilon-caprolactone) polymeric micelles encapsulating RIF functionalized by hydrolysed galactomannan (GalM-h), a polysaccharide of mannose and galactose both recognized by CLRs, to form a complex polymeric NP, coated with GalM-h/chitosan, with drug encapsulation efficacy of 12.9 times that of unfunctionalized NP. The presence of GalM-h was confirmed by an agglutination assay with concanavalin A, whilst its uptake and internalization into RAW 264.7 murine macrophages were observed by fluorescence microscopy, with the polysaccharide significantly improving the intracellular concentration of RIF [41].

Furthermore, on mannose-functionalized nanocarriers, [48] also demonstrated using chitosan-based bioadhesive nanoparticles loaded with clofazimine the effect of mannose functionalization, an excellent enhanced therapeutic efficacy in terms of inhibition and anti-mycobacterial activity by 46 fold as compared to free clofazimine. The nanosystem with a particle size ranging between 132–184 nm and an encapsulation efficacy of 95% demonstrated great in vitro release, showing a burst release, whereby 30–40% of the drug is released in the first 5 h followed by sustained release between 10 and 72 h, after which 70.98% of drugs in mannose-functionalized nanoparticles are released [7].

In the insight of a different ligand class from mannose, hyaluronic acid (HA) has been investigated, recognizable by its receptors such as CD44 and TLR-2 and -4, via MyD88 in order to activate NF-κB and induce proinflammatory cytokine gene expression [33]. The expression of these receptors on alveolar macrophages allows HA use for active targeting and drug delivery [20]. A targeted delivery system for RIF comprises the incorporation of HA into tocopherol/succinate (TS) micelles, increasing the in vitro uptake of RIF in murine alveolar macrophages (AMs) (MH-S cells) by phagocytosis, in a dose and energy-dependent manner. Concurrently, the micelles were taken up through CD-44 receptor-mediated endocytosis. The MH-S cells were activated, which in turn further improved the RIF-HA-TS uptake. The cytokine secreting activity demonstrated the enhanced secretion of Th1 cytokines due to the presence of the nanosystem in comparison to free drugs [18].

In [5], an oleic acid tuftsin-modified peptide (pTUF-OA) was synthesized to be used as a ligand unto an SLNP as shown in Fig. 5. The peptide is composed of threonine, lysine, proline and arginine amino acids and an oleic acid molecule. The SLNP is composed of a lipid phase of stearic and oleic acid and a surfactant mixture composed of Phospholipon 80H and Tween 80. In vitro pulmonary-simulated condition indicating an initial 10% burst release of drug for 2 h was followed by a slow sustained release. Cell viability of over 70% was established in J774 A.1 cells as

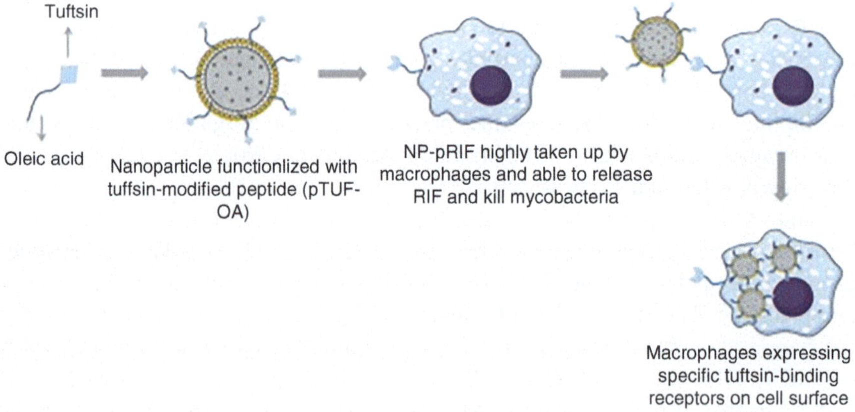

Fig. 5 Diagram showing pTUF-OA peptide functionalization unto RIF-loaded nanoparticles for active targeting of macrophages. (Figure obtained with permission from Ref. [5])

compared to a 60% cell viability of free RIF, improving the RIF treatment drawback with regard to toxicity. A twofold improved MIC of 0.48 μg/ml for RIF-loaded peptide NP was compared to 1 μg/ml for free RIF due to the enhanced penetration of NP across the mycobacteria cell wall. The enhanced internalization was attributed to the specific tuftsin-binding receptor, which could possibly be neuropilin-1 (Nrp1) [43] on the surface of the macrophages, higher contact surface and hydrophobic nature of the NP.

Other ligands with limited studies include O-stearoyl amylopectin, (O-SAP), and methylated bovine serum albumin (MBSA) that have a specific affinity for macrophages and selectivity for type 1 and 2 scavenger receptor, respectively. A RIF-loaded aerosolized egg phosphatidylcholine (PC)- and cholesterol (Chol)-based liposome coated with these ligands demonstrated rapid attainment of high concentration of drug in alveolar macrophages over an extended period of time as compared to free drugs. O-SAP-coated liposomes demonstrated better accumulation in macrophages, with 10.8% of O-SAP liposomes retained in the lungs after 24 h compared to 8.2% MBSA liposomes. Although the delivery system shows prospective use, the fast elimination and uptake by RES is a challenge that still needs to be overcome.

In extrapulmonary TB, fluoroquinolones with a low incidence of resistance, gatifloxacin (GAT) have been incorporated into nanoparticles and functionalized for the treatment of CNS tuberculosis. Marcianes *et al.* synthesized a PLGA 502 nanoparticle surface modified with polysorbate 80 or labrafil. Labrafil is a more lipophilic surfactant in comparison to polysorbate 80, therefore facilitates more access of nanoparticles to the CNS but has a low entrapment efficacy, giving polysobate 80 the advantage as a modifier. The distribution into the brain, lungs and liver were evaluated using rhodamine-loaded PLGA nanoparticles in male Wistar rats. In vivo fluorescence studies indicated a significant increase in rhodamine concentration in the cerebral cortex and hippocampus at 30 min and also after 60 min, indicating correspondence to the expected effect of gatifloxacin-loaded nanoparticles suggesting prolonged encapsulation of drug after delivery, as >80% of GAT was still encapsulated after 30 min, allowing enhanced drug residence time for the exertion of therapeutic efficacy [37]. The non-ionic surfactant potentially reduces cellular drug efflux [8]. In the cortex, rhodamine concentration was tenfold higher when incorporated into the nanosystem than in solution. The transportation of the nanocarrier across the BBB may be attributed to polysorbate 80, allowing adsorption of ApoA1, ApoB, and/or ApoE onto the surface of the low-density lipoprotein and other related receptors on the BBB as observed in polysorbate 80-coated poly butyl cyanoacrylate NPs, leading to endocytosis thereafter, transcytosis of the NP [28, 52]. No significant cytotoxicity was experienced with regard to cell viability and neuron survival. A greater distribution of NP was observed in the lung as compared to the liver after 30 min, which is advantageous in targeting the *M. tuberculosis* bacilli, from its origin. After 60 min, the concentration had decreased in both lungs and liver. In vitro studies of GAT release showed a biphasic release, with rapid release occurring within the first 10 h, followed by controlled release [36]. The Gat-loaded PLGA NPs functionalized with polysorbate 80 present a prospective system for the treatment of CNS TB.

Another instance of gatifloxacin use in surface-modified PLGA microparticles for active targeting to macrophages demonstrated the author's synthesis of PLGA microparticles using PLGA 502H and undergoing surface modification using labrafil. Labrafil actively increases the phagocytosis of PLGA 502H microparticles by macrophages as it makes the surface more lipophilic. Lipophilic surfaces are readily phagocytosed than hydrophilic surfaces since the same nanosystem without labrafil remained not phagocytosed after 5 h due to the formulation having a more hydrophilic surface. The internalization of the nanosystem occurred by the formation of a membrane-bound vesicle via protrusion of the cell membrane, causing invagination of the macrophage membrane and the possibility of phagosome development and phagocytosis. With a high encapsulation efficacy of 89.6%, the system was noticed in macrophages after 3 h and remained in macrophages for at least 48 h, showing suitability for the treatment of MDR-TB. In vitro studies showed a biphasic release of GAT, a rapid release in the first 3 days, followed by slow release [37].

Most recently, the enzyme mimetic activity of magnetic iron oxide nanoparticles has shown benefit in the delivery of anti-TB drugs. INH is a prodrug that is activated when exposed to KatG enzyme or catalase and peroxidase; for this reason, the absence of catalase and peroxidase or a KatG gene mutation, which is by large the most common mutations in INH-resistant MTBs is an important factor to take note of in treatment[76]. Magnetic iron oxide nanoparticles have intrinsic peroxidase-like and catalase-like enzymatic activity, showing a prospective combinatory effect against *M.tb*. A study incorporating INH in a lipoamino acid surface-modified magnetic nanoparticle with an average size of 12.93 nm demonstrated a reduced INH MIC from 1.26 μg/ml to 1.08 μg/ml when incorporated in the nanoparticle. In the use of unmodified nanoparticles, a more impressive MIC of 0.87 μg/ml was noticed, but due to the possible unintended effect of uncoated magnetic nanoparticles, it is preferable to coat it, since lipoamino acid may mimic the structure of the cell membrane [16] and prevent unintended interaction with plasma proteins in the body. Furthermore, the presence of a lipophilic structure enhances drug efficacy especially in the lung, the most prominent site for TB infection.

A study demonstrating the anti-*M.tb* effect of gallium (III) (Ga)-based compounds, types of drugs in development, when incorporated in nanoparticles was performed. In [10], various nanoparticles, comprising RIF, gallium incorporated in NPs made of pluronic F127, mannose/folic acid conjugated F127, dendrimer, and poloxamer 188, were synthesized. It was very interesting to note that Ga-loaded folate- or mannose-functionalized block copolymers provided sustained Ga release for 15 days whilst significantly inhibiting *M.tb* (H37Rv) microbial growth in THP-1 cells for the same period. Ga(III) concentration in folate-conjugated NPs was 2.5 and 10 folds compared to the mannose-NPs and free GA, respectively, by acting on folate receptor β (FRβ) [6]. In addition to the delivery and inhibition of *M.tb*, Ga was also found to block the phagosome maturation process by interrupting *M.tb*-mediated suppression of Gal3.

An active targeted system presents unique advantages and more studies, including extended in vivo studies, will confirm and assure the complete prospects of the delivery system. Furthermore, studies incorporating the delivery of combinatorial anti-TB drugs as performed in the standard regimen, or studies with the combination of anti-TB drugs and repurposed drugs in the active target nanosystems may have possible enhanced benefits to counter the progress of the disease. These strategies may be administered by various routes of administration, such as the pulmonary route for pulmonary TB, and IV or intraperitoneal for extrapulmonary tuberculosis [21].

6 Future Perspectives

The extensive studies over the years on nanotechnology have continuously gained grounds in its impressive ability to overcome physicochemical limitations, of solubility and stability, by addition virtue of its nano-size range, whilst enhancing the bioavailability of required anti-TB drugs. This size also contributes to enhancing drug permeation across membranes and cellular uptake, with its functionalization allowing specificity and selectivity to the required cell. For pulmonary TB, many studies on the administration of the medicines through the oral route for the administration of the free anti-TB drugs have been performed. More recently, pulmonary TB therapy, employing the use of nanosystems, is showing a prospective alternative to the oral route, with pulmonary administration of drugs generally gaining awareness due to its non-invasiveness, reduced side effects, and evasion of the hepatic first-pass metabolism, and the large alveoli surface area. Furthermore, since most TB is pulmonary, it will be administered directly to the required site of action [21].

Although the need to optimize the even distribution of nanocarriers across the lung remains a key parameter, another feature to consider is the aerodynamic behaviour of the nanoparticles in relation to the rapid clearance from the respiratory system. Methods to improve mucoadhesion and mucus penetration as present in polymers like chitosan and pluronic needs to be investigated [14]. In comparison to mucoadhesive formulations with identical particle size, the mucus-penetrating nanoparticles for the administration of baicalein permitted deeper penetration, cellular absorption, enhanced drug distribution in airways, and superior local distribution and bioavailability. As a result, there is a chance that pluronic and other mucus-penetrative polymers might be used to make nanosystems for pulmonary TB therapies.

More interestingly, the approach of active targeting using ligand-receptor pattern recognition demonstrates a promising preferential approach for internalization into Mtb reservoirs whilst optimizing the pharmacotherapy. Discovering and exploring other surface ligands apart from mannose and HA that may act as keys for pattern recognition on the receptors on Mtb reservoirs, such as toll-like receptors (TLRs), C-type lectin receptors (CLRs), NOD-like receptors (NLRs), that are normally used for the entrapment of the microbes need to be explored with the concept of drug internalization. With more promising in vivo data, a greater insight will allow preclinical studies to determine its possible use clinically. Only then will the prospect of acquiring a commercially viable nanoformulation become a possibility.

Overview Table

Type of nanocarrier	Drug	Particle size distribution	Surface modifier	Method employed	Targeting cells/organ	Ref.
Passive drug delivery systems						
Cetyl palmitate SLNP	Rifampicin	333–355 nm	Chitosan	Hot ultrasonication method	–	[70, 71]
Compritol® ATO 888	Rifampicin isoniazid	141.2 ± 13.5 nm 120.0 ± 0.7 nm	Polysorbate 80	Microemulsification method	–	[61]
Ethylene diamine poly(propylene imine) (EDA-PPI) 4th and 5th gen dendrimers	Rifampicin	–	PEG	Double Michael addition and exhaustive amidation reactions	–	[72] [13]
Glyceryl monostearate (GMS) and stearic acid (SA) containing SLNP	Rifampicin	30–200 nm	FITC labelled WGA	Emulsification followed by a solvent evaporation method	–	[51]
PLA micelles	Rifampicin	163–210 nm	Chitosan	Swern oxidation	–	[78]
PLGA nanoparticles	Rifampicin Isoniazid Pyrazinamide	350–400 nm	Lectin: Wheat germ agglutinin (WGA)	Two-step amide bond linkage using carbodiimide	–	[57]
PLGA nanoparticles	Moxifloxacin	50–200 nm	PEG-water soluble chitosan	Solvent-emulsion evaporation method	–	[42]
PLGA-PEG-PLGA triblock copolymer	Isoniazid	150–400 nm	Isoniazid	Water-oil-water double emulsification technique	–	[17]
Poly(ethylene glycol)-poly(aspartic acid) (PEG-PASP) copolymer	Rifampicin Isoniazid Pyrazinamide	239–293 nm	PEG	Condensation using EDC	–	[60]
Poly (ethylene glycol)–poly(L-lactide) (MPEG-PLLA) and poly(ethylene glycol)-poly(D-lactide) (MPEG-PDLA) block copolymers	Rifampicin	40–120 nm	Monomethoxy PEG	Ring-opening polymerization of L-lactide and D-lactide in the presence of monomethoxy PEG, respectively	–	[9]

Poly(ethylene oxide) monomethyl ether-block-poly(ε - caprolactone)	Rifampicin	20–60 nm	Förster resonance energy transfer (FRET) sensor	Ring-opening copolymerization	–	[69]
Poly(lactic-co-glycolic acid) (PLGA), poly(lactic-co-hydroxymethyl glycolic acid) (PLHMGA) and poly(lactic-co-benzyloxymethyl glycolic acid) (PLBMGA) copolymer	Bedaquiline	241–290 nm	mPEG	Ring-opening polymerization	–	[53]
Poly(ε-caprolactone-coglycolide)–poly(ethylene glycol)-poly(ε-caprolactone-co-glycolide)	Rifampicin	–	PEG	Ring-opening polymerization of ε-caprolactone and glycolide in the presence of PEG	–	[24]
Soy lecithin (LC) and DSPE (1,2-distearoyl-sn-glycero-3--phosphoethanolamine)	Isoniazid Ciprofloxacin	11.81 ± 1.2 nm 172.23 ± 2.31 nm	PEG	w-o-w double-emulsification-solvent-evaporation method	–	[3]
Active-targeting drug delivery system						
Chitosan-based bioadhesive nanoparticles	Clofazimine	132–184 nm	Mannose	Solvent evaporation method	Macrophages	[48]
Cholesteryl myristate with palmitic acid (PA set) or tripalmitin (TP set)	Rifampicin	0.72 ± 0.02 μm to 1.38 ± 0.19 μm	Methyl α-d-mannopyranoside	Melt emulsification technique	Respiratory tract, J774 murine macrophage	[38]
Ethylene diamine poly(propylene imine) (EDA-PPI) 5th gen dendrimers	Rifampicin	Less than 5 μm	Mannose	Double Michael addition and exhaustive amidation reactions	Alveolar macrophages	[35]
Graphene oxide nanocarrier	Rifampicin	50–300 nm	Mannose and PEG	EDC amide bond formation	Mucosal CD14+ macrophages	[49]
PLGA microparticles	Gatifloxacin	3–5 μm	Labrafil	Solvent evaporation-extraction method	Macrophages	[37]

PLGA nanoparticles	Gatifloxacin	194.9 ± 5.7 nm	Rhodamine and polysorbate 80	Acetone-water nanoprecipitation method	Brain cortex	[36]
PLGA nanoparticles	Rifampicin	3.2 μm	Mannitol (MAN) microspheres	Spray drying using four-fluid nozzle spray drier	NR8383 lungs macrophage cells	[44]
Pluronic F127	Rifampicin and gallium	~300 nm	Mannose	Solvent evaporation method and DCC amide bond linkage	THP-1 cells	[10]
Poly(epsilon-caprolactone)-poly(ethylene glycol)-poly(epsilon-caprolactone)	Rifampicin	263 and 340 nm bimodal distribution	Galactomannan	Ionotropic gelation	RAW 264.7 murine macrophages	[41]
Stearic and oleic acid, Phospholipon 80H and Tween 80	Rifampicin	–	Oleic acid-Tuftsin-modified peptide (pTUF-OA)	Peptide synthesis and purification through ion-exchange chromatography and microemulsion technique	Macrophage and microglial	[5]
Tocopherol/succinate (TS) micelles	Rifampicin	212–294.6 nm	Hyaluronic acid (HA)	–	Alveolar macrophages (AMs)	[18]
Tristearin, soya lecithin and stearyl amine.	Rifampicin	272 nm	Lactoferrin (Lf)	Solvent injection	Lungs	[59]
Witepsol® E85 forming the lipid phases and stearylamine (SA)	Isoniazid	500 nm	Mannose	Solvent emulsification-evaporation method	dTHP-1 cells	[12]

7 Conclusion

In order to get back on track with the SDGs and WHO goals to end TB by 2030 and 2035 respectively, nanotechnology systems need to be thoroughly scrutinized to obtain all possible benefits as carriers of anti-TB therapy. The improved clinical outcome will counter and concurrently treat MDR TB, therefore curbing the spread and severity of the disease. A combination of toxicity studies, biomaterial approval by regulatory affairs, and enhanced preclinical studies will bring us closer to the world of commercially available specific and selective nano-targeted therapies. This will allow dose reduction and dose frequency which will enhance patient compliance, whilst decreasing treatment cost. Moreover, simply incorporating nanoformulation with present conventional anti-TB drugs is more cost-effective, time-saving, and more beneficial to low and middle-income countries than the research and development of a new chemical entity with better therapeutic efficacy.

References

1. Abed N, Couvreur P. Nanocarriers for antibiotics: a promising solution to treat intracellular bacterial infections. Int J Antimicrob Agents. 2014;43(6):485–96.
2. Banik N, Hussain A, Ramteke A, Sharma HK, Maji TK. Preparation and evaluation of the effect of particle size on the properties of chitosan-montmorillonite nanoparticles loaded with isoniazid. RSC Adv. 2012;2(28):10519–28.
3. Bhardwaj A, Mehta S, Yadav S, Singh SK, Grobler A, Goyal AK, Mehta A. Pulmonary delivery of antitubercular drugs using spray-dried lipid–polymer hybrid nanoparticles. Artif Cells Nanomed Biotechnol. 2016;44(6):1544–55.
4. Brannon-Peppas L. Recent advances on the use of biodegradable microparticles and nanoparticles in controlled drug delivery. Int J Pharm. 1995;116(1):1–9.
5. Carneiro SP, Carvalho KV, Soares RDDOA, Carneiro CM, de Andrade MHG, Duarte RS, Dos Santos ODH. Functionalized rifampicin-loaded nanostructured lipid carriers enhance macrophages uptake and antimycobacterial activity. Colloids Surf B: Biointerfaces. 2019;175:306–13.
6. Chandrupatla DM, Molthoff CF, Lammertsma AA, van der Laken CJ, Jansen G. The folate receptor β as a macrophage-mediated imaging and therapeutic target in rheumatoid arthritis. Drug Deliv Transl Res. 2019;9(1):366–78.
7. Chaubey P, Mishra B. Mannose-conjugated chitosan nanoparticles loaded with rifampicin for the treatment of visceral leishmaniasis. Carbohydr Polym. 2014;101:1101–8.
8. Chen HH, Huang WC, Chiang WH, Liu TI, Shen MY, Hsu YH, Lin SC, Chiu HC. pH-responsive therapeutic solid lipid nanoparticles for reducing P-glycoprotein-mediated drug efflux of multidrug resistant cancer cells. Int J Nanomedicine. 2015;10:5035.
9. Chen L, Xie Z, Hu J, Chen X, Jing X. Enantiomeric PLA–PEG block copolymers and their stereocomplex micelles used as rifampin delivery. J Nanopart Res. 2007;9(5):777–85.
10. Choi SR, Britigan BE, Moran DM, Narayanasamy P. Gallium nanoparticles facilitate phagosome maturation and inhibit growth of virulent Mycobacterium tuberculosis in macrophages. PLoS One. 2017;12(5):e0177987.
11. Chono S, Tanino T, Seki T, Morimoto K. Efficient drug targeting to rat alveolar macrophages by pulmonary administration of ciprofloxacin incorporated into mannosylated liposomes for treatment of respiratory intracellular parasitic infections. J Control Release. 2008;127(1):50–8.
12. Costa A, Sarmento B, Seabra V. Mannose-functionalized solid lipid nanoparticles are effective in targeting alveolar macrophages. Eur J Pharm Sci. 2018;114:103–13.

13. Dineshkumar P, Panneerselvam T, Deepti Brundavani K, Selvaraj K, Vijayaraj Kumar P. Formulation of rifampicin loaded PEGylated 5.0 G EDA-PAMAM dendrimers as effective long-duration release drug carriers. Curr Drug Ther. 2017;12(2):115–26.
14. Dong W, Ye J, Zhou J, Wang W, Wang H, Zheng X, Yang Y, Xia X, Liu Y. Comparative study of mucoadhesive and mucus-penetrative nanoparticles based on phospholipid complex to overcome the mucus barrier for inhaled delivery of baicalein. Acta Pharm Sin B. 2020;10(8):1576–85.
15. Dye C. Global epidemiology of tuberculosis. Lancet. 2006;367(9514):938–40.
16. Fait ME, Hermet M, Comelles F, Clapés P, Alvarez HA, Prieto E, Herlax V, Morcelle SR, Bakás L. Microvesicle release and micellar attack as the alternative mechanisms involved in the red-blood-cell-membrane solubilization induced by arginine-based surfactants. RSC Adv. 2017;7(60):37549–58.
17. Gajendiran M, Gopi V, Elangovan V, Murali RV, Balasubramanian S. Isoniazid loaded core shell nanoparticles derived from PLGA–PEG–PLGA tri-block copolymers: in vitro and in vivo drug release. Colloids Surf B: Biointerfaces. 2013;104:107–15.
18. Gao Y, Sarfraz MK, Clas SD, Roa W, Löbenberg R. Hyaluronic acid-tocopherol succinate-based self-assembling micelles for targeted delivery of rifampicin to alveolar macrophages. J Biomed Nanotechnol. 2015;11(8):1312–29.
19. Garg T, Rath G, Murthy RR, Gupta UD, Vatsala PG, Goyal AK. Current nanotechnological approaches for an effective delivery of bioactive drug molecules to overcome drug resistance tuberculosis. Curr Pharm Des. 2015;21(22):3076–89.
20. Grassin-Delyle S, Abrial C, Salvator H, Brollo M, Naline E, Devillier P. The role of toll-like receptors in the production of cytokines by human lung macrophages. J Innate Immun. 2020;12(1):63–73.
21. Grotz E, Tateosian N, Amiano N, Cagel M, Bernabeu E, Chiappetta DA, Moretton MA. Nanotechnology in tuberculosis: state of the art and the challenges ahead. Pharm Res. 2018;35(11):1–22.
22. Haba Y, Kojima C, Harada A, Ura T, Horinaka H, Kono K. Preparation of poly (ethylene glycol)-modified poly (amido amine) dendrimers encapsulating gold nanoparticles and their heat-generating ability. Langmuir. 2007;23(10):5243–6.
23. Hussell T, Bell TJ. Alveolar macrophages: plasticity in a tissue-specific context. Nat Rev Immunol. 2014;14(2):81–93.
24. Jiang Z, You Y, Deng X, Hao J. Injectable hydrogels of poly (ε-caprolactone-co-glycolide)–poly (ethylene glycol)–poly (ε-caprolactone-co-glycolide) triblock copolymer aqueous solutions. Polymer. 2007;48(16):4786–92.
25. Kang PB, Azad AK, Torrelles JB, Kaufman TM, Beharka A, Tibesar E, DesJardin LE, Schlesinger LS. The human macrophage mannose receptor directs Mycobacterium tuberculosis lipoarabinomannan-mediated phagosome biogenesis. J Exp Med. 2005;202(7):987–99.
26. Karthikeyan R, Koushik OS, Kumar VP. Surface modification of cationic dendrimers eases drug delivery of anticancer drugs. Nanosci Nanotechnol. 2016;10:108.
27. Kaur R, Garg T, Goyal AK, Rath G. Development, optimization and evaluation ofelectrospun nanofibers: tool fortargeted vaginal delivery of antimicrobials against urinary tract infections. Curr Drug Deliv. 2015;22:328–34.
28. Kreuter J. Mechanism of polymeric nanoparticle-based drug transport across the blood-brain barrier (BBB). J Microencapsul. 2013;30(1):49–54.
29. Kumar A, Garg T, Sarma GS, Rath G, Goyal AK. Optimization of combinational intranasal drug delivery system for the management of migraine by using statistical design. Eur J Pharm Sci. 2015;70:140–51.
30. Kumar PV, Asthana A, Dutta T, Jain NK. Intracellular macrophage uptake of rifampicin loaded mannosylated dendrimers. J Drug Target. 2006;14(8):546–56.
31. Lam PL, Wong WY, Bian Z, Chui CH, Gambari R. Recent advances in green nanoparticulate systems for drug delivery: efficient delivery and safety concern. Nanomedicine. 2017;12(4):357–85.

32. Lee CC, MacKay JA, Fréchet JM, Szoka FC. Designing dendrimers for biological applications. Nat Biotechnol. 2005;23(12):1517–26.
33. Lee-Sayer SS, Dong Y, Arif AA, Olsson M, Brown KL, Johnson P. The where, when, how, and why of hyaluronan binding by immune cells. Front Immunol. 2015;6:150.
34. MacNeil A, Glaziou P, Sismanidis C, Maloney S, Floyd K. Global epidemiology of tuberculosis and progress toward achieving global targets—2017. Morb Mortal Wkly Rep. 2019;68(11):263.
35. Majoral JP, Zablocka M, Caminade AM, Balczewski P, Shi X, Mignani S. Interactions gold/phosphorus dendrimers. Versatile ways to hybrid organic–metallic macromolecules. Coord Chem Rev. 2018;358:80–91.
36. Marcianes P, Negro S, García-García L, Montejo C, Barcia E, Fernández-Carballido A. Surface-modified gatifloxacin nanoparticles with potential for treating central nervous system tuberculosis. Int J Nanomedicine. 2017;12:1959.
37. Marcianes P, Negro S, Barcia E, Montejo C, Fernández-Carballido A. Potential active targeting of gatifloxacin to macrophages by means of surface-modified PLGA microparticles destined to treat tuberculosis. AAPS PharmSciTech. 2020;21(1):1–14.
38. Maretti E, Costantino L, Rustichelli C, Leo E, Croce MA, Buttini F, Truzzi E, Iannuccelli V. Surface engineering of solid lipid nanoparticle assemblies by methyl α-d-mannopyranoside for the active targeting to macrophages in anti-tuberculosis inhalation therapy. Int J Pharm. 2017;528(1–2):440–51.
39. Mariappan TT, Singh S. Regional gastrointestinal permeability of rifampicin and isoniazid (alone and their combination) in the rat. Int J Tuberc Lung Dis. 2003;7(8):797–803.
40. Mignani S, Tripathi RP, Chen L, Caminade AM, Shi X, Majoral JP. New ways to treat tuberculosis using dendrimers as nanocarriers. Pharmaceutics. 2018;10(3):105.
41. Moretton MA, Chiappetta DA, Andrade F, Das Neves J, Ferreira D, Sarmento B, Sosnik A. Hydrolyzed galactomannan-modified nanoparticles and flower-like polymeric micelles for the active targeting of rifampicin to macrophages. J Biomed Nanotechnol. 2013;9(6):1076–87.
42. Mustafa S, Devi VK, Pai RS. Effect of PEG and water-soluble chitosan coating on moxifloxacin-loaded PLGA long-circulating nanoparticles. Drug Deliv Transl Res. 2017;7(1):27–36.
43. Nissen JC, Selwood DL, Tsirka SE. Tuftsin signals through its receptor neuropilin-1 via the transforming growth factor beta pathway. J Neurochem. 2013;127(3):394–402.
44. Ohashi K, Kabasawa T, Ozeki T, Okada H. One-step preparation of rifampicin/poly (lactic-co-glycolic acid) nanoparticle-containing mannitol microspheres using a four-fluid nozzle spray drier for inhalation therapy of tuberculosis. J Control Release. 2009;135(1):19–24.
45. Pabreja S, Garg T, Rath G, Goyal AK. Mucosal vaccination against tuberculosis using Ag85A-loaded immunostimulating complexes. Artif Cells Nanomed Biotechnol. 2016;44(2):532–9.
46. Pandey R, Sharma A, Zahoor A, Sharma S, Khuller GK, Prasad B. Poly (DL-lactide-co-glycolide) nanoparticle-based inhalable sustained drug delivery system for experimental tuberculosis. J Antimicrob Chemother. 2003;52(6):981–6.
47. Patra JK, Das G, Fraceto LF, Campos EVR, del Pilar Rodriguez-Torres M, Acosta-Torres LS, Diaz-Torres LA, Grillo R, Swamy MK, Sharma S, Habtemariam S. Nano based drug delivery systems: recent developments and future prospects. J Nanobiotechnol. 2018;16(1):1–33.
48. Pawde DM, Viswanadh MK, Mehata AK, Sonkar R, Poddar S, Burande AS, Jha A, Vajanthri KY, Mahto SK, Dustakeer VA, Muthu MS. Mannose receptor targeted bioadhesive chitosan nanoparticles of clofazimine for effective therapy of tuberculosis. Saudi Pharm J. 2020;28(12):1616–25.
49. Pi J, Shen L, Shen H, Yang E, Wang W, Wang R, Huang D, Lee BS, Hu C, Chen C, Jin H. Mannosylated graphene oxide as macrophage-targeted delivery system for enhanced intracellular M tuberculosis killing efficiency. Mater Sci Eng C. 2019;103:109777.
50. Piccaro G, Poce G, Biava M, Giannoni F, Fattorini L. Activity of lipophilic and hydrophilic drugs against dormant and replicating Mycobacterium tuberculosis. J Antibiot. 2015;68(11):711–4.

51. Pooja D, Tunki L, Kulhari H, Reddy BB, Sistla R. Characterization, biorecognitive activity and stability of WGA grafted lipid nanostructures for the controlled delivery of rifampicin. Chem Phys Lipids. 2015;193:11–7.
52. Ramge P, Unger RE, Oltrogge JB, Zenker D, Begley D, Kreuter J, Von Briesen H. Polysorbate-80 coating enhances uptake of polybutylcyanoacrylate (PBCA)-nanoparticles by human and bovine primary brain capillary endothelial cells. Eur J Neurosci. 2000;12(6):1931–40.
53. Ritsema JAS, Herschberg EMA, Borgos SE, Løvmo C, Schmid R, Te Welscher YM, Storm G, van Nostrum CF. Relationship between polarities of antibiotic and polymer matrix on nanoparticle formulations based on aliphatic polyesters. Int J Pharm. 2018;548(2):730–9.
54. Sanità G, Carrese B, Lamberti A. Nanoparticle surface functionalization: how to improve biocompatibility and cellular internalization. Front Mol Biosci. 2020;7:381.
55. Sarfraz M, Shi W, Gao Y, Clas SD, Roa W, Bou-Chacra N, Löbenberg R. Immune response to antituberculosis drug-loaded gelatin and polyisobutyl-cyanoacrylate nanoparticles in macrophages. Ther Deliv. 2016;7(4):213–28.
56. Sarkar K, Kumar M, Jha A, Bharti K, Das M, Mishra B. Nanocarriers for tuberculosis therapy: design of safe and effective drug delivery strategies to overcome the therapeutic challenges. J Drug Deliv Sci Technol. 2021;67:102850.
57. Sharma A, Sharma S, Khuller GK. Lectin-functionalized poly (lactide-co-glycolide) nanoparticles as oral/aerosolized antitubercular drug carriers for treatment of tuberculosis. J Antimicrob Chemother. 2004;54(4):761–6.
58. Shi X, Sun K, Baker JR Jr. Spontaneous formation of functionalized dendrimer-stabilized gold nanoparticles. J Phys Chem C. 2008;112(22):8251–8.
59. Shilpi S, Vimal VD, Soni V. Assessment of lactoferrin-conjugated solid lipid nanoparticles for efficient targeting to the lung. Prog Biomater. 2015;4(1):55–63.
60. Silva M, Lara AS, Leite CQF, Ferreira EI. Potential Tuberculostatic agents: micelle-forming copolymer poly (ethylene glycol)-poly (aspartic acid) prodrug with isoniazid. Archiv der Pharmazie Int J Pharm Med Chem. 2001;334(6):189–93.
61. Singh H, Bhandari R, Kaur IP. Encapsulation of rifampicin in a solid lipid nanoparticulate system to limit its degradation and interaction with isoniazid at acidic pH. Int J Pharm. 2013;446(1–2):106–11.
62. Singh J, Garg T, Rath G, Goyal AK. Advances in nanotechnology-based carrier systems for targeted delivery of bioactive drug molecules with special emphasis on immunotherapy in drug resistant tuberculosis–a critical review. Drug Deliv. 2016a;23(5):1676–98.
63. Singh N, Gautam SP, Singh HL, Dhiman A, Siddiqui G, Verma A. Isonizid loded dendrimer based nano carriers for the delivery of anti-tuberculosis. Indian Res J Pharmacol Sci. 2016b;3:519–29.
64. Singh S, Mariappan TT, Sharda N, Kumar S, Chakraborti AK. The reason for an increase in decomposition of rifampicin in the presence of isoniazid under acid conditions. Pharm Pharmacol Commun. 2000;6(9):405–10.
65. Smith I. Mycobacterium tuberculosis pathogenesis and molecular determinants of virulence. Clin Microbiol Rev. 2003;16(3):463–96.
66. Sosnik A, Carcaboso ÁM, Glisoni RJ, Moretton MA, Chiappetta DA. New old challenges in tuberculosis: potentially effective nanotechnologies in drug delivery. Adv Drug Deliv Rev. 2010;62(4–5):547–59.
67. Su Y, Bakker T, Harris J, Tsang C, Brown GD, Wormald MR, Gordon S, Dwek RA, Rudd PM, Martinez-Pomares L. Glycosylation influences the lectin activities of the macrophage mannose receptor. J Biol Chem. 2005;280(38):32811–20.
68. Thevenot J, Troutier AL, David L, Delair T, Ladavière C. Steric stabilization of lipid/polymer particle assemblies by poly (ethylene glycol)-lipids. Biomacromolecules. 2007;8(11):3651–60.
69. Trousil J, Filippov SK, Hrubý M, Mazel T, Syrová Z, Cmarko D, Svidenská S, Matějková J, Kováčik L, Porsch B, Konefał R. System with embedded drug release and nanoparticle degradation sensor showing efficient rifampicin delivery into macrophages. Nanomedicine. 2017;13(1):307–15.

70. Vieira AC, Chaves LL, Pinheiro M, Lima SAC, Ferreira D, Sarmento B, Reis S. Mannosylated solid lipid nanoparticles for the selective delivery of rifampicin to macrophages. Artif Cells Nanomed Biotechnol. 2018a;46(sup1):653–63.
71. Vieira AC, Chaves LL, Pinheiro S, Pinto S, Pinheiro M, Lima SC, Ferreira D, Sarmento B, Reis S. Mucoadhesive chitosan-coated solid lipid nanoparticles for better management of tuberculosis. Int J Pharm. 2018b;536(1):478–85.
72. Vijayaraj Kumar P, Agashe H, Dutta T, Jain NK. PEGylated dendritic architecture for development of a prolonged drug delivery system for an antitubercular drug. Curr Drug Deliv. 2007;4(1):11–9.
73. Wijagkanalan W, Kawakami S, Takenaga M, Igarashi R, Yamashita F, Hashida M. Efficient targeting to alveolar macrophages by intratracheal administration of mannosylated liposomes in rats. J Control Release. 2008;125(2):121–30.
74. Williams K, Minkowski A, Amoabeng O, Peloquin CA, Taylor D, Andries K, Wallis RS, Mdluli KE, Nuermberger EL. Sterilizing activities of novel combinations lacking first- and second-line drugs in a murine model of tuberculosis. Antimicrob Agents Chemother. 2012;56(6):3114–20.
75. World Health Organization. Implementing the end TB strategy: the essentials, No. WHO/HTM/TB/2015.31. World Health Organization; 2015.
76. WHO. Treatment guidelines for isoniazid-resistant tuberculosis. 2018 at https://www.who.int/tb/publications/2018/FAQ_TB_policy_recommendations_guidelines.pdf. Accessed on 15 Nov 2021.
77. World Health Organization Global tuberculosis report (2020). Licences: CC BY-NC SA 3.0 IGO.
78. Wu Y, Li M, Gao H. Polymeric micelle composed of PLA and chitosan as a drug carrier. J Polym Res. 2009;16(1):11–8.
79. Yamori S, Ichiyama S, Shimokata K, Tsukamura M. Bacteriostatic and bactericidal activity of antituberculosis drugs against Mycobacterium tuberculosis, Mycobacterium avium-Mycobacterium intracellular complex and Mycobacterium kansasii in different growth phases. Microbiol Immunol. 1992;36(4):361–8.
80. Zahoor A, Sharma S, Khuller GK. Inhalable alginate nanoparticles as antitubercular drug carriers against experimental tuberculosis. Int J Antimicrob Agents. 2005;26(4):298–303.
81. Zhang Z, Lu J, Wang Y, Pang Y, Zhao Y. Prevalence and molecular characterization of fluoroquinolone-resistant Mycobacterium tuberculosis isolates in China. Antimicrob Agents Chemother. 2014;58(1):364–9.

Tuberculosis and Drug Delivery System: Clinical Trials in TB

Shahnaz Alom, Farak Ali, Dimpee Choudhury, Apurba Gohain, Abdul Baquee Ahmed, and Md Kamaruz Zaman

Abstract Tuberculosis (TB) is a communicable disease caused by the bacterium *Mycobacterium tuberculosis* which is the second most lethal infectious disease after AIDS. The global problem of multidrug-resistant tuberculosis is nearing epidemic proportions, and it is a leading killer of young adults worldwide. It is endemic in the majority of developing countries and has resurfaced in both developed and emerging countries with high rates of HIV infection. This chapter reviews the clinical trials and drug delivery approaches for tuberculosis. A number of innovative implant, microparticulate, and other carrier-based drug delivery systems containing the main anti-tuberculosis medicines have been developed with the goal of improving patient outcomes by either targeting the location of tuberculosis infection or reducing dose frequency. To fully realize the promise of drug development, it will take creativity, perseverance, collaboration, and resources. A delicate balance must be struck between preserving medications from developing resistance and ensuring that regimens are low cost, readily available, and widely accepted by healthcare systems and clinicians.

S. Alom
Department of Pharmacology, Girijananda Chowdhury Institute of Pharmaceutical Science, Tezpur, Assam, India

F. Ali
Department of Pharmaceutical Chemistry, Girijananda Chowdhury Institute of Pharmaceutical Science, Tezpur, Assam, India

D. Choudhury
Department of Botany, Abhayapuri College, Abhayapuri, Assam, India

A. Gohain
Department of Chemistry, Assam University Silchar, Silchar, Assam, India

A. B. Ahmed
Department of Pharmaceutics, Girijananda Chowdhury Institute of Pharmaceutical Science, Tezpur, Assam, India

M. K. Zaman (✉)
Department of Pharmaceutical Sciences, Dibrugarh University, Dibrugarh, Assam, India
e-mail: kzaman71@dibru.ac.in

R. Shegokar, Y. Pathak (eds.), *Tubercular Drug Delivery Systems*,
https://doi.org/10.1007/978-3-031-14100-3_14

Keywords Tuberculosis · Drug delivery · Mycobacterium · Lungs · Pulmonary · Clinical trials

1 Introduction

Tuberculosis (TB) is a common and extremely infectious chronic granulomatous bacterial illness, where mortality is very high among young adults across the world. It is caused by the bacillus *Mycobacterium tuberculosis*, which most commonly affects the lungs (pulmonary tuberculosis), where chances of infection may spread to other body parts also (extrapulmonary TB). The illness generally spreads into the air, when persons with pulmonary tuberculosis exhale germs, such as by coughing. TB is still a serious worldwide health issue that affects millions of individuals each year. The illness can strike anybody, anywhere, although the majority of persons who contract it (about 90%) are adults, the male-to-female ratio is 2:1, and annual case rates range from fewer than 50 to more than 5000 per 1 million populations. Each year, 30 nations with a high TB burden account for about 90% of all cases. The rate per 100,000 population is used to describe the incidence of all types of tuberculosis, the incidence of infected cases, and death. The majority of cases (5–6 million) affect persons between the ages of 15 and 49. The chance of developing active illness varies depending on the length of time since infection, age, and host immunity; nonetheless, a newly infected young child's lifetime risk of disease has been estimated at 10% [1]. Figure 1 depicts a schematic description of TB.

The region of Southeast Asia has the highest number of instances, accounting for 33% of all incident cases worldwide. In 2003, however, the estimated incidence per capita in sub-Saharan Africa was about twice that of Southeast Asia, ranging from 290 to 350 cases per 100,000 people [1]. After the human immunodeficiency virus, it is the second largest cause of mortality from an infectious illness globally (HIV). Noncompliance with recommended regimens has become one of the primary issues in tuberculosis treatment today, owing to the fact that TB therapy requires constant, frequent multiple drug therapy. The introduction of long-duration medication formulations that release antitubercular drugs in a gradual and sustained way should enhance treatment adherence and effectiveness [2]. For patients with pulmonary tuberculosis, a 6-month regimen of four first-line medications is now recommended: isoniazid, rifampicin, ethambutol, and pyrazinamide where 85% chances of successful therapy are there. This medication is beneficial in both HIV-positive and HIV-negative people. The liver and kidneys are overworked and progressively deteriorate as a result of the cumulative effects of the medications over such a lengthy period of time. The most of the standard anti-TB medications are taken orally and undergo first-pass metabolism in the liver, resulting in adverse effects [3]. Moreover, the recent rise of resistant TB strains and the scarcity of novel anti-TB medications pose a potential danger to TB prevention and treatment. The exposure of mycobacteria to sub-therapeutic levels of antibiotic is one of the factors for the formation of resistant TB strains. Treatment failure is mostly caused by patient noncompliance

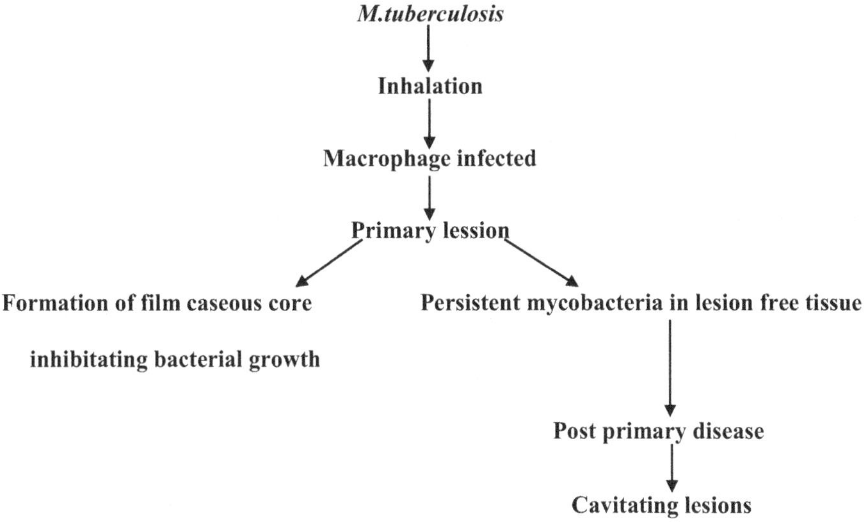

Fig. 1 Progression of tuberculosis (TB)

with the treatment regimen and multidrug-resistant tuberculosis (MDR-TB). MDR-TB therapy necessitates the use of second-line medicines, which are less effective and well tolerated. In order to prevent resistant TB, each case of tuberculosis must be adequately treated, and patient compliance must be improved. The progression of TB can be briefly described with the following schematic diagram (Fig. 1).

Pulmonary delivery systems have been extensively studied for the treatment of pulmonary-related disorders in order to avoid systemic adverse effects and improve therapeutic efficacy by delivering the medicine directly to the site of action. Inhalable dry powders, in particular, were created for the aerosol administration of anti-TB medications to the lungs. These powders, which include anti-TB drugs, can be administered straight to the lungs. They are made to go straight to the alveolar macrophages, where the bacteria grow. Local administration may result in high medication concentrations in the lungs, thereby decreasing treatment time and avoiding multidrug resistance (MDR). This chapter aims to examine clinical studies for medication delivery techniques for TB. The bacterium that causes TB is *Mycobacterium tuberculosis*. Other 'atypical' mycobacteria, such as *M. kansasii*, can cause illness with a similar clinical and pathologic presentation. In terms of organ distribution, *M. avium-intracellulare infection* (MAI) observed in immunocompromised hosts (especially in people with AIDS) is not primarily a pulmonary infection (mostly in organs of the mononuclear phagocyte system) [4]. Tuberculosis is becoming a global issue. War, famine, homelessness, and a lack of medical treatment are all factors that contribute to the rising prevalence of tuberculosis among the poor. Because tuberculosis is easily transmitted from person to person, an increase in TB in any sector of the population poses a threat to everyone [5]. This means that relevant public health interventions, including as screening, immunization (where necessary), and treatment, must be implemented and maintained. An increase in instances will be aided by a lack of public health measures. Inadequate treatment increases the establishment of TB-resistant strains.

2 Patterns of Infection

There are two main TB disease patterns:

Primary TB: It is a type of tuberculosis that occurs in youngsters. A tiny subpleural granuloma is the first site of infection, which is followed by granulomatous hilar lymph node infection. The Ghon complex is made up of all of these things. In nearly all cases, the granulomas dissolve and the infection does not spread further.

Secondary TB: It is a reactivation (or reinfection) of a previous infection that occurs predominantly in adults when their health deteriorates. The granulomatous inflammation is far more extensive and florid. The higher lobes of the lungs are usually the most damaged, and cavitation can ensue.

When infection resistance is very low, a 'miliary' pattern of dissemination can develop, with a slew of minute millet seed (1–3 mm) granulomas appearing in the lungs or other organs.

When tuberculosis spreads outside of the lungs, it can cause a variety of unusual results with recognizable patterns:

Skeletal Tuberculosis Tuberculous osteomyelitis, often known as Pott's disease, affects the thoracic and lumbar vertebrae, followed by the knee and hip. With compressive fractures (kyphosis) and extension to soft tissues, including a psoas 'cold' abscess, there is significant necrosis and skeletal destruction.

Genital Tract Tuberculosis Tuberculous salpingitis and endometritis are caused by tuberculosis spreading to the fallopian tube, causing granulomatous salpingitis, which can flow into the endometrial cavity, resulting in granulomatous endometritis, irregular monthly bleeding, and infertility. Tuberculosis most commonly affects the prostate and epididymis in men, causing non-tender induration and infertility.

TB in the urinary tract: A 'sterile pyuria' with WBCs in the urine but a negative routine bacterial culture may indicate renal tuberculosis. If not treated, the renal parenchyma would gradually deteriorate. With ureteral stricture, drainage to the ureters might cause irritation.

Tuberculosis in the CNS A meningeal pattern of spread can develop, with high protein, low glucose, and lymphocytosis in the cerebrospinal fluid. Because the base of the brain is frequently implicated, cranial nerve symptoms may be present. A solitary granuloma, also known as a 'tuberculoma', can occur and cause seizures in some people.

Tuberculosis in the Gastrointestinal Tract This is now uncommon due to widespread pasteurization of milk, which has eliminated *Mycobacterium bovis* infections. *M. tuberculosis* organisms coughed up in sputum, on the other hand, may be swallowed and spread throughout the GI tract. Circumferential ulcerations with small intestine stricture are the most common lesions. Because of the abundance of lymphoid tissue and the slower pace of passage of luminal substances, ileocecal involvement is more common.

Tuberculosis of the Adrenals Tuberculosis of the adrenals is frequently bilateral, resulting in enlargement of both adrenals. Addison's disease is caused by the destruction of the cortex.

Scrofula Tuberculous lymphadenitis of the cervical lymph nodes can result in a mass of hard, matted nodes right beneath the jaw. Overlying skin may have chronic draining fistulous tracts. Children may have this problem, and *Mycobacterium scrofulaceum* can be cultured.

Tuberculosis of the Heart The pericardium is the most common site of tuberculosis of the heart. A granulomatous pericarditis develops as a result, which can be haemorrhagic. If the fibrosis is significant and prolonged, it might develop to calcification and constrictive pericarditis.

3 Microscopic Findings

Microscopically, TB infection causes granulomatous inflammation with epithelioid macrophages and Langhans giant cells, lymphocytes, plasma cells, possibly a few PMNs, fibroblasts with collagen, and caseous necrosis in the core. A type IV hypersensitivity reaction triggers the inflammatory response [1]. This can be used as a starting point for a TB skin test diagnosis. The organisms will appear as slender red rods in an acid-fast stain (Ziehl-Neelsen or Kinyoun's acid fast stains). The organisms will be easier to screen with an auramine stain when seen under fluorescence microscopy, and more organisms will be visible. Sputum is the most typical item tested; however, histologic staining can also be done on tissues or other bodily fluids. Below is the microscopic view of TB infection caused by granulomatous inflammation (Fig. 2)

4 Drug Delivery Systems in TB

Drug administration to the lungs provides a number of benefits, including quick and prolonged drug delivery, high effectiveness, no first-pass metabolism, and the ability to induce both local and systemic effects. The enormous surface area of the lungs, the weak absorption barrier, and the high vascularity all contribute to improved medication delivery via the lungs [7]. The use of aerosolized medications in the treatment of localized disease conditions in the lungs is a well-established modality [8]. Some important modes of drug delivery are discussed here. Liposomes are bilayer vesicles with a membranous lipid bilayer that completely encloses an aqueous volume. Because they may be made with phospholipids endogenous to the lungs as surfactants, they are the most intensively explored methods for regulated delivery of pharmaceuticals to the lungs. They have the ability to entrap both hydrophilic and hydrophobic medicines. The supply of oxygen to the lungs has been enhanced and even tested in animals and humans [9]. Medicines encapsulated in liposomes have higher bactericidal action than free drugs, notably in the treatment of monocytes and macrophages [10]. Nanocapsules are the vesicular systems where the medication is encapsulated within the center volume of these vesicular networks, which is surrounded by an embryonic continuous polymeric sheath. These are a combination of polymer nanocapsules and liposomes. Liposomes are made with organic solvents and are leaky and unstable in biological fluids. Lung nanocapsules, on the other hand, are made with a solvent-free, soft energy method and have a high level of stability (with physical stability up to 18 months) [11, 12]. Nanoparticles are colloidal formulations made up of synthetic or semisynthetic polymers. The drug molecules are trapped in the polymer matrix as particulates enmesh or solid solution, or they may be physically or chemically attached to the particle surface. Inhalation is the most common way for airborne nanoparticles to be delivered [13]. Microparticles are spherical particles having a core material that range in size from 50 nm to 2 mm. They are usually injected intraperitoneally or directly into the target organs, and due to their size, they offer a long-term drug depot. They necessitate the diversity of colloidal carrier bio-disposition's normal course, that is, passive accumulation. Intracellular targeting with nanoparticles can be generated by derivatized polymers, which orient projecting their hydrophilic portion exposed to aqueous bulk while the hydrophobic segment is sheltered, resulting in a hydrophilic surface that evades identification by the reticulo-endothelial system [14]. Because of its favourable pharmacokinetics, nano-particulate drug delivery may be able to overcome a wider range of drug resistance. Niosomes are the structures created by the self-assembly of non-ionic amphiphiles in an aqueous environment, culminating in a closed bilayer structure, known as non-ionic surfactant-based vesicles (niosomes). The assembly is seldom spontaneous [15], and it almost always requires some form of energy input, such as physical agitation or heat. As a result, the hydrophobic regions of the molecule are protected from the aqueous solvent, while the hydrophilic head groups have the most interaction with it. Rifampicin encapsulated in Span 85 (sorbitan tri-oleate)-based niosomes in the 8–15 mm size

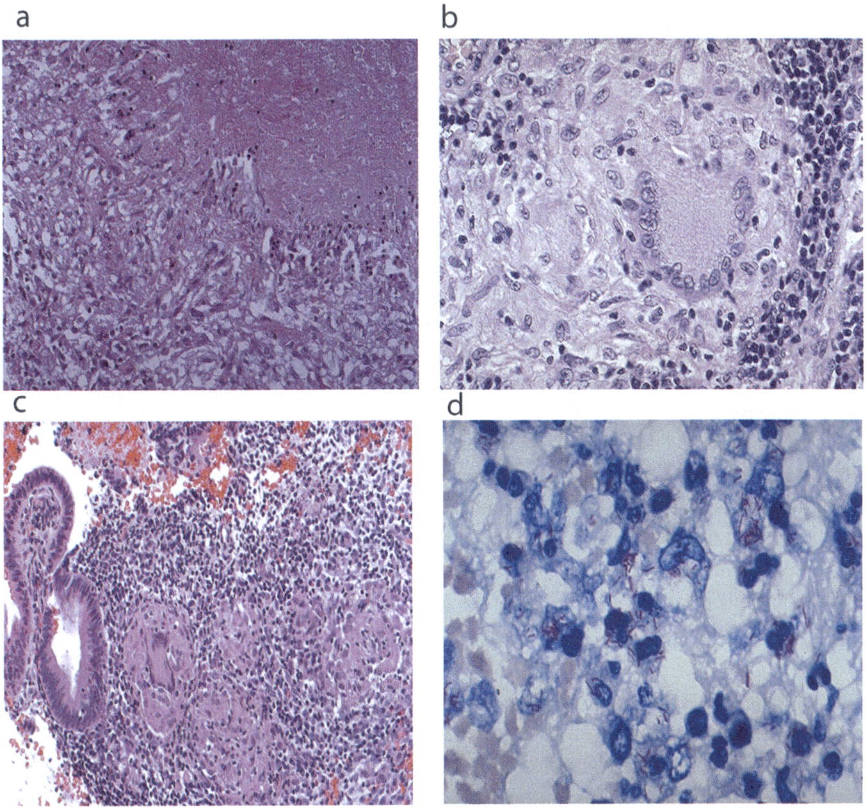

Fig. 2 Microscopic view of TB infection caused by granulomatous inflammation. (**a**) The pink, amorphous region in the center of this granuloma in the upper right, surrounded by epithelioid cells on the left and lower areas of this photomicrograph, should be noted. Caseous necrosis appears like this on a microscopic level. (**b**) A Langhans big cell (a committee of macrophages with the nuclei placed at the periphery) and elongated epithelioid cells, which are changed from macrophages, are illustrated here. (**c**) Granulomas are not necessarily well-formed, although epithelioid cells should be present. The presence of giant cells is common. Lymphocytes, plasma cells, and neutrophils are among the other inflammatory cell components. Collagenization is frequently a sign of a healing process. Granulomas can become calcified as they recover. (**d**) The presence of mycobacteria in tissue and cytologic preparations is determined using an acid-fast stain. The thin red rod-like organisms should be noted Ref. [6]

range were observed to aggregate in the lungs of mice indicating that enhanced anti-tuberculosis therapy may be possible [16].

Biodegradable Microspheres

These drug carriers may be made in a variety of particle sizes, which is important for *in vivo* deposition of particulate carriers. Drugs can be readily included with relatively high efficiency, and variable drug release rates may be obtained by manipulating the synthetic synthesis technique. *In vitro* and *in vivo*, they are more physicochemically stable. Entrapped drugs have a longer duration of action and a slower

Table 1 Table showing numerous delivery systems and their modes of distribution, and the proportion of drug put into the system

Sl. no.	Delivery system	Mode of delivery	Drug loading
1	Liposomes conventional	Nebulization	Rifampicin 22% Isoniazid 14%
2	Liposomes ligand appended	Nebulization	Rifampicin 40%
3	Liposome	Nebulization	Amikacin 15–21%
4	Liposome	Instillation	Amikacin 40%
5	Liposome	Nebulization	Rifampicin 69.3%
6	Liposome	Nebulization	Rifampicin 85.3%
7	Microparticles PLG	Nebulization/Insufflation	Rifampicin 30%
8	Microparticles PLA dry powders	Inhalation	Rifampicin11% Isoniazid 4%
9	Microparticles DPP	Insufflation	PAS, 95%
10	Microparticles	Inhalation	Rifampicin 20.8 ± 2.4%
11	Nanoparticles PLG	Nebulization	Rifampicin 60–70%
12	Lectin PLG	Nebulization	Rifampicin 60–70%
13	Solid lipid nanoparticles	Nebulization	Rifampicin 40–50%
14	Microspheres	Dry powder inhaler	Capreomycin sulphate 6.2 ± 2%

rate of release. Easy formulation is made possible by the increased stability. Following systemic injection, a variety of biodegradable microspheres were shown to be non-toxic, biodegradable, and non-immunogenic [17]. After intravenous administration, model microspheres coated with a polaxamine-980 block copolymer showed extended circulation half-life in the arterial compartment, little or no RES absorption, and significant deposition levels in the lungs [18]. The tabular representation is as follows (Table 1).

5 Clinical Studies and Novel Medication Regimens for Tuberculosis

A potential TB medication pipeline has emerged in the last five years. Several candidates with potent anti-*Mycobacterium tuberculosis* activity are in the lead optimization stages, as well as phase II and phase III clinical studies.

5.1 Fluoroquinolones

Fluoroquinolones are the mainstay of treatment for MDR-TB, and their efficacy has been demonstrated in many studies [19]. They also have a potential role in reducing the treatment duration in drug-susceptible TB [20]. Guidelines for the treatment of

MDR-TB recommend late-generation fluoroquinolones (levofloxacin and moxifloxacin). The finest fluoroquinolones, on the other hand, are a contentious topic. Moxifloxacin has lower minimum inhibitory concentrations (MICs) and stronger bactericidal activity than levofloxacin, according to many experimental and animal studies [21]. However, high dosages of levofloxacin (1000 mg/day) have been shown in some studies to have outstanding early bactericidal activity (EBA) comparable to moxifloxacin [4]. There were no differences in sputum culture conversion at 3 months of treatment (88.3% in the levofloxacin group vs. 90.5% in the moxifloxacin group, $P > 0.05$) or the proportions of adverse drug reactions in a recently reported clinical trial comparing the efficacy of moxifloxacin (400 mg/day) and levofloxacin (750 mg/day) in MDR-TB treatment. The efficacy and safety of both medicines in the early stages of MDR-TB treatment were demonstrated in this study.

Aside from the development of new medications, drug resistance prevention is an essential concern in MDR-TB care because fluoroquinolone resistance is a significant risk factor for poor outcomes [5]. Only 7 of the 15 countries with their own national lower respiratory tract infection and community-acquired pneumonia guidelines had information on the potential risk of developing fluoroquinolone resistance in patients with misdiagnosed TB, according to a recent study conducted in 24 European countries. However, a meta-analysis of six studies found that individuals who were exposed to fluoroquinolones before their TB diagnosis had a threefold increased chance of acquiring fluoroquinolone-resistant TB [22].

5.2 *Rifapentine*

Rifapentine was tested as a once-weekly TB treatment in combination with isoniazid because of its long half-life and lower MIC [23]. Patients with cavitary pulmonary TB or HIV infection, on the other hand, had a significant failure rate with the regimen. Furthermore, in pulmonary TB, 5 days/week of rifapentine treatment in combination with isoniazid, ethambutol, and pyrazinamide was not more effective than a standard rifampin regimen [24], despite another study showing that a 3-month rifapentine regimen cured TB and prevented its relapse in mouse models. Because the efficacy of rifapentine is related to peak concentration [25], the disparity in these results between mice and human investigations could be explained by an inadequate peak concentration. To determine the safety and activity, more research is needed. More research is needed to determine the safety, activity, and pharmacokinetics of the drug. A systemic and conclusive research is required to determine the safety, activity and pharmacokinetic profile of the drug. In animal studies, it was found that, the rifapentine exhibit longer half-life and more potent anti-*M. tuberculosis* activity than rifampin, as a result, it can be opted for the treatment of latent tuberculosis. Research revealed that, a 3-month course of these drugs would be helpful for treating latent *M. tuberculosis* because weekly rifapentine and isoniazid are efficacious in the continuation phase of tuberculosis treatment in individuals with a low bacillary burden. Direct observation revealed a similar result. In terms of preventing tuberculosis, 3 months of rifapentine with isoniazid treatment was just as effective as 9 months of isoniazid treatment alone, and it had a higher treatment completion rate. It will be crucial to monitor safety over the long run [26].

5.3 Bedaquiline

A diarylquinoline, bedaquiline, is a strong novel medication that inhibits the proton pump in adenosine triphosphate (ATP) synthesis, resulting in insufficient ATP synthesis. In a murine model, it possesses a very low MIC against *M. tuberculosis* and robust bactericidal action that outperforms isoniazid and rifampin [27]. A conventional 2-month treatment regimen with bedaquiline produced high culture conversion rates, fast sputum culture conversion, and little acquired resistance to companion medicines in phase II studies for newly diagnosed MDR-TB [28].

When an effective treatment regimen cannot be offered, the WHO and the US Centers for Disease Control and Prevention have advised that bedaquiline be used as an interim recommendation for MDR-TB adult patients. However, due of an elevated risk of death and QT prolongation, there are still safety concerns about this medicine.

5.4 Nitroimidazoles

Delamanid and PA-824 are novel nitroimidazoles that inhibit mycolic acid formation in phase II and phase III clinical trials. Delamanid exhibited high EBA in patients with pulmonary tuberculosis and potent action against drug-susceptible and drug-resistant *M. tuberculosis* in both in vitro and in vivo experiments [29]. In a subsequent study of the efficacy and safety of delamanid in patients with MDR-TB, researchers found that 2 months of treatment with two different drug doses (100 and 200 mg twice daily) in combination with a standard MDR-TB regimen resulted in a significant increase in culture conversion compared to placebo (45 percent in the 100 mg group vs. 42% in the 200 mg group vs. 29% in the placebo group). In the subsequent open-label extension experiment for MDR and XDR-TB, a mortality advantage was also demonstrated; specifically, mortality rates in patients receiving delamanid for at least 6 months and 2 months or less were 1% and 8.3%, respectively. QT prolongation, on the other hand, was recorded substantially more frequently in delamanid-treated patients than in placebo-treated individuals.

PA-824 has a low MIC that is comparable to isoniazid. In a recent phase II clinical trial, the mean 14-day EBA of this regimen was comparable to that of standard treatment in patients with drug-susceptible pulmonary TB. In February 2013, an 8-week serial sputum colony count trial utilizing this regimen was completed, although the results have yet to be published [30].

5.5 *Clofazimine*

Clofazimine has been used to treat leprosy for a long time; however, it was first created as an anti-TB medicine. According to meta-analyses [31], the drug has lately emerged for the treatment of drug-resistant tuberculosis. The most common reported side effects were gastrointestinal disturbances and skin pigmentation, and a clofazimine-containing regimen exhibited satisfactory treatment outcomes in 62–65% of cases [30]. A clofazimine-containing regimen consisting of gatifloxacin, ethambutol, and pyrazinamide obtained an 88% relapse-free cure rate among 206 patients in observation research on the efficiency of standardized MDR-TB regimens [32]. A clofazimine-containing regimen resulted in a reduced bacillary load after 2 months of treatment and negative conversion after 5 months of treatment in a recent mouse research.

6 Meropenem and Clavulanate in Combination Therapy

M. tuberculosis develops extended-spectrum b-lactamase, which makes it intrinsically resistant to b-lactam medicines like meropenem (BlaC). Because clavulanate inhibits BlaC, meropenem and clavulanate have significant anti-*M. tuberculosis* activity *in vitro* [33]. A meropenem-clavulanate-containing regimen has been shown in recent studies to have high smear and culture conversion in MDR/XDR-TB.

6.1 *SQ109*

SQ109 is a 1,2-ethylenediamine derivative of ethambutol that exhibits action against both drug-susceptible and drug-resistant tuberculosis by inhibiting protein synthesis in *M. tuberculosis*. In vitro, it displays synergistic effects with bedaquiline and preferential interactions with PNU-100480 (sutezolid). SQ109 is now undergoing phase II studies [34].

6.2 *Benzothiazinones*

Benzothiazinones are a new class of antimycobacterial drugs that inhibit the formation of decaprenylphospho-arabinose, the precursor of the arabinans in the mycobacterial cell wall, and have the strongest antimycobacterial efficacy against *M. tuberculosis*. BTZ043 is effective against 240 *M. tuberculosis* clinical isolates, including drug-susceptible MDRTB and XDR-TB [35]. There were no antagonistic interactions detected in a study of BTZ043 and rifampin, isoniazid, ethambutol, TMC207, PA-824, moxifloxacin, meropenem with or without clavulanate, and

SQ109, and most of the interactions were additive. When BTZ043 was coupled with bed aquiline, synergic effects were observed [36].

6.3 Novel Regimens

In clinical trials, numerous novel regimens for drug-susceptible and drug-resistant tuberculosis are being tested. These efforts have been made to reduce the time of chemotherapy in drug-susceptible and drug-resistant TB patients, as well as to improve drug-resistant TB efficacy.

7 Treatment Regimens for TB That Are Less Time-Consuming

Treatment shortening could lead to a breakthrough in TB control, as protracted treatment duration causes many patients to drop out early. Patients with noncavitary disease whose sputum cultures converted to negative after 2 months had significantly higher rates of relapse in the 4-month arm than in the standard 6-month arm in a randomized equivalence trial evaluating whether treatment duration could be shortened to a 4-month regimen using standard doses of rifampicin in patients with a low risk of relapse (with noncavitary disease whose sputum cultures converted to negative after 2 months) [37]. The failure to remove dormant bacilli may be to blame for the increased relapse rate in the shorter treatment group. As a result, successful treatment regimens that reduce treatment time must emphasize the introduction of new medications with unique mechanisms of action.

Fluoroquinolones could play a significant role in this regard. However, a recent randomized trial found that, the 4-month thrice-weekly regimens of gatifloxacin or moxifloxacin with isoniazid, rifampin and pyrazinamide (2GHRZ3/2MHR3 or 2MHRZ3/2MHR3) had a higher relapse rate of 15% and 11%, respectively, compared to 6% in the standard 6-month thrice weekly treatment (2EHRZ3/4HR3) of long run tuberculosis. Three noninferiority trials are currently evaluating the use of fluoroquinolones to reduce treatment duration to 4 months by employing a daily dose strategy at least during the intensive phase rather than an intermittent dosing regimen. OFLOTUB (A Multicentre, Randomized, Control Trial of Ofloxacin-Containing, Short-Course Regimen for the Treatment of Pulmonary Tuberculosis; NCT00216385) compares the standard 6-month regimen to a 2-month intensive phase in which gatifloxacin replaces ethambutol, followed by a 2-month maintenance phase in which gatifloxacin, isoniazid, and rifampin. Follow-up data has been collected; however; the results are yet to be released [38]. REMox TB (NCT00864383) compares normal 6-month therapy to two study regimens [2 months of moxifloxacin, isoniazid, rifampin, and pyrazinamide followed by 2 months of moxifloxacin, isoniazid, and pyrazinamide (2MHRZ/2MHR) or 2 months of ethambutol, moxifloxacin, rifampin, and pyra]. The RIFAQUIN trial (ISRCTN44153044) is comparing the standard 6-month regimen with two study regimens [2 months of daily ethambutol, moxifloxacin, rifampin, and pyrazinamide followed by 2 months of twice weekly moxifloxacin and rifapentine (2EMRZ/2PM2) or 4 months of once-weekly moxifloxacin and rifapentine (2EMRZ/4MP1)]

8 New Tuberculosis Treatment Regimens

In most patients, WHO recommendations on the treatment of MDR-TB propose an intense treatment period of 8 months and a total treatment duration of 20 months [19]. However, treatment success rates for MDR-TB were unsatisfactory as a result of this programmatic management [5]. An observational analysis of standardized treatment regimens for MDR-TB found that they are extremely effective and have few side effects. In particular, an 88% cure rate was achieved using gatifloxacin in combination with ethambutol, pyrazinamide, and clofazimine (5-month maintenance phase), supplemented with kanamycin, prothionamide, and isoniazid during an intensive phase of 4 months or until sputum smear conversion (4KCGEHZP/5GEZC) [32]. Despite the fact that this observational trial had substantial limitations, such as a lack of similar patients treated with WHO-recommended regimens, the fact that this regimen significantly decreased the duration of MDR-TB treatment and raised the cure rate is encouraging. The STREAM trial (The Evaluation of a Standardized Treatment Regimen of Anti-Tuberculosis Drugs for Patients with Multidrug-Resistant Tuberculosis; ISRCTN78372190) is currently underway, and it is comparing WHO-recommended individualized regimens with 4KCMEHZP/5MEZC (moxifloxacin instead of gatifloxacin).

9 Fixed-Dose Combinations

Since the 1980s, fixed-dose combination (FDC) tablets have been manufactured to improve patient compliance with anti-TB treatment. FDC pills combine two or more anti-TB medications. FDCs have been frequently utilized because they were thought to prevent physician error when prescribing medications and patient error when taking only one medicine [39]. The effectiveness of FDCs has been studied extensively through observational research and clinical trials. The outcomes, however, have been inconsistent. To see if FDCs are useful in treating tuberculosis, a systematic review and meta-analysis was recently conducted. The use of FDCs for the purpose of improving treatment outcomes is not supported by this systematic review of available data. FDCs, on the other hand, were linked to a higher risk of illness or treatment failure. However, the outcomes have not always been consistent. A systemic review and meta analysis were just completed to see, if FDC's are beneficial in the treatment of tuberculosis or not, as the limitations of FDC's are found to be higher than separate drug regimens. Moreover, there is no such concrete evidence of improving patients compliance as well as treatment satisfaction through FDC's. This systematic review of current evidence on TB [40] does not believe that FDCs should be used for this purpose. Treatment outcomes are improving. FDCs, on the other hand, were related with an increase in the risk of treatment failure or a return of the disease.

10 Future Prospects

According to Wade Hampston Frosts, 'age and earlier exposure do not confer immunity against TB as they do against many acute infections'. Because pulmonary TB is the most common form of tuberculosis and *Mycobacterium tuberculosis* dwells in alveolar macrophages, giving anti-tubercular drugs via the respiratory system is a potential promise. Due to the difficulties of diagnosis, MDR, and treatment adherence, the goal of eliminating the transmission of the causal organism is now out of reach. The inability to control a treatable disease has several causes. The HIV/AIDS pandemic [41] cases and case fatality have grown more quickly, and drug resistance rates are high in several countries (especially to rifampicin, which is too expensive to be included in many of the regimens). Inactivation of the drug, changed cell wall permeability or drug efflux, drug titration owing to target overproduction, and target modification are the main routes of resistance. Intrinsic drug delivery principles may be used to segregate the distribution of pure pharmaceuticals in order to create an oral system as cost effective, low dose intake (for new potent drugs), robust supply chain and access to remotely spread population are the key factors.

11 Conclusion

Tuberculosis has a devastating effect on underdeveloped countries. In this environment, drug-resistant tuberculosis poses a significant obstacle to efficient tuberculosis control. The goal is to discover a solution by developing or manufacturing better and more effective medications that minimize treatment time, reduce drug toxicity, and have a higher bioavailability. The objective is to discover a way to stop the spread of the causal organism, but owing to the difficulty of identification, antibiotic resistance, and patients' low compliance with therapy, this is a complex, multifaceted, and thorny task. A number of new anti-TB potential medications are under development, but they have several serious limitations, including a lack of thorough research and a high price tag. For the past 40 years, a standard TB treatment regimen has been used with established high efficacy and safety, but improvements are needed to reduce treatment duration and increase efficacy in drug-resistant

TB. These modifications in TB treatment could be enabled by recent advances in TB medication research. However, there are still a number of issues that need to be addressed before these new drugs can be used, including dose optimization and the development of new regimens that not only combine new drugs but also add new drugs to standard regimens, which could improve treatment efficacy while reducing adverse events and drug interactions. Additionally, enhanced efforts should be undertaken to decrease drug resistance throughout TB treatment.

Conflicts of interest Authors declares that there is no any competing interest.

References

1. Frieden TR, Sterling TR, Munsiff SS, Watt CJ, Dye C. Tuberculosis. In: Lancet. Elsevier; 2003. p. 887–99.
2. Zumla A, Chakaya J, Centis R, D'Ambrosio L, Mwaba P, Bates M, et al. Tuberculosis treatment and management-an update on treatment regimens, trials, new drugs, and adjunct therapies. Lancet Respiratory Med Elsevier. 2015;3:220–34.
3. Chaisson RE, Schecter GF, Theuer CP, Rutherford GW, Echenberg DF, Hopewell PC. Tuberculosis in patients with the acquired immunodeficiency syndrome. Clinical features, response to therapy, and survival. Am Rev Respir Dis. 1987;136(3):570–4.
4. Johnson JL, Hadad DJ, Boom WH, Daley CL, Peloquin CA, Eisenach KD, et al. Early and extended early bactericidal activity of levofloxacin, gatifloxacin and moxifloxacin in pulmonary tuberculosis. Int J Tuberc Lung Dis. 2006;10(6):605–12.
5. Orenstein EW, Basu S, Shah NS, Andrews JR, Friedland GH, Moll AP, et al. Treatment outcomes among patients with multidrug-resistant tuberculosis: systematic review and meta-analysis. Lancet Infect Dis. 2009;9(3):153–61.
6. Ehlers LP, Bianchi MV, Argenta FF, Lopes BC, Taunde PA, Wagner PGC, et al. Mycobacterium tuberculosis var. tuberculosis infection in two captive black capuchins (Sapajus nigritus) in Southern Brazil. Braz J Microbiol. 2020;51(4):2169–73.
7. Ruge CC, Kirch J, Lehr CM. Pulmonary drug delivery: from generating aerosols to overcoming biological barriers-therapeutic possibilities and technological challenges. Lancet Respiratory Med Elsevier. 2013;1:402–13.
8. Zeng XM, Martin GP, Marriott C. The controlled delivery of drugs to the lung. Int J Pharm. 1995;124:149–64.
9. Kellaway IW, Farr SJ. Liposomes as drug delivery systems to the lung. Adv Drug Deliv Rev. 1990;5:149–61.
10. Couvreur P, Fattal E, Andremont A. Liposomes and nanoparticles in the treatment of intracellular bacterial infections. Off J Am Assoc Pharm Sci Pharm Res. 1991;8:1079–86.
11. Heurtault B, Saulnier P, Pech B, Proust JE, Benoit JP. A novel phase inversion-based process for the preparation of lipid nanocarriers. Pharm Res. 2002;19(6):875–80.
12. Byron PR. Prediction of drug residence times in regions of the human respiratory tract following aerosol inhalation. J Pharm Sci. 1986;75(5):433–8.
13. Hoet PHM, Brüske-Hohlfeld I, Salata OV. Nanoparticles – known and unknown health risks. J Nanobiotechnol BioMed Central. 2004;2:12.
14. Gref R, Domb A, Quellec P, Blunk T, Müller RH, Verbavatz JM, et al. The controlled intravenous delivery of drugs using PEG-coated sterically stabilized nanospheres. Adv Drug Deliv Rev Elsevier. 1995;16:215–33.
15. Lasic DD. On the thermodynamic stability of liposomes. J Colloid Interf Sci. 1990;140:302–4.

16. Uchegbu IF, Vyas SP. Non-ionic surfactant based vesicles (niosomes) in drug delivery. Int J Pharm. 1998;172(1–2):33–70.
17. Zeng XM, Martin GP, Marriott C. The controlled delivery of drugs to the lung. Int J Pharm. 1995;124(2):149–64.
18. Illum L, Davis SS, Müller RH, Mak E, West P. The organ distribution and circulation time of intravenously injected colloidal carriers sterically stabilized with a blockcopolymer – poloxamine 908. Life Sci. 1987;40(4):367–74.
19. Falzon D, Jaramillo E, Schünemann HJ, Arentz M, Bauer M, Bayona J, et al. WHO guidelines for the programmatic management of drug-resistant tuberculosis: 2011 update. Eur Respir J. 2011;38(3):516–28.
20. Dorman SE, Johnson JL, Goldberg S, Muzanye G, Padayatchi N, Bozeman L, et al. Substitution of moxifloxacin for isoniazid during intensive phase treatment of pulmonary tuberculosis. Am J Respir Crit Care Med. 2009;180(3):273–80.
21. Ahmad Z, Tyagi S, Minkowski A, Peloquin CA, Grosset JH, Nuermberger EL. Contribution of moxifloxacin or levofloxacin in second-line regimens with or without continuation of pyrazinamide in murine tuberculosis. Am J Respir Crit Care Med. 2013;188(1):97–102.
22. Migliori GB, Langendam MW, D'Ambrosio L, Centis R, Blasi F, Huitric E, et al. Protecting the tuberculosis drug pipeline: stating the case for the rational use of fluoroquinolones. Eur Respiratory J Eur Respiratory Soc. 2012;40:814–22.
23. Benator D, Bhattacharya M, Bozeman L, Burman W, Catanzaro A, Chaisson R, et al. Rifapentine and isoniazid once a week versus rifampicin and isoniazid twice a week for treatment of drug-susceptible pulmonary tuberculosis in HIV-negative patients: a randomised clinical trial. Lancet. 2002;360(9332):528–34.
24. Dorman SE, Goldberg S, Stout JE, Muzanyi G, Johnson JL, Weiner M, et al. Substitution of Rifapentine for Rifampin during intensive phase treatment of pulmonary tuberculosis: study 29 of the tuberculosis trials consortium. J Infect Dis. 2012;206(7):1030–40.
25. Mitchison DA. Pharmacokinetic/pharmacodynamic parameters and the choice of high-dosage rifamycins. Int J Tuberc Lung Dis. 2012;16(9):1186–9.
26. Sterling TR, Villarino ME, Borisov AS, Shang N, Gordin F, Bliven-Sizemore E, et al. Three months of Rifapentine and isoniazid for latent tuberculosis infection. N Engl J Med. 2011;365(23):2155–66.
27. Wallis RS, Jakubiec W, Kumar V, Bedarida G, Silvia A, Paige D, et al. Biomarker-assisted dose selection for safety and efficacy in early development of PNU-100480 for tuberculosis. Antimicrob Agents Chemother. 2011;55(2):567–74.
28. Diacon AH, Pym A, Grobusch M, Patientia R, Rustomjee R, Page-Shipp L, et al. The Diarylquinoline TMC207 for multidrug-resistant tuberculosis. N Engl J Med. 2009;360(23):2397–405.
29. Matsumoto M, Hashizume H, Tomishige T, Kawasaki M, Tsubouchi H, Sasaki H, et al. OPC-67683, a nitro-dihydro-imidazooxazole derivative with promising action against tuberculosis in vitro and in mice. PLoS Med. 2006;3(11):2131–44.
30. Dawson R, Diacon A. PA-824, moxifloxacin and pyrazinamide combination therapy for tuberculosis. Expert Opin Investig Drugs. 2013;22(7):927–32.
31. Dey T, Brigden G, Cox H, Shubber Z, Cooke G, Ford N. Outcomes of clofazimine for the treatment of drug-resistant tuberculosis: a systematic review and meta-analysis. J Antimicrob Chemother. 2013;68(2):284–93.
32. Van Deun A, Maug AKJ, Salim MAH, Das PK, Sarker MR, Daru P, et al. Short, highly effective, and inexpensive standardized treatment of multidrug-resistant tuberculosis. Am J Respir Crit Care Med. 2010;182(5):684–92.
33. Hugonnet JE, Tremblay LW, Boshoff HI, Barry CE, Blanchard JS. Meropenem-clavulanate is effective against extensively drug-resistant Mycobacterium tuberculosis. Science (80-). 2009;323(5918):1215–8.

34. Sacksteder KA, Protopopova M, Barry CE, Andries K, Nacy CA. Discovery and development of SQ109: a new antitubercular drug with a novel mechanism of action. Future Microbiol NIH Public Access. 2012;7:823–37.
35. Pasca MR, Degiacomi G, Lopes Ribeiro ALDJ, Zara F, De Mori P, Heym B, et al. Clinical isolates of Mycobacterium tuberculosis in four European hospitals are uniformly susceptible to benzothiazinones. Antimicrob Agents Chemother. 2010;54(4):1616–8.
36. Lechartier B, Hartkoorn RC, Cole ST. In vitro combination studies of benzothiazinone lead compound BTZ043 against mycobacterium tuberculosis. Antimicrob Agents Chemother. 2012;56(11):5790–3.
37. Johnson JL, Hadad DJ, Dietze R, Maciel ELN, Sewali B, Gitta P, et al. Shortening treatment in adults with noncavitary tuberculosis and 2-month culture conversion. Am J Respir Crit Care Med. 2009;180(6):558–63.
38. Merle CSC, Sismanidis C, Sow OB, Gninafon M, Horton J, Lapujade O, et al. A pivotal registration phase III, multicenter, randomized tuberculosis controlled trial: design issues and lessons learnt from the Gatifloxacin for TB (OFLOTUB) project. Trials. 2012;61:13.
39. Blomberg B, Spinaci S, Fourie B, Laing R. The rationale for recommending fixed-dose combination tablets for treatment of tuberculosis. Bull World Health Organ. 2001;79:61–8.
40. Albanna AS, Smith BM, Cowan D, Menzies D. Fixed-dose combination antituberculosis therapy: a systematic review and meta-analysis. Eur Respir J. 2013;42(3):721–32.
41. Nagelkerke NJD, De Vlas SJ, Mahendradhata Y, Ottenhoff THM, Borgdorff M. The search for a tuberculosis vaccine: an elusive quest? Tuberculosis. 2006;86(1):41–6.

Potential of Herbal Drugs for Treatment of Tuberculosis

Vishwa Patel, Dipal Gandhi, Hetanshi Patel, and Niyati Acharya

Abstract Worldwide, TB is estimated as the 13th leading cause of death and the second leading infectious disease after COVID-19. In 2020, an estimated 10 million people fell ill with tuberculosis (TB) worldwide. The current first-line drug treatment against tuberculosis has various limitations, of poor efficiency in complicated TB cases, side effects and lengthy drug treatment regime. Antibiotics like rifampicin, ethambutol, isoniazid, and pyrazinamide are currently used to treat tuberculosis, but multiple drug resistance is a major problem associated with it. Many traditional practices and medicinal plants have been in use as adjuvant for the treatment and management of TB.

This chapter gives an overview of medicinal plants and some of the bioactives found to be reported with a prominent effect against TB. The mechanism behind the interaction of anti-TB drugs and herbal constituents and their novel delivery approaches are also discussed.

Keywords Tuberculosis · Ethnopharmacology · Bioactive phytoconstituents · Herbal remedies · NDDS

1 Introduction

Tuberculosis (TB) is an infectious disease caused by the bacillus *Mycobacterium tuberculosis.* It is a chronic specific inflammatory infectious disease caused by *M. tuberculosis* in humans. Tuberculosis usually attacks the lungs but it can also affect any parts of the body [2, 49]. TB disease can be traced back to *Mycobacterium tuberculosis*. Lungs are the main line of attack by *M. tuberculosis* and symptoms are often presented as severe coughing, fever and chest pain [40]. The current first-line

V. Patel · D. Gandhi · H. Patel · N. Acharya (✉)
Institute of Pharmacy, Nirma University, Ahmedabad, India
e-mail: niyati.acharya@nirmauni.ac.in

R. Shegokar, Y. Pathak (eds.), *Tubercular Drug Delivery Systems*,
https://doi.org/10.1007/978-3-031-14100-3_15

drug treatment against tuberculosis has various limitations, such as efficiency towards complicated TB cases, drug-related side effects and lengthy drug therapy (6–12 months) [77]. The first-line cocktail drugs used in the treatment of TB infections include isoniazid and rifampicin, which are usually supplemented with pyrazinamide and ethambutol in the first 2 months [5, 78]. Most cases of TB are pulmonary and acquired by person to person transmission of airborne droplets of organisms. Oropharyngeal and intestinal TB contracted by drinking dairy milk contaminated with *M. bovis* are rarely seen nowadays and Products made from unpasteurized cow's milk have been linked to a number of infectious diseases, and importing them from nations where *M. bovis* is a common cattle bacteria increases the risk of transmission [31].

According to the World Health Organization (WHO), approximately 10 million people were infected with TB in 2020 and 1.5 million died from the disease in the same year. WHO declared TB a global public health emergency. About one-third of the world's population (>2 billion), are infected with TB bacilli. Ten percent of the people infected with TB bacilli will become sick with active TB in their lifetime. According to WHO report, global population with burden of disease caused by TB from 1990–2011 was 6948 million and total number of MDR cases from 2005 to 2011 were 61690.3 In 2011, there were an estimated 8.7 million incident cases of TB (range, 8.3 million–9.0 million) globally. Highest numbers of incidents were reported in Asia (59%) and Africa (26%). Estimates of the burden of TB disease among children have also been carried out [45].

The burden of TB is highest in Asia and Africa. In 2011, the largest number of cases was reported from India, China, South Africa, Indonesia and Pakistan. India and China alone accounted for 26% and 12% of global cases, respectively. An estimated 1.4 million (range 1.3–1.6 million) of the 8.7 million TB incident cases reported in 2011 died from the disease, translating to a mortality rate of 20 fatalities per 100,000 persons [45]

1.1 Tuberculosis: Initiation, Etiology, Symptoms, Pathophysiology

M. tuberculosis is the most common cause but other than that it includes: *M. avium intracellulare, M. kansasii, M. scrofulaceum, M. marinum, M.ulcerence, M. fortuitum,* and *M. chelonae* [2, 31]. TB is transmitted through the air by a person with active TB disease of the lungs and less frequently transmitted by ingestion of *M. bovis* found in unpasteurized milk products or auto-ingestion and inoculation (in skin tuberculosis). Symptoms include persistent cough sometimes with chronic obstructive pulmonary disease (COPD) (Fig. 1).

Pathophysiology of tuberculosis [32] is divided into three Acts. Act I is the War of Attrition wherein MTB tries to multiply while the host attempts to contain them within granulomas, with no or little immunity, there is greater lymphatic or

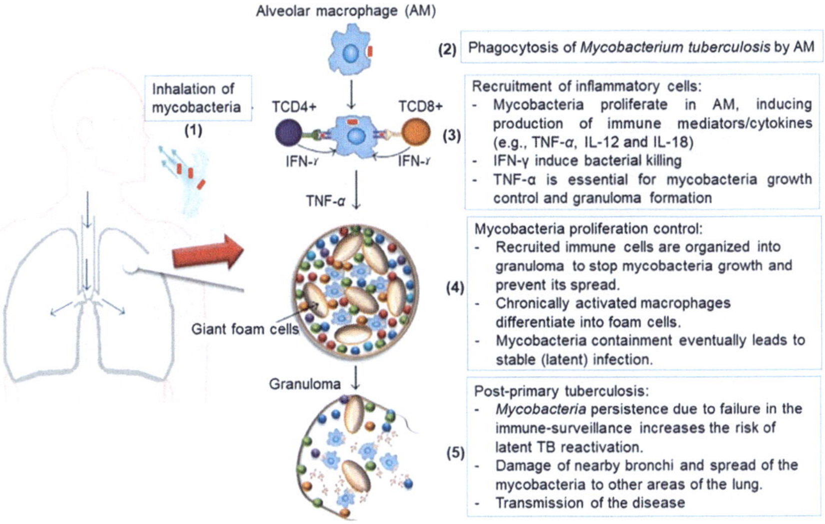

Fig. 1 Illustrative description of TB pathogenesis in five consecutive steps: (1) *Mycobacterium* entry, (2) interactions with alveolar macrophages, (3) recruitment and stimulation of immune cells, (4) *Mycobacterium* containment in granuloma, (5) active TB disease [97]

hematogenous spread. Control through cell mediated Immunity. In Act II – The Sneak Attack – post-primary bronchogenic TB begins asymptomatically in the apices of the lung, at some distance from the site initial infection, and it is part of latent TB since there are no clinical symptoms. In Act III- the fallout, is responsible for nearly all clinical post primary disease with formation of a cavity surrounded by epithelioid cells and fibrosis which produces granulomatous inflammation. Cavities form when caseous stage encompasses the pneumonia softens, fragments and is coughed out of the body leaving a hole. Pneumonia that is not coughed out remains to induce inflammation (Fig. 2).

1.2 M. tuberculosis *Drug Development and Phage-Based Therapy: Potential Anti-TB Treatment*

The triple combination of bedaquiline, pretomanid and linezolid has been used as a potential "game-changer" with high efficacy towards extensively drug-resistant TB. This multidrug target regimen has been inhibiting mycobacterial ATP synthase, mycolic acid synthesis and energy production [93]. The discovery of a potent anti-TB drug, TB47, a pyrazolo[1,5- a] pyridine-3-carboxamide, has new hope in the fight against TB. The compound, TB47, was revealed to be effective against 37 MDR-TB clinical strains. This compound has indicated that to be highly anti-mycobacterial, inhibiting the growth of *M. bovis, M. ulcerans, M. marinum, M. smegmatis* and *M. abscessus* at very low concentrations of TB47 displayed

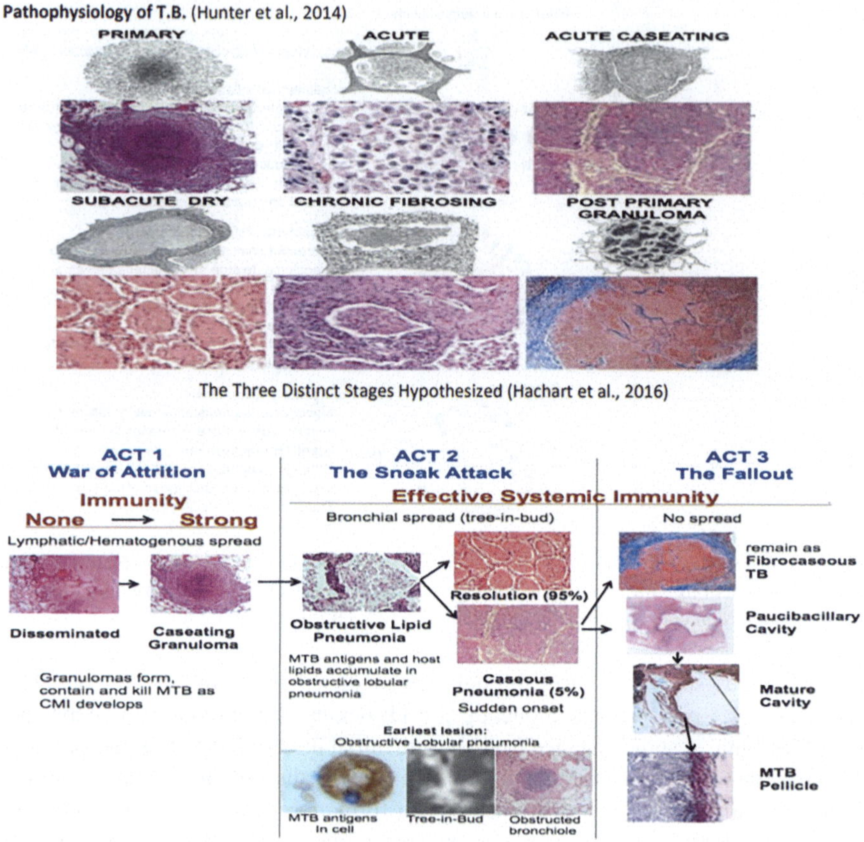

Fig. 2 Pathophysiology of TB

potent bactericidal activity in comparison to rifampicin. TB47 directly depends on the respiratory cytochrome bcc complex and is a potential antitubercular agent that synergistically inhibits *M. tuberculosis* [86].

The drugs now used to treat tuberculosis were discovered over a two-decade period (1944–1965), during which a relatively intensive search was conducted in various industrial and nonindustrial laboratories. There are multiple reasons for the decline in interest in research into new anti-tuberculosis drugs, including incorrect assumptions, investment of time, and issues with how the pathogen should be handled. In addition, developing an anti-TB drug requires more human resources than developing other antimicrobial agents. Finally, and perhaps most importantly, tuberculosis is prevalent in developing countries with limited economic resources, and industrial laboratories are reluctant to invest in research for new products to be used in those areas, where another major drawback is the lack of patent protection. Medicinal plants are an everlasting gift from nature that has been utilized to treat a variety of ailments in humans since the dawn of time. Herbs, either alone or in combination, are effective at reducing drug-related side effects. In absence of effective therapeutic drugs for TB, hope is built on plant-based natural products due to their chemical diversity and important role as phyto-drugs as well as herbs and bioactive which can be further utilized to develop novel targeted formulation.

2 Herbs Used in the Treatment of Tuberculosis

TB is a highly contagious viral disease that has been declared a global health emergency by the World Health Organization, with over one-third of the world's population latently infected with *M. tuberculosis*. The treatment for tuberculosis is divided into two phases: intensive and maintenance [79].

Tuberculosis (TB) is a disease that has afflicted humans since the dawn of humanity. The increasing rise in multidrug-resistant clinical isolates of *Mycobacterium* TB has necessitated the rapid discovery and development of novel anti-tuberculosis leads. Because of the toxicity and side effects of allopathic drugs, herbal therapy is becoming more popular [4].

Antibiotics such as rifampicin, ethambutol, isoniazid, and pyrazinamide are currently used to treat tuberculosis, but the rise of multiple drug resistant (MDR) and extensively drug-resistant (XDR) *Mycobacterium* strains is a common concern with anti-TB medications [70].

Traditional medicines (TM) play an important role in the health care industry around the world. The popularity of TM is growing in industrialized countries for a variety of reasons, one of which is that ethnopharmacological research has proven the effectiveness of these TM. Nearly half of all medications approved by the FDA in the United States in the previous 20 years have been natural product derivatives, including natural plant products. An estimated 70,000 plant species are used for medical reasons among the 435,000 species identified globally. As a result, picking plants based on ethnobotanical knowledge can improve the chances of discovering new anti-TB chemicals [94].

The traditional anti-TB medicinal plants are classified into 90 families, with 230 genera and 277 species. Fabaceae (21 species in 18 genera), Asteraceae (20 species in 16 genus), Euphorbiaceae (14 species in 11 genus), Lamiaceae (13 species in 11 genus), Rutaceae (14 species in 10 genus), Combretaceae (9 species in 4 genus), Piperaceae (9 species in 1 genus), Zingiberaceae (8 species in 3 genus), Annonaceae (7 species in 6 genus), Apiaceae (7 in 7 genus). A total of 40 plant families are only mentioned once. There are up to six anti-TB plant species in the *Terminalia* genus, which belongs to the Combretaceae family, and roughly nine anti-TB plant species in the *Piper* genus [94].

GIT symptoms, hepatotoxicity, ototoxicity, nephrotoxicity, skin rashes, fever, peripheral neuritis, and rarely psychotic alterations are all side effects of anti-TB medications. The plant extract prevents cellular damage and can greatly restore normal hepatic enzyme levels in the event of any hepatic irregularities. Glycosides, flavonoids, triterpenes, and the phenolic group of chemicals isolated from plants have been shown to have hepatoprotective properties in numerous studies. These phytochemicals have antioxidant and free radical scavenging activities, which reduce catalase, superoxide dismutase, and glutathione levels in the liver and prevent excessive lipid peroxidation [50].

The herbs can help to decrease high blood enzymes, total bilirubin, and protein levels. They can help repair enzymatic antioxidants' abnormal activity and liver damage induced by anti-TB medications. Traditional TB treatments include *Apium graveolens, Apium indica, A. paniculata, Ficus religiosa, Fumaria indica, Glycyrrhiza glabra, Syzygium aromaticum, W. somnifera, Tinospora cordifolia,* and other herbs with hepatoprotective qualities. Alkaloids, flavonoids, diterpenoids, tannins, lipids, sterols, and other compounds found in these plants have antibacterial and liver protective properties. Several anti-TB plants have been found to have antimicrobial action against MTB-H37RV [50].

Phytochemicals such as alkaloids, flavonoids, carbohydrates, glycosides, saponins, tannins, and terpenoids are abundant in *Moringa oleifera* leaves. Hepatic damage caused by INH, PZA, and RIF, *M. oleifera* leaf extract can restore normal liver activity. The enzymes ALS, AST, alkaline phosphatase, lipid peroxidation, and bilirubin in the serum appear to enhance recovery from liver injury and restore regular functioning [50].

In several ways, phytochemicals are useful in the treatment of anti-TB drug-induced toxicity. They primarily act on a few key mechanisms that are responsible for drug toxicity. Cytochrome P450, free radicals, and reactive oxygen species are all targets for phytochemicals. Flavonoids, tannins, and carotenoids, for example, have an effect on reactive oxygen species. The activity of Cytochrome P450 is also affected by organosulphur compounds and flavonoids, which affects drug metabolism [50].

The detailed description of the plants, family, plant parts used, extract type and in vitro activity with minimum inhibitory concentration value (MIC), information on active chemicals (if any), and ethnomedicine and Ayurvedic usage have been mentioned in Table 1 Surprisingly, the majority of plant species have a significant ethnopharmacological relationship with traditional knowledge [30].

2.1 Multitargeting Potential of Natural Products in TB Treatment

Medicinal plants have traditionally been used in the treatment of pulmonary diseases such as TB, yet most of these have not been fully elucidated. Plant-derived compounds with low toxicity and high activity towards *M. tuberculosis* have previously been identified against *M. tuberculosis* species and these should be tested in combination with the current first-line drugs. In the last decade, 25 newly approved drugs were derived from natural products, with 31 additional drugs either at or past Phase III clinical trials. In the last decade, a number of researchers have reported extracts from edible plants such as aromatic ginger, roselle, celebes pepper and

Table 1 List of medicinal plants used as antitubercular agent along with its mechanism of action

Name of plant	Common name	Family	Part of plants	Extract/constituents	Pharmacological action	References
Acalypha indica	Acalypha	Euphorbiaceae	Stem bark and leaves	Petroleum ether crude extract	In vitro for their activity against two MDR isolates (DKU-156 and JAL-1236)	
Adhatoda vasica	Vasaka	Acanthaceae	Leaves	Bromhexine and ambroxol (benzylamines)	pH-dependent growth-inhibitory effect on *M. tuberculosis*	[29]
Allium sativum	Garlic	Amaryllidaceae	Cloves	Allicin, ajoene	80 mg/ml of garlic oil almost completely inhibited the growth of *M. tuberculosis* $H_{37}Rv$ (almost 97% reduction in colony count) as against 0.03 mg/ml of rifampicin which showed similar inhibition of growth of *M. tuberculosis* $H_{37}Rv$	[89]
Diospyros anisandra		Ebenaceae	Stem bark	Monomeric naphthoquinone (plumbagin)	Two strains of MTB (H37Rv); cytotoxicity of naphthoquinones was estimated against two mammalian cells, vero line and primary cultures of human peripheral blood mononuclear (PBMC) cells, and their selectivity index	[88]
Kaempferia galanga	Sand ginger, cekur	Zingiberaceae	Rhizome	Ethyl p-methoxycinnamate (EPMC)	Inhibit MDR strains of *M. tuberculosis*	[46]

(continued)

Table 1 (continued)

Name of plant	Common name	Family	Part of plants	Extract/constituents	Pharmacological action	References
Artemisia capillaris	Yin Chen Hao	Compositae	Aerial parts	Ursolic acid (UA) and hydroquinone (HQ)	Inhibited the growth of both susceptible and resistant strains of *M. tuberculosis*	[40]
Euclea natalensis	Natal guarri	Ebenaceae	Root	Diospyrin (binaphthoquinoid)	Against the H37Rv strain and drug-resistant strains	[47]
Caesalpinia pulcherrima	Peacock flower	Fabaceae	Root	Cassane furanoditerpenoids	Antitubercular activity assessed against *M. tuberculosis* H37Ra using MABA (microplate Alamar blue assay)	[63]
Phyllanthus niruri	Meniran	Phyllanthaceae	Leaves	Aqueous extract	Possesses an in vitro immunomodulatory activity on tuberculosis patients	[64]
Glycyrrhiza glabra	Licorice	Fabaceae	Root	Glabridin	Active against both the strains of *mycobacterium*. Additionally, glabridin was more active against gram-positive strains than gram-negative	[26]
Prunus armeniaca	Siberian apricot	Rosaceae	Fruit	Aqueous and ethanolic extract	In vitro activity in cup plate method	[41]
Alstonia scholaris	Devil's tree	Apocynaceae	Leaves, fruit	Indole alkaloids	In vitro antituberculosis activity (89% inhibition against *M. tuberculosis* Hsub37Rv at 50 ug/mL) using microplate Alamar blue assay	[52]

Name of plant	Common name	Family	Part of plants	Extract/constituents	Pharmacological action	References
Argyreia speciosa	Elephant creeper	Convulvulaceae	Roots	Flavonoid sulfates	*M. tuberculosis* H37 Rv sensitive strain by in vitro and in vivo assays	[33]
Bridelia micrantha	Coastal golden tree	Phyllanthaceae	Bark	*n*-hexane fraction	Fraction showed 20% inhibition of MTB $H_{37}Ra$ and almost 35% inhibition of an MTB isolate resistant to all first-line drugs at 10μg/mL	[28]
Caesalpinia pulcherrima	Peacock flower	Caesalpiniaceae	Roots	Furano diterpenoid 6 beta-cinnamoyl-7 beta-hydroxyvouacapen-5 alpha-ol	Strong antitubercular activity with a minimum inhibitory concentration (MIC) of 6.25 microg/mL	[63]
Ipomea leptophylla	Bush morning glory	Convolvulaceae	Leaves, stem	Resin glycosides	In vitro activity against *M. tuberculosis*	[10]
Laggera pterodonta	Ko Kuna Sigi	Asteraceae	Leaves/aerial part	Crude methanol extract	Screened against *M. bovis* (BCG strains). The two extracts were found to be active at minimum inhibitory concentrations (MIC) of 625μg/ml	[17]
Mallotus philippinensis	Kumkum tree	Euphorbiaceae	Dried flowers organic extract	Mallotophilippen F, rottlerin, isoallorottlerin, isorottlerin dimethyl chromene	Active against both *M. tuberculosis* and *M. avium*	[34]

(continued)

Table 1 (continued)

Name of plant	Common name	Family	Part of plants	Extract/constituents	Pharmacological action	References
Morinda citrifolia	Indian mulberry	Rubiaceae	Leaves	Ethanolic and hexane extracts	E-phytol, a mixture of the two ketosteroids, and the epidioxysterol derived from campesta-5,7,22-trien-3beta-ol all show pronounced antitubercular activity	[80]
Pelargonium reniforme	Geraniums	Geraniaceae	Roots	Root extracts	Acetone, chloroform and ethanol extracts of *P. reniforme* showed activity against *M. tuberculosis* exhibiting a minimum inhibitory concentration of 5x 10^3 mg/L	[48]
Senecio chionophilus	Umbrella hairgrass	Asteraceae	Roots	Sesquiterpenoids	Antitubercular potential against *M. tuberculosis* in a microplate Alamar Blue Assay	[24, 25]
Trichosanthes dioica	Pointed gourd	Cucurbitaceae	Leaves	Extract	In vitro activity against *M. smegmatis*	Kumar N, Singh S, Manvi, Gupta R. Trichosanthes dioica Roxb.: An overview. *Pharmacogn Rev*. 2012;6(11):61–67. https://doi.org/10.4103/0973-7847.95886
Carum carvi	Persian cumin	Apiaceae	Whole plant	Ethanolic extract	Acts as a bioenhancer and modifies the kinetics of Antitubercular Treatment (ATT) favorably	[11]

Name of plant	Common name	Family	Part of plants	Extract/constituents	Pharmacological action	References
Moringa oleifera	Drumstick	Moringaceae	Leaves	Hydroalcoholic root	Enhance the recovery from hepatic damage induced by antitubercular drugs	[65]
Anacyclus pyrethrum	Akarkara	Asteraceae	Root		Possesses promising hepatoprotective activity against INH- and RIF-induced hepatic damage in experimental rats	
Ziziphus oenoplia	Makai (Hindi) Jackal jujube (English)	Rhamnaceae	Root	Ethanolic extract	Potent hepatoprotective action against INH + RIF induced hepatic damage	[71]
Actiniopteris radiata	Nemaliadugu	Actiniopteridaceae	Whole plant	Chloroform extract	Significant antitubercular activity	[53]
Terminalia chebula	Chebulic myrobalan	Combretaceae	Fruit	Alcoholic extract	Hetero-protective action against anti-TB induced hepatotoxicity	[82]
Solanum xanthocarpum	Kantakari	Solanaceae	Fruit	Extract	Protective action against anti-TB induced hepatotoxicity	[35]
Sterculia setigera	Kukkuki	Malvaceae	Leaves		Active against *M. tuberculosis,* and the anti-TB activity	[7]
Arctium lappa	Burdock	Asteraceae	Aerial parts	Loliolide	Antitubercular remedies	[96]
Tussilago farfara	Coughwort	Asteraceae	Aerial parts	Terpenoids	Antitubercular remedies	[96]
Laurus novocanariensis	Lauras	Lauraceae	Ripe fruit (oil)	Methanolic extract	Antimicrobial activity against *M. tuberculosis* H37Rv	
Triumfetta rhomboidea	Burr bush	Tiliaceae	Whole plant	Acetone, chloroform and ethanol extracts	Significant effects against *M. tuberculosis*	[90]

(continued)

Table 1 (continued)

Name of plant	Common name	Family	Part of plants	Extract/constituents	Pharmacological action	References
Pelargonium reniforme	Geranium	Geraniaceae	Roots	Acetone, chloroform and ethanol extracts	Inhibitory activity against the drug-sensitive strain of *M. tuberculosis*	[48]
Pelargonium sidoides	Kalwerbossie	Geraniaceae	Roots	Methanol extract	Inhibitory activity against the drug-sensitive strain of *M. tuberculosis*	[48]
Piper sarmentosum	Wild pepper	Piperaceae	Root, stem, leaves and fruit	Ethanol extract	All the extracts have exhibited anti-TB activity	[36]
Artocarpus integrifolia	Jack fruit	Moraceae	Root bark	Methanol extract	Anti-TB and anti-diabetic effects	[37]
Tinospora sinensis	Gurjo	Menispermaceae	Leaves	Ethyl acetate and N-hexane extract	Ethyl acetate and n-hexane extracts demonstrated the promising ($p < 0.05$) antitubercular activity with inhibition (%) of 92% and 86%	[8]
Vitex negundo	Nirgundi	Verbenaceae	Leaf extract	Iridoid glycosides, isomeric flavanones and flavonoids	Hepatoprotective activity	[83]
Byrsonima crassa	Golden spoon	Malpighiaceae	Leaves and bark (chloroform extract)	Triterpenes:A-Amyrin, B-Amyrin and their acetates, lupeol, oleanolic acid, ursolic acid and A-amyrinone	Triterpenes exhibited minimum inhibitory concentrations (MICs) of 31.25–312.25μg/mL	[39]
Galenia africana	Brakkraalbossie	Asteraceae	Leaves	Flavonoids	Flavonoids exhibited moderate antituberculosis activity	[54]

Name of plant	Common name	Family	Part of plants	Extract/constituents	Pharmacological action	References
Quinchamalium majus		Santalaceae	Roots	Triterpenes from dichloromethane extracts	Growth inhibition of *M. tuberculosis* and green monkey Vero cells by constituents of *Q. majus* and related analogues in MABA	[24, 25]
Erythrina indica	Indian coral tree	Papilionaceae	Leaves	Ethanolic extract	Hepatoprotective activity	[56]
Alpinia officinarum	Galangal	Zingiberaceae	Rhizome	Extracts	Hepatoprotective activity	[16]
Leucas marrubioides	Horehound Leucas	Lamiaceae	Root	Petroleumether extract	Active against *M. tuberculosis* H73Rv strain using Microplate Alamar Blue Assay (MABA) within MIC range of 0.2–100μg/m	[22]
Tinospora cordifolia	Guduchi	Menispermaceae	Stem and fruit	Combination effects was studied at different dose levels	Protective effects against antitubercular drug-induced hepatic damage	[69]
Phyllanthus emblica	Indian gooseberry	Euphorbiaceae				
Garcinia indica	Kokum butter tree	Clusiaceae/guttiferae	Fruit rind	Aqueous extract	Hepatoprotective effects in antitubercular drug-induced liver injury	[68]
Mucuna pruriens	Cowitch	Fabaceae	Leaves	Hydroalcoholic extract	Hepatoprotective against antitubercular drug-induced hepatic damage	[62]
Thalia multiflora	Alligator flag	Marantaceae	Aerial parts	Crude extract and sterols	Showed ability to inhibit the growth of *M. tuberculosis* H37Rv (ATCC 27294) at 50μg/mL	[23]

(continued)

Table 1 (continued)

Name of plant	Common name	Family	Part of plants	Extract/constituents	Pharmacological action	References
Cnidoscolus chayamansa	Tree spinach	Euphorbiaceae	Leaf	Sesquiterpenes	Hepatoprotective effect against hepatotoxicity induced by anti-TB drugs	[67]
Pteris ensiformis	Sword brake	Pteridaceae	Whole plant	Steroidal saponins	Exhibited antitubercular activity against *M. tuberculosis* H37Rv in vitro.	[12]
Asparagus africanus lam	African asparagus	Asparagaceae	Shoots	Steroidal saponins	Antimicrobial used for the treatment of TB.	[58]
Plumeria bicolor	Nosegay	Apocynaceae	Tree bark	Iridoids and terpene	Plumericin was very active against all strains. Higher activity against the MDR strains	
Thonningia sanguinea Vahl.	Ground pineapple	Balanophoraceae	Fruits	Brevifolin carboxylic acid gallic acid	Antioxidative and radical scavenging activities and lipid peroxidation inhibitory activity	[73]
Tiliacora triandra	Bai yanang	Menispermaceae	Roots	20-Nortiliacorinine and tiliacorine	Active against MDR-MTB strains with median inhibitory concentration (MIC); cytotoxicity in normal MRC-5 cell line was also determined	
Pluchea indica Linn	Indian camphorweed	Asteraceae	Flowers, leaves	Aqueous extract	Inhibitory activities against H37Rv strain at the MIC of 800 μg/ml each	[13]
Uvaria rufa Blume	Susung-Kalabaw	Annonaceae	Fruits, flowers	Methanolic extract	Activity against H37Rv strain was observed at the MIC ranging from 33.1 to >100 microgram/ml in MABA	[84]

banana leaves as potent MDR-TB inhibitors. Examples of some antitubercular drugs with multitarget properties include SQ109 and its derivatives, which have been reported to inhibit MmpL3 and MmpL11, transporter proteins involved in cell wall biosynthesis, and MenA and MenG, involved in menaquinone biosynthesis. Furthermore, the inhibitory capabilities of the SQ109 derivatives against a panel of various Gram-positive and Gram-negative bacteria, including the *M. tuberculosis,* fungi and the *P. falciparum* parasite, have been reported [51, 85].

3 Potential of Phytoconstituents

1. *Acalypha indica*

Acalypha indica Linn.(Family: Euphorbiaceae) It has been traditionally used for diuretic, anthelmintic, respiratory problems, rheumatoid arthritis, to cure scabies and other skin infection [57]. A. *indica* has different pharmacological activities such as anti-inflammatory, anti-bacterial, anti-fungal, hepatoprotective, anthelmintic, anti-fertility, antiulcer activity, antioxidant and anticancer activity [21]. A.*indica* revealed the presence of β-sitosterol and its β-D-glucoside were isolated from the leaves and twigs of *A. indica.* Several secondary metanolites are found to be present such as Potassium brevifolin carboxylate, 1-O-galloyl-β-D-glucose, 1,2,3,6-tetra-D-galloyl-β-D-glucose, corilagin, geraniin, acaindinin, acetonylgeraniin A, euphormisin M2, repandusinic acid A, and chebulagic acid, as well as two flavonoid glycosides quercetin 3-O-β-D-glucoside and rutin. Chrysin and galangin were isolated from the whole plant extract. Acalphin, an acyanogenic glycoside, was also isolated from the same plant [42]. Kaempferol glycosides, mauritianin, clitorin, nicotiflorin and biorobin were isolated from the flowers and leaves of *A. indica* [44]. The phytochemicals such as phenolics and flavonoids are important components of the plant and some of their biological activities could be imputed to the presence of these constituents [72]. The crude extracts of *A. indica* have shown activity against *M. tuberculosis* H37Rv strain. Antitubercular potential of various fractions of methanol extract of A. indica were performed and checked on drugresistant strain. Three fractions such as ethyl acetate, aqueous fraction and n-butanol fraction were analyzed for the activity. The prepared fractions found to contain phytochemicals viz. flavonoids, alkaloids, saponins and steroids. The importance and effect of these phytocinstituents were studied. Flavonoids show activity by damaging cytoplasmic membrane with the generation of hydrogen peroxide, inhibition of nucleic acid synthesis and inhibition of ATP synthase. This might be the reason for the mechanism of action of saponins and flavonoids present which may inhibit *M. tuberculosis.* Hence, the ethyl acetate and aqueous fractions will be carried out to separate the phytochemicals responsible for the anti-TB activity. The antitubercular activity of the fractions is carried out on *M. tuberculosis* H37Rv strain, by Microplate Alamar Blue Assay (MABA) method [59].

2. *Allium sativum*

Allium sativum (Family: Liliaceae) therapeutic effects are due to specific oil and water-soluble organosulfur compounds, which are responsible for the typical odor and flavor of garlic. They exhibit different antibacterial, antifungal, antiseptic, antiviral, expectorant, antihistamine properties [18]. Researchers have discovered that *Allium sativum* exhibits in vitro activities such as (i) antimicrobial activity against both Gram-positive and Gram-negative bacteria, including species of *Escherichia coli, Salmonella, Staphylococcus and Streptococcus, Klebsiella, Proteus, Bacillus, Clostridium, Helicobacter pylori*, and even Mycobacterium tuberculosis (Mtb); (ii) antifungal activity – particularly against Candida albicans – (iii) antiparasitic and (iv) antiviral action [18]. It has in vitro a strong anticancer effect as well as anti-inflammatory, immunomodulatory and antioxidant properties [3]. It also possesses antihyperlipidemic, anthelmintic, antihypercholesterolemic, antihypertensive activity. The high potential of garlic extract was revealed by the ability to inhibit the growth of *M. tuberculosis* H37Rv and *M. tuberculosis* TRC-C1193, susceptible and resistant to isoniazid (first-line antituberculosis medication), respectively. The minimum inhibitory concentration (MIC) [1] the garlic extracts effectiveness against clinical isolates of MDR-TB are of scientific importance. Water extract of *Allium sativum* was found to have activity against two MDR *M. tuberculosis* isolates that were found to be resistant against rifampicin and isoniazid [20].

3. *Diospyros anisandra*

Effects of *Diospyros anisandra* (Family: Ebenaceae) were reported against *M. tuberculosis* H37Rv strain sensitive to streptomycin, isoniazid, rifampin, ethambutol, and pyrazinamide, and *M. tuberculosis*. Antimycobacterial activity was evaluated by the Microplate Alamar Blue Assay (MABA). Leaves, root and stem bark of *Diospyros anisandra* were screened against two strains of *M. tuberculosis*, one resistant and one susceptible to antibiotics, using the Microplate Alamar Blue Assay test. The lipophilic fractions of the root and bark showed significant inhibitory activity against both strains, with the hexane fraction of the bark showing the strongest activity against the resistant strain and a significant antimicrobial activity against *Staphylococcus aureus, Bacillus subtilis, Candida albicans, Aspergillus niger*, and *Colletotrichum gloeosporioides*. The methanolic extract of stem bark of *D. anisandra* showed significant antimycobacterial and antimicrobial activities, inhibiting the growth of a resistant strain of *M. tuberculosis* [9, 14].

4. *Glycyrrhiza glabra*

The antitubercular activity of acetone extract of *G. glabra*, was evaluated against *M. tuberculosis* H37Rv (ATCC 27294). The in vitro antitubercular activity was determined by the Resazurin Microtiter Plate Assay (REMA) and colony count method. The antitubercular activity of acetone extract of *G. glabra* was determined in human macrophage U937 cell lines and was compared against that of the standard drugs isoniazid (INH), rifampicin (RIF) and ethambutol (ETH). *G. glabra* extract showed significant activity against *M. tuberculosis*, when evaluated by REMA and colony count methods and in U937 human macrophage cell lines infected with *M. tuberculosis* H37Rv [60]. Liposomal dry powder (using freeze-dryer method) for inhalation (LDPI) formulation containing licorice extract for use in tuberculosis treatment. The in vivo lung deposition of the LDPI was evaluated in Swiss Albino mice to validate the in vitro ACI (Anderson cascade impactor) and TSI (twin stage impinger). The lung deposition of the LDPI also provided an idea about the release of drugs in lung milieu. The in vivo pharmacodynamic studies of the LDPI were also conducted in mice infected with *M. tuberculosis* H37Rv [91].

5. *Mallotus philippinensis*

Mallotus philippinensis, (Family: Euphorbiaceae) species are known to contain different natural compounds such as phenols, diterpenoids, steroids, flavonoids, cardenolides, triterpenoids, coumarins, isocoumarins, and many more. Specially, phenolics such as bergenin, mallotophilippinens, rottlerin, and isorottlerin have been isolated, identified, and reported interesting biological activities such as antimicrobial, antioxidant, antiviral, cytotoxicity, antioxidant, anti-inflammatory, immunoregulatory activity protein inhibition against cancer cells [19]. Bioassay-directed fractionation of the organic extract of *Mallotus philippinensis* gave five compounds the most active of which against *M. tuberculosis* was a new compound, 8-cinnamoyl-5,7-dihydroxy-2,2-dimethyl-6- geranyl chromene for which the name mallotophilippen F is suggested. Compound 8-cinnamoyl-2,2-dimethyl-7-hydroxy-5-methyl chromone was isolated from a natural source for the first time, while the remaining three compounds, rottlerin (3), isoallorottlerin, iso rottlerin and the so-called "red compound," 8-cinnamoyl-5,7-dihydroxy-2,2,6- trimethyl-chromene, had been isolated previously from this plant [34].

4 NDDS Approaches for TB

Drug delivery and drug targeting systems are currently used to minimize drug degradation and loss, to prevent harmful side effects and to increase drug bioavailability and the fraction of the drug accumulated in the required zone. Among drug carriers, one can name soluble polymers, microparticles made of insoluble or biodegradable natural and synthetic polymers, microcapsules, cells, cell ghosts, lipoproteins, liposomes and micelles. The carriers can be made slowly degradable, stimuli-reactive (e.g., pH- or temperature-sensitive) and even targeted (e.g., by conjugating them with specific antibodies against certain characteristic components of the area of interest). Targeting is the ability to direct the drug-loaded system to the site of interest.

Two major mechanisms can be distinguished for addressing the desired sites for drug release: (i) passive and (ii) active targeting. An example of passive targeting is the preferential accumulation of chemotherapeutic agents in solid tumors as a result of the enhanced vascular permeability of tumor tissues compared with healthy tissue. A strategy that could allow active targeting involves the surface functionalization of drug carriers with ligands that are selectively recognized by receptors on the surface of the cells of interest. Since ligand–receptor interactions can be highly selective, this could allow a more precise targeting of the site of interest. Controlled drug release and subsequent biodegradation are important for developing successful formulations [15].

Potential release mechanisms involve (i) the desorption of surface-bound/adsorbed drugs; (ii) diffusion through the carrier matrix; (iii) diffusion (in the case of nano capsules) through the carrier wall; (iv) carrier matrix erosion and (v) a combined erosion/diffusion process. The mode of delivery can be the difference between a drug's success and failure, as the choice of a drug is often influenced by the way the medicine is administered. Sustained (or continuous) release of a drug involves polymers that release the drug at a controlled rate due to diffusion out of the polymer or by degradation of the polymer over time. Pulsatile release is often the preferred method of drug delivery, as it closely mimics the way by which the body naturally produces hormones such as insulin. It is achieved by using drug-carrying polymers that respond to specific stimuli (e.g., exposure to light, changes in pH or temperature) [74].

Other approaches to drug delivery are on crossing particular physical barriers, such as the blood–brain barrier, in order to better target the drug and improve its effectiveness; or on finding alternative and acceptable routes for the delivery of protein drugs other than via the gastrointestinal tract, where degradation can occur.

Presently novel drug delivery systems have been widely utilized only for allopathic drugs, but they have their own limitations; hence, turning to safe, effective and time-tested Ayurvedic herbal drug formulation would be a preferable option [75].

4.1 Potential of Novel Drug Delivery for Herbal Drugs

Drug delivery system used for administering the medicine to the patient is traditional and out-of-date, resulting in reduced efficacy of the drug. In case of herbal extracts, there is a great possibility that many compounds will be destroyed in the highly acidic pH of the stomach. Other components might be metabolized by the liver before reaching the blood. As a result, the required amount of the drug may not reach the blood. If the drug does not reach the blood at a minimum level, which is known as 'minimum effective level' then there will be no therapeutic effect.

Phytopharmaceuticals are pharmaceuticals using traditional compounds derived from botanicals instead of chemicals. Natural ingredients are more easily and more readily metabolized by the body. Therefore, they produce fewer, if any, side effects and provide increased absorption in the bloodstream resulting in more thorough and effective treatments. Pharmaceuticals made from chemical compounds are prone to adverse side effects. The human body will have a tendency to reject certain chemical compounds which do not occur naturally. These rejections occur in the form of side effects; some as mild as minor headaches, and others as severe as to be potentially lethal. It is important to note while phytopharmaceuticals produce fewer to no side effects, chemical interactions with other prescription drugs can occur.

Furthermore, as they are single and purified compounds, they can be easily standardized making it easier to incorporate them in modern drug delivery systems compared to herbs [61].

Lipid-based drug delivery systems have been shown their potential in controlled and targeted drug delivery. Pharmacosomes are amphiphilic phospholipids complexes of drugs bearing active hydrogen that bind to phospholipids. They impart better biopharmaceutical properties to the drug, resulting in improved bioavailability. Phytosomes are novel compounds comprising of lipophilic complexes of components of plant origin like *Silybum marianum, Ginkgo biloba, ginseng* and so on, with phospholipid [76]. They are also called as phyto lipids delivery system. They have high lipophilicity and improved bioavailability and therapeutic properties. These are advanced forms of herbal extract that have improved pharmacokinetic and pharmacological parameters, whose result can advantageously be used in treatment of acute liver diseases, either metabolic or infective origin. Phytosomes are produced by a patent process in which individual components of herbal extract like flavonolignans and terpenoids are bound on a molecular level to the phospholipids like phosphatidylcholine through a polar end. Phytosomes are used as a medicament and have wide scope in cosmetology. Phytosomes form a bridge between the conventional delivery system and novel delivery system. If the herbs themselves or the purified phytopharmaceuticals or phytosomes are incorporated in novel drug delivery systems, we can get the benefits of both [6].

4.2 Bioenhancers Based on Mechanism of Action

Cuminum cyminum (black cumin), *Carvum carvi*, genistein, sinomenine, naringin, quercetin act as inhibitors of p-glycoprotein (p-gp) efflux pump and other pumps. Quercetin, naringin, gallic acid and its esters also act as inhibitors of CYP-450 enzyme and its isoenzymes. Niaziridin from drumstick pods, glycyrrhizin from licorice, gingerol and shogaol from ginger act by inhibiting regulators of gastrointestinal functions to facilitate better absorption [66].

Black Cumin is reported to contain the chemical constituents namely 3″, 5-dihydroxyflavone-7-O-ß-D glucuronide-4″-ß-O-D′Glucopyranoside. Anti-TB phytochemicals are abietane, ethyl-p-methoxycinnamate, ergosterol peroxide, mono-O-methyl curcumin isoxazole, 7-methyl juglone, 12-dimethyl multicaulis, 12-methyl-5- dehydroacetylhorminone, tryptanthrin, etc. [27].

Many bioactives such as Alliin, aloin, octyl-β-d-glucopyranoside, oleanolic acid, and phytol [43] were evaluated against two standard front-line anti-TB drugs, isoniazid (ISN) and ethambutol (EMB), to decipher their potential anti-tuberculosis efficacy, targeting four of the mycobacterial receptor proteins/enzymes (arabinosyltransferase C, protein kinase A, glutamine synthetase, and proteasomal ATPase).

Phytol metabolite from *Leucas volkensii* was found to have inhibitory activity against *M. tuberculosis* H37Rv and is believed to be a better therapeutic agent for the treatment of TB.

Alliin (S-allyl-L-cysteine sulfoxide), the most abundant sulfur compound in *Allium sativum* L., had been a potent cardioprotective and neuroprotective agent having antidiabetic, anticholesteremic, and anticarcinogenic effects. Allicin (diallyl thiosulfinate) is a sulfur-containing, volatile, oxygenated chemical derived from garlic (*Allium sativum*). The combination of garlic extract-derived AgNP synthesis and conjugation with INH might pave the approach for effective MDR–TB treatment. Fresh garlic extract was combined with isoniazid hydrazide (INH), a commonly used antibiotic to treat tuberculosis, to create a green synthesis of silver nanoparticles (AgNPs). The in vitro drug release study was done at 37 ± 0.5 °C under using sodium phosphate buffer solution or simulated lung fluid from buffer medium at pH 7.2. The chosen compounds were compared to anthranilate phosphoribosyltransferase (trpD) from *Mycobacterium tuberculosis* in a molecular docking study. It had a higher docking score than trpD, suggesting that it could aid in antitubercular activity [55, 81].

Glycyrrhitinic acid, from licorice, targets the medication to macrophages, a unique technique was used to produce bioactive licorice extract with significant antitubercular action in the form of mannose-linked gelatin nanoparticles. The mannosylated formulation showed increased cellular absorption in vitro via the mannose receptor route, resulting in active drug targeting to macrophages. In vivo, the mannosylated formulation was able to maintain therapeutically substantial drug levels in a reasonable manner. In vivo pharmacodynamic tests of mannosylated gelatin nanoparticles in a murine tuberculosis model revealed potential antitubercular efficacy in vivo. The proposed formulation was found to be promising, safe, and effective, and it has the potential to move forward with further exploratory research in order to develop it for use in tuberculosis treatment. A novel method for delivering biologically active licorice extract with significant antitubercular activity straight to the lungs via liposomal dry powder inhalation. As a result, licorice extract liposomal dry powder for inhalation (LDPI) has the potential to be investigated as an effective antitubercular treatment or as a supplement to currently available antitubercular drugs.

Curcumin, a polyphenol, has been demonstrated to target several signaling molecules while also displaying cellular activity, supporting its multiple health advantages. Inflammatory disorders, metabolic syndrome, pain, and the management of inflammatory and degenerative eye conditions have all been proven to benefit from it. Curcumin, the bioactive component of turmeric also known as "Indian Yellow Gold," has been shown to be effective in the treatment of a variety of chronic inflammatory and infectious disorders. Curcumin incorporated in nanoparticles dramatically shortened the time required for antibiotic therapy to achieve sterile immunity, lowering the risk of drug-resistant strains of bacteria. Curcumin nanoparticles inhibited the growth of the MTB H37Rv strain by at best 1-log in the mice and accelerated the clearance of the MTB from the lung and spleen of BALB/c mice by promoting an antitubercular response, which in turn reduced the duration of therapy. Curcumin nanoparticles restored the isoniazid (INH)-induced suppression in antigen-specific cytokine, the proliferation of T cells suppressed by INH, and reduced hepatotoxicity in mice induced by antitubercular antibiotics. As a result, adjuvant therapy with nanoformulated curcumin may be effective in the treatment of tuberculosis and other disorders [87].

Piperine was recently used as a bioenhancer in a formulation comprising lower concentrations of RIF and INH for the treatment of drug-susceptible pulmonary TB, which was efficacious and resulted in good clinical cure rates while lowering treatment time and side effects. The efflux pump inhibitor activity of PIP was determined by bromide accumulation assay and cytotoxicity carried out in VERO cells and J774. A1 macrophages. Piperine (PIP) and Streptomycin (SM) synergism was found in the *M. tuberculosis* reference strain, with equivalent effects in susceptible and MDR clinical isolates. In *M. tuberculosis* H37Rv and clinical isolates with varied resistance profiles, RIF + PIP and SM + PIP combinations showed synergism but had a low effect in improving efficiency [38].

Aloin, a major compound of *A. vera* latex, is a well-known laxative agent, generally existing as a mixture of two diastereomers, aloin A and aloin B, also referred to as barbaloin and isobarbaloin, respectively.

Quercetin, a flavonoid with excellent activity against *M. tuberculosis* H37Rv, was found to have a very robust and considerable growth inhibitory function. In vivo lung deposition studies of LDPI was carried out in mice. In tuberculosis-infected mice, the combination of quercetin and polyvinylpyrrolidone (PVP) was observed to prevent the spread of necrosis to unaffected organs [91].

Phyllanthus emblica (amla). The major bioactive compounds of *P. emblica*, including octyl-β-D-glucopyranoside, have been well known for curing effects against various diseases, such as fever, cough, piles, constipation, anorexia, hemorrhoids, skin diseases, asthma, biliousness, respiratory disorders, tumors, and cancer.

Oleanolic acid isolated from *Lantana hispida* showed a potential inhibitory potential against *M. tuberculosis* H37RV.

In addition to above bioactives used, few more bioactives have been studied as antitubercular agents such as Phenazines obtained biosynthetically by many species of the *Actinobacteria phylum,* Clofazimine extracted from *Buellia canescens*, Piperidines derivative of piperine obtained from black pepper, Mycins sourced from *Actinobacteria* spp., Thiolactomycin isolated from *Nocardia* spp., Rifapentine isolated from *Amycolatopsis rifamycinica*, Quinolones *Pseudomonas spp.*, *Escherichia spp.* and many other bacterial species, Antimicrobial peptides polypeptides produced by living organisms, Lariatins and few marine *Streptomyces* CN3–982 and many more natural products as well as phytoconstituents inspired from natural sources showed to be promising effect as anti-mycobacterial agents with clinical potential [92].

5 Conclusion

Nature provides numerous medicinal plants that can be used to treat human illnesses. Plants are the primary source of molecules for the development of new drugs, which highlights transnational industries' interest in discovering substances derived from plants, especially since the vast majority of species have yet to be chemically or biologically researched specially in terms of antitubercular agents. As a result, combining anti-TB medications' target-specific characteristics with the many beneficial properties of medicinal plants could be a viable solution. The present chapter gives an overview of medicinal plants found to be reported to impart a prominent effect against TB. The mechanism behind the interaction of anti-TB drugs and herbal constituents has received little attention. Thus, novel targets can also be set to overcome the disease using herbs as well as bioactives that can also be further explored for modern targets for preparation of formulations. As a result, the current chapter summarizes findings from the literature on plants as anti-TB agents.

References

1. Arditti FD, Rabinkov A, Miron T, Reisner Y, Berrebi A, Wilchek M, Mirelman D. Apoptotic killing of B-chronic lymphocytic leukemia tumor cells by allicin generated in situ using a rituximab-alliinase conjugate. Mol Cancer Ther. 2005;4(2):325–32.
2. Adami AJ, Cervantes JL. The microbiome at pulmonary alveolar niche and its role in *Mycobacterium tuberculosis* infection. Tuberculosis. 2015;95(6):651–8.
3. Ankri S, Mirelman D. Antimicrobial properties of allicin from garlic. Microbes Infect. 1999;1(2):125–9.
4. Arya V. A review on anti-tubercular plants. Int J Pharmtechnol Res. 2011;3:872–80.
5. Abedinzadeh M, Gaeini M, Sardari S. Natural antimicrobial peptides against *Mycobacterium tuberculosis*. J Antimicrob Chemother. 2015;70(5):1285–9.
6. Amit J, Sunil C, Vimal K, Anupam P. Phytosomes: a revolution in herbal drugs. The pharma review. New Delhi: Kongposh Publications Pvt. Ltd; 2008. p. 24–8.
7. Babalola IT, Adelakun EA. Compendium of medicinal plants for the ethnotherapeutic management of tuberculosis and other respiratory diseases. J Pharmacogn Phytochem. 2018;7(3):1983–94.
8. Badavenkatappa SG, Peraman R. In vitro antitubercular, anticancer activities and IL-10 expression in HCT-116 cells of *Tinospora sinensis* (Lour.) Merr. leaves extract. Nat Prod Res. 2021;35(22):4669–74.
9. Borges-Argáez R, Canche-Chay CI, Peña-Rodríguez LM, Said-Fernández S, Molina-Salinas GM. Antimicrobial activity of *Diospyros anisandra*. Fitoterapia. 2007;78(5):370–2.
10. Barnes CC, Smalley MK, Manfredi KP, Kindscher K, Loring H, Sheeley DM. Characterization of an anti-tuberculosis resin glycoside from the prairie medicinal plant *Ipomoea leptophylla*. J Nat Prod. 2003;66(11):1457–62.
11. Choudhary N, Khajuria V, Gillani ZH, Tandon VR, Arora E. Effect of *Carum carvi*, a herbal bioenhancer on pharmacokinetics of antitubercular drugs: a study in healthy human volunteers. Perspect Clin Res. 2014;5(2):80.
12. Chen JJ, Wang TC, Yang CK, Liao HR, Sung PJ, Chen IS, Cheng MJ, Peng CF, Chen JF. New pterosin sesquiterpenes and antitubercular constituents from *Pteris ensiformis*. Chem Biodivers. 2013;10(10):1903–8.

13. Cho JJ, Cho CL, Kao CL, Chen CM, Tseng CN, Lee YZ, Liao LJ, Hong YR. Crude aqueous extracts of *Pluchea indica* (L.) Less. inhibit proliferation and migration of cancer cells through induction of p53-dependent cell death. BMC Complement Altern Med. 2012;12(1):1–1.
14. Dini C, Fabbri A, Geraci A. The potential role of garlic (Allium sativum) against the multi-drug resistant tuberculosis pandemic: a review. Ann Ist Super Sanita. 2011;47:465–73.
15. Devi VK, Jain N, Valli KS. Importance of novel drug delivery systems in herbal medicines. Pharmacogn Rev. 2010;4(7):27.
16. Eram S, Mujahid M, Bagga P, Ahsan F, Rahman MA. Hepatoprotective evaluation of Galanga (*Alpinia Officinarum*) rhizome extract against antitubercular drugs induced hepatotoxicity in rats. Int J Geogr Inf Syst. 2020;26(2):113–25.
17. Egharevba OH, Oladosu P, Okhale ES, Ibrahim I, Folashade KO, Okwute KS, Okogun IJ. Preliminary anti-tuberculosis screening of two Nigerian Laggera species (*Laggera pterodonta* and *Laggera aurita*). J Med Plant Res. 2010;4(12):1235–7.
18. Block E, John Dane A, Thomas S, Cody RB. Applications of direct analysis in real time mass spectrometry (DART-MS) in allium chemistry. 2-Propenesulfenic and 2-Propenesulfinic acids, diallyl trisulfane S-oxide, and other reactive sulfur compounds from crushed AnkGarlic and other alliums. J Agric Food Chem. 2010;58(8):4617–25.
19. Gangwar M, Goel RK, Nath G. *Mallotus philippinensis* Muell. Arg (Euphorbiaceae): ethnopharmacology and phytochemistry review. Biomed Res Int. 2014;2014:213973.
20. Gupta R, Thakur B, Singh P, Singh HB, Sharma VD, Katoch VM, Chauhan SV. Anti-tuberculosis activity of selected medicinal plants against multi-drug resistant *Mycobacterium tuberculosis* isolates. Indian J Med Res. 2010;131(6):809.
21. Godipurge SS, Biradar JS, Mahurkar N. Phytochemical and pharmacological evaluation of *Acalypha indica* Linn in experimental animal models. Int J Pharmacogn Phytochem Res. 2014;6(4):973–9.
22. Gowrish A, Vagdevi H, Rajashekar H. In vitro antioxidant and antitubercular activity of *Leucas marrubioides* Desf. root extracts. J Appl Pharm Sci. 2015;5(02):137–42.
23. Gutierrez-Lugo MT, Wang Y, Franzblau SG, Suarez E, Timmermann BN. Antitubercular sterols from *Thalia multiflora* Horkel ex Koernicke. Phytother Res. 2005;19(10):876–80.
24. Gu JQ, Wang Y, Franzblau SG, Montenegro G, Timmermann BN. Constituents of *Quinchamalium majus* with potential antitubercular activity. Z Naturforsch C. 2004a;59(11–12):797–802.
25. Gu JQ, Wang Y, Franzblau SG, Montenegro G, Timmermann BN. Constituents of *Senecio chionophilus* with potential antitubercular activity. J Nat Prod. 2004b;67(9):1483–7.
26. Gupta VK, Fatima A, Faridi U, Negi AS, Shanker K, Kumar JK, Rahuja N, Luqman S, Sisodia BS, Saikia D, Darokar MP. Antimicrobial potential of *Glycyrrhiza glabra* roots. J Ethnopharmacol. 2008;116(2):377–80.
27. NabiQazi G, Bedi KL, Johri RK, Tikoo MK, Kumar A, Sharma SC, Abdullah ST, Suri OP, Gupta BD, Suri KA, Satti NK, Khajuria RK, Singh S, Khajuria A, KrishanKapahi B. Bioavailability/bioefficacy enhancing activity of cuminmcyminum and extracts and fractions thereof. U.S. Patent US 7514105; 2009.
28. Green E, Obi LC, Samie A, Bessong PO, Ndip RN. Characterization of n-Hexane subfraction of *Bridelia micrantha* (Berth) and its antimycobacterium activity. BMC Complement Altern Med. 2011;11(1):1–5.
29. Grange JM, Snell NJ. Activity of bromhexine and ambroxol, semi-synthetic derivatives of vasicine from the Indian shrub *Adhatoda vasica*, against mycobacterium tuberculosis in vitro. J Ethnopharmacol. 1996;50(1):49–53.
30. Gautam R, Saklani A, Jachak SM. Indian medicinal plants as a source of antimycobacterial agents. J Ethnopharmacol. 2007;110(2):200–34.
31. Pamela HB. Tuberculosis pathogenesis and transmission. Oakland Country Michiga Health Division; 2016. p. 6, 8,1 2, 14, 20–28
32. Hunter RL, Actor JK, Hwang SA, Karev V, Jagannath C. Pathogenesis of post primary tuberculosis: immunity and hypersensitivity in the development of cavities. Ann Clin Lab Sci. 2014;44(4):365–87.

33. Habbu PV, Mahadevan KM, Shastry RA. Antimicrobial activity of the flavonoid sulphates and other fractions of *Argyreia speciosa*. Indian J Exp Biol. 2009;47:121–8.
34. Hong Q, Minter DE, Franzblau SG, Arfan M, Amin H, Reinecke MG. Anti-tuberculosis compounds from *Mallotus philippinensis*. Nat Prod Commun. 2010;5(2):211–7. https://doi.org/10.1177/1934578X1000500208.
35. Hussain T, Gupta RK, Sweety K, Khan MS, Hussain MS, Arif MD, Hussain A, Faiyazuddin MD, Rao CV. Evaluation of antihepatotoxic potential of *Solanum xanthocarpum* fruit extract against antitubercular drugs induced hepatopathy in experimental rodents. Asian Pac J Trop Biomed. 2012;2(6):454–60.
36. Hussain K, Ismail Z, Sadikun A, Ibrahim P. Antioxidant, anti-TB activities, phenolic and amide contents of standardised extracts of *Piper sarmentosum* Roxb. Nat Prod Res. 2009;23(3):238–49.
37. Hema MR, Ramaiah M, Vaidya VP, Shivakumar BS, Begum A. Evaluation of antidiabetic and antitubercular activities of methanol extract of root bark of *Artocarpus integrifolia*. Res J Pharm, Biol Chem Sci. 2017;2(4):886–93.
38. Hegeto LA, Caleffi-Ferracioli KR, Nakamura-Vasconcelos SS, de Almeida AL, Baldin VP, Nakamura CV, Siqueira VLD, Scodro RB, Cardoso RF. In vitro combinatory activity of piperine and anti-tuberculosis drugs in *Mycobacterium tuberculosis*. Tuberculosis. 2018;111:35–40.
39. Higuchi CT, Pavan FR, Leite CQ, Sannomiya M, Vilegas W, Leite SR, Sacramento LV, Sato DN. Triterpenes and antitubercular activity of *Byrsonima crassa*. Química nova. 2008;31:1719–21.
40. Jyoti MA, Nam KW, Jang WS, Kim YH, Kim SK, Lee BE, Song HY. Antimycobacterial activity of methanolic plant extract of *Artemisia capillaris* containing ursolic acid and hydroquinone against *Mycobacterium tuberculosis*. J Infect Chemother. 2016;22(4):200–8.
41. Jaya S, Siddheswaran P, Kumar KS, Karthiyayini T. Anti-tubercular activity of fruits of *Prunus armeniaca* (L.). Int J Pharm Bio Sci. 2010;1(2):1885.
42. Kurandawad JM, Lakshman HC. Diversity of the endophytic fungi isolated from *Acalypha indica* Linn – a promising medicinal plant. Int J Sci Res Publ. 2014;4:2250–3153.
43. Kumar M, Singh SK, Singh PP, Singh VK, Rai AC, Srivastava AK, Shukla L, Kesawat MS, Kumar Jaiswal A, Chung SM, Kumar A. Potential anti-*Mycobacterium tuberculosis* activity of plant secondary metabolites: insight with molecular docking interactions. Antioxidants. 2021;10(12):1990.
44. Kumar SS, Kumar CV, Vardhan AV. Hepatoprotective activity of *Acalypha indica* linn against thioacetamide induced toxicity. Int J Pharm Sci. 2013;5:356–9.
45. Kaufman S. Introduction. Semin Immunol. 2014;26(6):429–30.
46. Lakshmanan D, Werngren J, Jose L, Suja KP, Nair MS, Varma RL, Mundayoor S, Hoffner S, Kumar RA. Ethyl p-methoxycinnamate isolated from a traditional anti-tuberculosis medicinal herb inhibits drug resistant strains of *Mycobacterium tuberculosis* in vitro. Fitoterapia. 2011;82(5):757–61.
47. Lall N, Meyer JJ. Inhibition of drug-sensitive and drug-resistant strains of *Mycobacterium tuberculosis* by diospyrin, isolated from *Euclea natalensis*. J Ethnopharmacol. 2001;78(2–3):213–6.
48. Mativandlela SP, Lall N, Meyer JJ. Antibacterial, antifungal and antitubercular activity of (the roots of) *Pelargonium reniforme* (CURT) and *Pelargonium sidoides* (DC) (Geraniaceae) root extracts. S Afr J Bot. 2006;72(2):232–7.
49. Makhoba XH, Viegas C Jr, Mosa RA, Viegas FPD, Pooe OJ. Potential impact of the multi-target drug approach in the treatment of some complex diseases. Drug Des Devel Ther. 2020;14:3235.
50. Mangwani N, Singh PK, Kumar V. Medicinal plants: adjunct treatment to tuberculosis chemotherapy to prevent hepatic damage. J Ayurveda Integr Med. 2020;11(4):522–8.
51. Molina-Salinas GM, Uc-Cachón AH, Peña-Rodríguez LM, Dzul-Beh AD, Escobedo García-Medrano RM. Bactericidal effect of the leaf extract from musa spp. (AAB group, silk sub-

group), cv."manzano" against multidrug-resistant *Mycobacterium tuberculosis*. J Med Food. 2019;22(11):1183–5.
52. Macabeo A, Read R, Brophy J, Krohn K. Activity of the extract and indole alkaloids from *Alstonia scholaris* against *Mycobacterium tuberculosis*. Philipp Agric Sci. 2008;91(3):348–51.
53. Munna S, Basha SC, Reddy PR, Pramod N, Kumar YP, Basha GM. Antitubercular activity of *Actiniopteris radiata* linn. Phytochem Anal. 2014;10:12.
54. Mativandlela SP, Muthivhi T, Kikuchi H, Oshima Y, Hamilton C, Hussein AA, Van Der Walt ML, Houghton PJ, Lall N. Antimycobacterial flavonoids from the leaf extract of *Galenia africana*. J Nat Prod. 2009;72(12):2169–71.
55. Mohamed JMM, Alqahtani A, Kumar TVA, Fatease AA, Alqahtani T, Krishnaraju V, Ahmad F, Menaa F, Alamri A, Muthumani R, Vijaya R. Superfast synthesis of stabilized silver nanoparticles using aqueous *Allium sativum* (garlic) extract and isoniazid hydrazide conjugates: molecular docking and in-vitro characterizations. Molecules. 2021;27(1):110.
56. Mujahid M, Hussain T, Siddiqui HH, Hussain A. Evaluation of hepatoprotective potential of *Erythrina indica* leaves against antitubercular drugs induced hepatotoxicity in experimental rats. J Ayurveda Integr Med. 2017;8(1):7–12.
57. Muzammil MS, Manikandan M, Jafar A, Sakthivel P, Geetha S, Malarkodi R. Anti-inflammatory studies on *Acalypha indica* L. leaves by membrane stabilization. Indian J Nat Prod Resour. 2014;5(2):195–7.
58. N'guessan JD, Bidié AP, Lenta BN, Weniger B, Andre P, Guédé-Guina F. In vitro assays for bioactivity-guided isolation of antisalmonella and antioxidant compounds in *Thonningia sanguinea* flowers. Afr J Biotechnol. 2007;6(14):1685–9.
59. Nahrstedt A, Kant JD, Wray V. Acalyphin, a cyanogenic glucoside from *Acalypha indica*. Phytochemistry. 1982;21(1):101–5.
60. Nair SS, Pharande RR, Bannalikar AS, Mukne AP. In vitro antimycobacterial activity of acetone extract of *Glycyrrhiza glabra*. J Pharm Pharmacogn Res. 2015;3(4):80–6.
61. Norman GB. Herbal drugs and phytopharmaceuticals. A handbook for practice on a scientific basis. 2nd ed. New York/Stuttgart: Medpharm Scientific Publishers/CRC Press; 2001. p. 230–48.
62. Obogwu MB, Akindele AJ, Adeyemi OO. Hepatoprotective and in vivo antioxidant activities of the hydroethanolic leaf extract of *Mucuna pruriens* (Fabaceae) in antitubercular drugs and alcohol models. Chin J Nat Med. 2014;12(4):273–83.
63. Promsawan N, Kittakoop P, Boonphong S, Nongkunsarn P. Antitubercular cassane furanoditerpenoids from the roots of *Caesalpinia pulcherrima*. Planta Med. 2003;69(08):776–7.
64. Putri DU, Rintiswati N, Soesatyo MH, Haryana SM. Immune modulation properties of herbal plant leaves: *Phyllanthus niruri* aqueous extract on immune cells of tuberculosis patient-in vitro study. Nat Prod Res. 2018;32(4):463–7.
65. Pari L, Kumar NA. Hepatoprotective activity of *Moringa oleifera* on antitubercular drug-induced liver damage in rats. J Med Food. 2002;5(3):171–7.
66. Prasad R, Singh A, Gupta N, Tarke C. Role of bioenhancers in tuberculosis. Int J Health Sci Res. 2016;2016:3076.
67. Pérez-González MZ, Macías-Rubalcava ML, Hernández-Ortega S, Siordia-Reyes AG, Jiménez-Arellanes MA. Additional compounds and the therapeutic potential of *Cnidoscolus chayamansa* (McVaugh) against hepatotoxicity induced by antitubercular drugs. Biomed Pharmacother. 2019;1(117):109140.
68. Panda VS, Ashar HD, Sharan A. Antioxidant and hepatoprotective effects of *Garcinia indica* fruit rind in antitubercular drug-induced liver injury in rats. Botanics Targets Therapy. 2013;27(3):29–37.
69. Panchabhai TS, Ambarkhane SV, Joshi AS, Samant BD, Rege NN. Protective effect of *Tinospora cordifolia*, *Phyllanthus emblica* and their combination against antitubercular drugs induced hepatic damage: an experimental study. Phytother Res. 2008;22(5):646–50.
70. Rajandeep K, Pushpinder K, Harpreet K. A review on antitubercular plants. J Pharm Innov. 2011;1:11–22.

71. Rao CV, Rawat AK, Singh AP, Singh A, Verma N. Hepatoprotective potential of ethanolic extract of *Ziziphus oenoplia* (L.) Mill roots against antitubercular drugs induced hepatotoxicity in experimental models. Asian Pac J Trop Med. 2012;5(4):283–8.
72. Ranju G, Niranjan S, Kumar PS, Kumar PV, Kumar PS. In vitro anthelmintic activity of *Acalypha indica* leaves extracts. Int J Res Ayurveda Pharm. 2011;2(1):247–9.
73. Sultana N, Lee NH. Antielastase and free radical scavenging activities of compounds from the stems of *Cornus kousa*. Phytother Res. 2007;21(12):1171–6.
74. Sungthongjeen S, Puttipipatkhachorn S, Paeratakul O, Dashevsky A, Bodmeier R. Development of pulsatile release tablets with swelling and rupturable layers. J Control Release. 2004;95(2):147–59.
75. Sahoo SK, Labhasetwar V. Nanotech approaches to drug delivery and imaging. Drug Discov Today. 2003;8(24):1112–20.
76. Semalty A, Semalty M, Rawat BS, Singh D, Rawat MS. Pharmacosomes: the lipid-based new drug delivery system. Expert Opin Drug Deliv. 2009;6(6):599–612.
77. Sejlitz T, Neujahr HY. Phenol hydroxylase from yeast: a model for phenol binding and an improved purification procedure. Eur J Biochem. 1987;170(1–2):343–9.
78. Sharma A, Hill A, Kurbatova E, van der Walt M, Kvasnovsky C, Tupasi TE, Caoili JC, Gler MT, Volchenkov GV, Kazennyy BY, Demikhova OV. Estimating the future burden of multidrug-resistant and extensively drug-resistant tuberculosis in India, the Philippines, Russia, and South Africa: a mathematical modelling study. Lancet Infect Dis. 2017;17(7):707–15.
79. Sharifi-Rad J, Salehi B, Stojanović-Radić ZZ, Fokou PV, Sharifi-Rad M, Mahady GB, Sharifi-Rad M, Masjedi MR, Lawal TO, Ayatollahi SA, Masjedi J. Medicinal plants used in the treatment of tuberculosis-ethnobotanical and ethnopharmacological approaches. Biotechnol Adv. 2020;15(44):107629.
80. Saludes JP, Garson MJ, Franzblau SG, Aguinaldo AM. Antitubercular constituents from the hexane fraction of *Morinda citrifolia* Linn. (Rubiaceae). Phytother Res. 2002;16(7):683–5.
81. Subbanna S, Gopenath TS, Basalingappa KM. Biogenic nanoparticles from *Allium sativum* and its bioactives applications. Eur J Mol Clin Med. 2020;7(08):2020.
82. Tasduq SA, Singh K, Satti NK, Gupta DK, Suri KA, Johri RK. *Terminalia chebula* (fruit) prevents liver toxicity caused by sub-chronic administration of rifampicin, isoniazid and pyrazinamide in combination. Hum Exp Toxicol. 2006;25(3):111–8.
83. Tandon VR, Khajuria V, Kapoor B, Kour D, Gupta S. Hepatoprotective activity of *Vitex negundo* leaf extract against anti-tubercular drugs induced hepatotoxicity. Fitoterapia. 2008;79(7–8):533–8.
84. Thang TD, Luu HV, Tuan NN, Hung NH, Dai DN, Ogunwande IA. Constituents of essential oils from the leaves and stem barks of *Uvaria rufa* and *Uvaria cordata* (Annonaceae) from Vietnam. J Essent Oil-Bear Plants. 2014;17(3):427–34.
85. Thakur JP, Gothwal PP. Edible plants as a source of antitubercular agents. J Pharmacogn Phytochem. 2015;4(1):228–34.
86. Tang J, Wang B, Wu T, Wan J, Tu Z, Njire M, Wan B, Franzblauc SG, Zhang T, Lu X, Ding K. Design, synthesis, and biological evaluation of pyrazolo [1, 5-a] pyridine-3-carboxamides as novel antitubercular agents. ACS Med Chem Lett. 2015;6(7):814–8.
87. Tousif S, Singh DK, Mukherjee S, Ahmad S, Arya R, Nanda R, Ranganathan A, Bhattacharyya M, Van Kaer L, Kar SK, Das G. Nanoparticle-Formulated Curcumin Prevents Posttherapeutic Disease Reactivation and Reinfection with Mycobacterium tuberculosis following Isoniazid Therapy. Front Immunol. 2017 Jun 30;8:739. https://doi.org/10.3389/fimmu.2017.00739. PMID: 28713372; PMCID: PMC5491555.
88. Uc-Cachón AH, Borges-Argáez R, Said-Fernández S, Vargas-Villarreal J, González-Salazar F, Méndez-González M, Cáceres-Farfán M, Molina-Salinas GM. Naphthoquinones isolated from *Diospyros anisandra* exhibit potent activity against pan-resistant first-line drugs *Mycobacterium tuberculosis* strains. Pulm Pharmacol Ther. 2014;27(1):114–20.
89. Viswanathan V, Phadatare AG, Mukne A. Antimycobacterial and antibacterial activity of *Allium sativum* bulbs. Indian J Pharm Sci. 2014;76(3):256.
90. Vadivel E, Arondekar SG. Antitubercular and anthelmintic activity of *Triumfetta rhomboidea*. J Pharm Pharm Sci (IRJPMS). 2018;1(3):46–8.

91. Viswanathan V, Pharande R, Bannalikar A, Gupta P, Gupta U, Mukne A. Inhalable liposomes of *Glycyrrhiza glabra* extract for use in tuberculosis: formulation, in vitro characterization, in vivo lung deposition, and in vivo pharmacodynamic studies. Drug Dev Ind Pharm. 2019;45(1):11–20.
92. Quan D, Nagalingam G, Payne R, Triccas JA. New tuberculosis drug leads from naturally occurring compounds. Int J Infect Dis. 2017;56:212–20.
93. Wellington S, Hung DT. The expanding diversity of *Mycobacterium tuberculosis* drug targets. ACS Infect Dis. 2018;4(5):696–714.
94. Xu Y, Liang B, Kong C, Sun Z. Traditional medicinal plants as a source of antituberculosis drugs: a system review. Biomed Res Int. 2021;2021:9910365.
95. Ying-Tsun M, Chuang JI, Lin J, Hsu F. Phenolic from *Acalypha indica*. J Chin Chem Soc. 1997;44:499–502.
96. Zhao J, Evangelopoulos D, Bhakta S, Gray AI, Seidel V. Antitubercular activity of *Arctium lappa* and *Tussilago farfara* extracts and constituents. J Ethnopharmacol. 2014;155(1):796–800.
97. Zuniga J, Torres-García D, Santos-Mendoza T, Rodriguez-Reyna TS, Granados J, Yunis EJ. Cellular and humoral mechanisms involved in the control of tuberculosis. Clin Dev Immunol. 2012;2012:193923.

Correction to: Drug Delivery by Micro, Nanoemulsions in Tuberculosis

Dignesh Khunt, Bhupendra G. Prajapati, Mehul Prajapti, Manju Misra, Sagar Salave, Jayvadan K. Patel, and Ravish J. Patel

Correction to:
Chapter 9 in: R. Shegokar, Y. Pathak (eds.), *Tubercular Drug Delivery Systems*, https://doi.org/10.1007/978-3-031-14100-3_9

In the original version of this book, in chapter "Drug Delivery by Micro, Nanoemulsions in Tuberculosis" the author name "Sagar Salave" was incorrectly set as "Sagar Salva". This has now been rectified.

The updated original version of this chapter can be found at
https://doi.org/10.1007/978-3-031-14100-3_9

R. Shegokar, Y. Pathak (eds.), *Tubercular Drug Delivery Systems*,
https://doi.org/10.1007/978-3-031-14100-3_16

Index

R. Shegokar, Y. Pathak (eds.), *Tubercular Drug Delivery Systems*, https://doi.org/10.1007/978-3-031-14100-3

The manufacturer's authorised representative in the EU is Springer Nature Customer Service Centre GmbH, Europaplatz 3, 69115 Heidelberg, Germany. If you have any concerns regarding our products, please contact ProductSafety@springernature.com

Printed and bound by CPI Group (UK) Ltd, Croydon, CR0 4YY
01/10/2026
02225893-0009